ALSO BY JOHN HAMILTON:

Cytomegalovirus and Immunity in

Monographs in Virology, Vol 12, Basel: Karger, 1982 and

Nearly two hundred scientific publications, abstracts, and chapters

THE HISTORY OF INFECTIOUS DISEASES AT DUKE UNIVERSITY IN THE TWENTIETH CENTURY

JOHN D. HAMILTON, MD

ISBN: 978-1-4834-2374-6 (sc)
ISBN: 978-1-4834-2376-0 (hc)
ISBN: 978-1-4834-2375-3 (e)

Library of Congress Control Number: 2014922708

Cover logo created and used by the adult Division of
Infectious Diseases between 1980 and 2010.

Lulu Publishing Services rev. date: 1/20/2015

ACKNOWLEDGMENTS

For my wife, Carol, whose support and ideas have made this book possible; for Patricia Spivey, whose editorial advice has been invaluable; for the Duke University Archives and the knowledgeable and expert archivists who oversee the collection; for Virginia Carden, whose expertise on citation management proved to be critical; for Susan Reeves, whose facility with all things visual was incredibly helpful; for Cara Ragusa, the cover artist; for the over one hundred faculty whose recollections of published and unpublished events clarified a murky history; and for readers of earlier drafts, I am truly grateful.

In the continual remembrance of a glorious past, individuals and nations find their noblest inspiration.
—Sir William Osler (1849–1919)
"The Leaven of Science,"
Aequanimitas, with Other Addresses

CONTENTS

PHOTOGRAPHS

FOREWORD

The Duke University Medical Center is widely recognized for excellence, and its Division of Infectious Diseases enjoys a reputation for outstanding academic achievements. These include innovative advancements in the clinical, educational, and research realms. So it is fitting that one of its former division chiefs would place those achievements in historical perspective.

Dr. John Hamilton, a leading figure in viral infections, has penned an elegant and accessible treatise on the history of infectious diseases at Duke University. Key leaders behind the dynamic changes at Duke are briefly profiled, and their contributions are chronicled against the backdrop of social forces that were shaping the American landscape at each time.

In the era leading up to the creation of Duke University, infections dominated the leading causes of death and disability. Smallpox, typhoid fever, cholera, and other infections in the nineteenth and early twentieth centuries often arrived in epidemic fashion, cutting short the lives of children and young adults. Hamilton illustrates the incremental effect of wars on increasing the infection rates during this period. He then outlines the responses to these challenges in North Carolina: the creation of the State Medical Society, the mandate for reporting communicable diseases, and the development of both the state reference laboratory for tuberculosis and the two sanatoriums for treatment by 1910.

Duke University was established in 1924 and the Duke Hospital in 1930—in the exciting era of the recognition of the germ theory of disease, the period of rational experimentation in science, and the growth of public health and sanitation. Yet Duke was also challenged by the Wall Street economic crash, the prominent activities of the Ku Klux

Klan in North Carolina, and the unprecedented population growth in Durham, related to the success of tobacco farming and cigarette manufacturing.

Hamilton's signature contributions to understanding the initial spread and control of hepatitis B and HIV and the development of effective programs in employee health and hospital infection control are modestly stated in the book. Duke infectious diseases faculty were also quickly responsive to the rise of bacterial resistance to available antibiotics and were among the earliest to create antibiotic stewardship programs—years before that term was popularized. The subsequent development of international programs, statewide surveillance of hospital-acquired infections, and specific approaches to vulnerable, immune-suppressed patients followed.

In separate chapters on tuberculosis, microbiology, vaccines, and others, Hamilton details the unique activities of the Infectious Diseases faculty. His writing is terse yet interesting and inclusive but unlabored. He clearly focuses on the critical responses to important problems in infectious diseases. One cannot help but recognize the high standards of the faculty, the tradition for respect they have had for their colleagues, their sense of urgency in addressing the challenges they faced, and their caring for patients' welfare. Along the way, there were the social and medical biases affecting behavior nationally, the politics of academic medicine that shaped priorities, and the people who rose above it all to make a difference.

Richard P. Wenzel, MD, MSc
Professor and Former Chairman
VCU Department of Internal Medicine
Richmond, Virginia
February 2014

PREFACE

This book is about the history of infectious diseases in the twentieth century at Duke and its affiliate, the Durham VA. It is not just about the diseases themselves but more so about the physicians, researchers, and the programs supporting these individuals, followed over the course of the century. The text itself is not technical, but the references cited for each individual can elaborate on the actual science if the reader chooses to delve into it. The diseases discussed in the book are those with a high prevalence or severity and with substantial links to Duke over the course of the century. To explain the rationale for James B. "Buck" Duke's investment in what became Duke, the initial chapters briefly describe the state of medicine, medical education, public health, and the disease portfolio before the twentieth century in the world, the state, and the city of Durham. The Duke University Medical School and Hospital did not exist before Mr. Duke's bequest in 1924, but from the turn of the century, the events leading up to the creation of Duke were of considerable importance, including the Influenza Pandemic of 1917–18, World War I, and the Great Depression. In order to retain the style of the time in direct quotes, I have not edited portions of the text.

I have been closely connected with Duke since 1970 and was an active faculty member from then until 2010, Chief of the Durham VA Infectious Diseases Section from 1971 to 1994, and Chief of the Duke Adult Infectious Diseases Division from 1994 to 2010. During that time, I was heavily involved in the clinical, teaching, and basic and clinical research missions in the Department of Medicine.

I am fortunate to have had direct experience with many individual faculty and events but also to have interviewed over one hundred current and former faculty members as well as have had complete access to the Medical Center Archives that contain the personal and professional papers belonging to departed or deceased faculty. I am grateful for the support and contributions of all.

John D. Hamilton, MD
Professor of Medicine, Emeritus

CHAPTER 1

MEDICINE BEFORE JAMES B. DUKE'S BEQUEST IN 1924

> Humanity has but three great enemies; *fever, famine and war*; of these by far the greatest, by far the most terrible, is *fever*.
>
> —Sir William Osler[1]

Osler's statement about epidemics has ample support in books by Stephen H. Gehlbach,[2] Irwin W. Sherman,[3] and William H. McNeill,[4] among many others. Surprising to us today, however, is how little was known about the cause of epidemics. Prior to the seventeenth century, theories of disease causation were often founded on religious principles of good and evil and on notions of the body as a balanced entity, with health being a state of equilibrium and illness being an imbalance.[5] Francis Bacon challenged these and other theories in 1605, stating, "The logic now in use serves rather to fix and give stability to their errors which have their foundation in commonly received notions than to help the search after truth. So it does more harm than good."[5] Barry expands on this, writing, "But if the first failing of medicine, a failing that endured virtually unchallenged for two millennia and then only gradually eroded over the next three centuries, was that it did not probe nature through experimentation, that it simply observed and reasoned from observation to a conclusion, that failing was finally about to be challenged."[5] The impending challenge apparently reflected the imposition of science on the understanding and practice of medicine occurring in the eighteenth, nineteenth, and twentieth centuries.

Several authors make a convincing case that the study of anatomic pathology, in fact, preceded and led the way to the introduction of what is called *discovery science*, when Vesalius performed dissections of the human body and when Benivieni and later Morgagni performed postmortem examinations to correlate patient symptoms with morphological findings.[6] In his textbook, Florey perhaps best summarizes the transition to discovery science: "By the end of the eighteenth century, gross pathologic anatomy was established as a firm basis for medical science." He goes on to say that pathologists such as Rokitansky (1804–1878) and Virchow (1821–1905) added substance, with the former performing over thirty thousand postmortem examinations and the latter theorizing that tissues constituting the human body are made of cells (the cell theory). The earlier discovery of the microscope by van Leeuwenhoek (1632–1723) made these advances possible.[6] But although van Leeuwenhoek appears to have seen bacteria, he described them as "little animals."[7]

It is beyond the scope of this book to provide a detailed history of the development of medicine in North Carolina prior to 1800. Two excellent volumes published by the North Carolina Medical Society do just that.[8,9] Suffice it to say there was no organized or scientifically based medicine or medical education before the mid-1850s in North Carolina, in spite of ongoing endemic and epidemic tuberculosis (TB), smallpox, influenza, malaria, typhoid fever, and cholera (albeit they were not named as such, because the germ theory of disease was not yet widely understood or embraced). There were numerous proprietary medical schools that, by standards fifty years later, were woefully inadequate and not affiliated with hospitals. And although there were medical libraries, they were few in number, located in the northeast for the most part, and not scientifically substantive. Nevertheless, a great deal was happening in the world, particularly in the area of discovery science.

The first International Sanitary Conference convened in Paris in 1851. It consisted of European physicians and politicians concerned about the serious and widespread epidemics around the world.[10] That this conference was held in Europe and attended by Europeans reflected their more advanced understanding of medical science at that time. Subsequent conferences were held over the next fifty years, with the goal of establishing strategies to prevent or limit epidemics, in spite of limited or nonexistent understanding of their causes. Whatever the limitations of those conferences, they served as the founding network of what later became organizations committed to improving the health of the public.

Somewhat similar organizations existed in the United States in the form of boards of health, medical associations, and medical societies. The American Medical Association (AMA), for example, was founded in 1847.[8,9] The first state board of health in the United States was established in Massachusetts in 1799. It focused less on broad discussions of the causes of epidemics and more on practical measures, such as assuring pure water, sewer systems, elimination of standing water, burial of the dead, quarantine, and fumigation. The North Carolina Medical Society was established initially in 1799, ceased functioning in 1804, and was revived in 1849. It argued against establishing a new medical college in North Carolina and began publishing the *Transactions of the North Carolina Medical Society* and, not long after, the *Journal of the North Carolina Medical Society*.[8,9]

Although the society's meetings ceased during the Civil War (1861–1866), medicine in North Carolina became increasingly important because of the exceptionally high number of infected battlefield wounds in Confederate soldiers. Medical problems, however, were not confined to the military. A yellow fever epidemic and other communicable diseases affected civilians, who also suffered from severe shortages of food.

After the war, the society lobbied the state legislature in 1877 to establish a board of medical examiners for regulating both the practice of medicine and the State Board of Health.[8,9] Ironically, the first president of the board, Dr. S. S. Satchwell, died of typhoid fever in 1892. Subsequent influential leaders, however, including Drs. Thomas Woods and Richard Lewis, focused on health education as a priority mission, as well as on the implementation of public health regulations.[11] The regulations were not enthusiastically endorsed by practitioners, and a further publication outlined the rationale for the new regulations.[12] Failure to comply with the regulations could result in a penalty. Dr. Watson Rankin, appointed as the first full-time state health officer in 1909, was instrumental in establishing mandatory disease reporting, which provided data to the newly formed North Carolina Bureau of Vital Statistics. He was also instrumental in establishing the State Laboratory of Hygiene, which focused primarily on testing the water.

Although others suggested the concept that germs cause disease, Louis Pasteur and Robert Koch are generally credited with proof of the concept and with creating the foundations for the disciplines bacteriology and microbiology.[7] Pasteur's contributions were many. Among them, he found that germs are responsible for the process of fermentation and putrefaction, he created rabies and anthrax vaccines, and he dismissed what was called the *doctrine of spontaneous generation*, all in the later part of the nineteenth century. Koch, for his part, demonstrated that *Bacillus anthracis* is the cause of anthrax infection and bacterial spores explain latency. He isolated *Mycobacterium tuberculosis* in 1882, demonstrated different susceptibility to bacterial infections by different animals, and created his now-famous Koch's postulates.

With the more general acceptance of the germ theory of disease in the last quarter of the nineteenth century, there came the development of strategies for disease prevention, such as quarantine. Discovery science increasingly focused on expanding our understanding of the array of disease-causing microorganisms—including viruses, fungi, and rickettsia—and their routes of transmission. It is somewhat surprising that the germ theory of disease was not embraced earlier, given examples such as the removal of the Broad Street Pump, which terminated the cholera epidemic in London in 1848, and the later identification of the *Cholera bacillus* organism in 1884.

Indeed, discovery science exploded.[13] Before the twentieth century, discoveries were made by the following scientists: Edward Jenner (smallpox vaccination), Louis Pasteur (the germ theory of disease, among other discoveries), Paul Ehrlich (humoral immunity), Elie Metchnikoff (cellular immunity), Ignaz Semmelweiss (contagion), Edward Trudeau (TB management), William Farr (epidemiology), Walter Reed (yellow fever transmission), John Snow (cholera transmission), Robert Koch (discovery of *Mycobacterium tuberculosis*), Joseph Lister (antisepsis), Howard Ricketts (discovery of rickettsias), Hans Gram (bacterial stains), William Halstead (rubber surgical gloves), Emil von Behring (serum therapy), William Welch (medical education based in science), and Wilhelm Roentgen (X-rays). Additional insights led to the acceptance of the germ theory of disease, the recognition that public health measures are essential in the prevention of disease, the identification and importance of portions of the immune system, the recognition of the importance of antisepsis and asepsis, the increased use of diagnostic technologies with real utility, and the notion that treatments of diseases may actually be feasible. Woven into this revolution in understanding the cause of disease was the impetus given repeatedly by war and the onslaught of infectious complications of wounds and of crowding, poor sanitation, disastrous practices, and marginal nutrition.

After 1850, medicine and medical education began to accept and embrace the advances, but by no means was it instantaneous. This is understandable given the lack of organization, the primitive modes of communication, and the slow acceptance that medicine could be based on scientific experiments usefully and reliably. Advocates of earlier notions of disease causation were not easily won over. By another measure, however, as judged by the number of recipients of the Nobel Prize in physiology and medicine between the inception of the prize in 1901 and the opening of Duke Hospital in 1930, there were fifteen recipients in those twenty-nine years who had done their work on infectious diseases before 1900.[14] Evidently, scientists recognized these contributions even if others did not.

In 1878, Congress authorized the US Marine Hospital to collect reports of select diseases, first from US consuls overseas and then, in 1893, from states and municipalities.[15] Congress directed the surgeon general in 1902 to provide standardized forms for data collection and to compile and publish the data. By 1912, state and territorial health authorities reported ten diseases from nineteen states, the District of Columbia, Hawaii, and Puerto Rico. The Centers for Disease Control (CDC) later assumed this responsibility.

Fortunately, scientists and practitioners of medicine eventually embraced the extraordinary achievements of the nineteenth century, for the same endemic and epidemic diseases still existed at the turn of century. Disease reporting and cause of death were not required in North Carolina before 1909, when the Sanitary Code of the North Carolina Board of Health was established and required the reporting of scarlet fever, diphtheria, epidemic cholera, typhus, typhoid fever, rubella, plague,

TB, chicken pox, and whooping cough with a penalty or a fine for each case not reported. Later, in the first quarter of the century, a more systematic collection of the data indicated a substantial number of cases in North Carolina,[16] especially gonorrhea and syphilis in Durham.[17] Of note, there was excess morbidity and mortality associated with those diseases among African Americans, with 14 percent fatality for all forms of TB disease in African Americans versus 9 percent in whites.[16,17]

Medical education in the United States, except arguably in the northeast, was rudimentary compared with that in Europe. Until the mid-1800s, there were few medical schools in the United States, and none truly based on scientifically sound teaching. This is not surprising, especially in a state such as North Carolina, which in 1867 had a population of one million, 97 percent of whom lived on farms. Of the so-called medical schools that existed, lectures constituted the sole form of education, and for the most part, they were based on observations by the few faculty members who delivered the lectures. Laboratories for both testing and instruction essentially did not exist. Prior to 1850, the number of proprietary schools for the training of doctors flourished, but students were relatively few. Almost none of the schools were connected to hospitals, libraries were rare, and scientific publications were essentially nonexistent. North Carolinians who sought an MD degree went north, most often to the University of Pennsylvania. Other than those who attended schools that conferred the MD degree, individuals who merely described themselves as doctors did not need such a formal degree or even a bona fide apprenticeship, although the latter was not an uncommon predecessor to practicing medicine. As more MDs were produced, which allowed them to reside in closer proximity to each other than before, discontent arose among that cadre. They considered the others who called themselves "doctors" as charlatans who were simply taking advantage of a gullible populace. Indirectly, that discontent led the MDs to call for more standardized educational requirements. Another development that reinforced the rise in professional medical education was the publication in 1910 of Flexner's report to the Carnegie Foundation, referred to as *The Flexner Report*, which summarized the state of medical education in the United States and Canada.[18] That report, overseen by Abraham Flexner, was a highly influential critique of the current system and a guide for the future. Regarding the existence of proprietary medical schools, Flexner is quoted as saying: "These enterprises—for the most part they can be called schools or institutions only by courtesy—were frequently set up regardless of opportunity or need; in small towns as readily as in large, and at times almost in the heart of wilderness." Specifically, the *Report* revealed the following facts [my enumeration]:

1) For the twenty-five years past, there has been an enormous over-production of uneducated and ill-trained medical practitioners.
2) Over-production of ill-trained men is due in the main to the existence of a very large number of commercial schools.
3) Until recently the conduct of a medical school was a profitable business, for the methods of instruction were mainly didactic. As the need for laboratories has become more keenly felt, the expenses of an efficient medical school have been greatly increased.
4) The existence of many unnecessary and inadequate medical schools has been defended by the argument that a poor medical school is justified (falsely) in the interest of the poor boy.
5) A hospital under complete educational control is as necessary to a medical school as is a laboratory of chemistry or pathology.

Because of the report, which rated medical schools, many poorly rated schools were forced to close, and those that remained were required to enhance their instructional capabilities. Many North Carolina medical schools were among those forced out of business, but those affiliated with the University of North Carolina and Wake Forest University continued on, somewhat erratically, as preparatory or two-year schools until they became four-year schools in 1952 and 1941, respectively. The first woman admitted to a medical school in the United States occurred in 1849 at Case Western Reserve in Ohio, and the first woman admitted to the North Carolina Medical Society occurred in 1872.[8,9] Admission of blacks to either North Carolina school was substantially more complicated, and full unrestricted admission to either medical school or medical society did not occur until the mid-1900s.

Simultaneously, one of the premier universities and medical schools, The Johns Hopkins University and School of Medicine, opened in Baltimore in 1876 and 1883, respectively. Mr. Hopkins bequeathed $3.5 million for this purpose to replicate some already distinguished institutions in Germany that used scientific principles as the basis for instruction. William Henry Welch, at the age of thirty-four, was recruited to lead the medical school in 1884. He built an exemplary school that was founded on scientific teaching and emphasized research laboratories, and he recruited a talented faculty. By the end of the first quarter of the twentieth century, Hopkins had elevated its reputation to that of peer European institutions. Dr. Welch would go on to become a central figure in all of medicine in the United States through his contacts with prominent persons in the field, including persons at Duke and those in politics, foundations, and industry. He was heavily involved in organizing a response to the influenza epidemic at Hopkins in 1918. He died of prostate cancer in 1934.

The first quarter of the twentieth century was a period marked by tremendous growth in Durham's population, owing in large measure to the tobacco industry, growing from five thousand persons in 1885 to 45,000 in 1927.[19] There was not, however, a commensurate improvement of the infrastructure to support the growth, which resulted in serious disparities between the haves and the have-nots.[17,20] Those disparities were most evident between the whites and the blacks. Race relations were poor, and the Ku Klux Klan activities became prominent. The governor of North Carolina, William W. Holden, was impeached by a Democratic-controlled state senate as a result of his efforts to prevent Klan activities in 1871.[21] The leading reportable cause of death in the 1920s was TB, with rates of 10 whites per 1,000 and 29.9 blacks per thousand in 1920.[17] Other common causes of death were pellagra (niacin deficiency) and kidney disease, although malaria, typhoid, smallpox, hookworm, and yellow fever continued as major causes of ill health. One in three black infants died before the age of one, and 64 percent of blacks died before the age of forty. A survey revealed that the difference in mortality between whites and blacks was attributable to poor housing, lack of sewerage, years of poor nutrition, and general medical neglect. Infrastructure to support those afflicted with these diseases consisted

8

of a patchwork of charities that included the Red Cross and Salvation Army, among others.

There were disparities not only in the prevalence and outcome of diseases between whites and blacks but also in the health-care provider pool and accessible infrastructure.[22] The two major medical associations—the American Medical Association (AMA), founded in 1847 for white physicians, and the National Medical Association (NMA), founded in 1895 for black physicians—were completely segregated, and admission privileges to hospitals were linked to membership in one medical association or another. This had the effect of isolating the black professionals, especially in the southern states, including North Carolina. Rush Medical School was the first to award a medical degree to an African American, in 1847, and the Massachusetts Medical Society was the first to admit an African American, in 1854.

Despite the Civil War and subsequent passage of the Thirteenth, Fourteenth, and Fifteenth Amendments to the Constitution, professional segregation persisted, championed even by some white physicians. An important landmark Supreme Court case[23] legitimized a policy of "Jim Crow segregation." Medical education for minorities was compromised as well when *The Flexner Report* in 1910 recommended the closure of all but two historically black medical schools, Howard and Meharry. Subsequent developments addressed more evident discriminating policies, such as (a) the Hill-Burton Act, which allowed federal funds to be used for construction of "separate but equal" facilities and which reinforced segregation; (b) *Brown v. Board of Education* in 1954, which found segregation in public schools unconstitutional; (c) the Civil Rights Act in 1964; and (d) Medicare and Medicaid in 1965, which banned segregation in hospitals. Curiously, the AMA as an organization resisted integration despite many of its members favoring it.

Between 1900 and 1930, advances in medicine in North Carolina were slow. To be sure, there was increasing recognition that diseases had tangible causes. Pathologists therefore continued to focus on postmortem examinations, and subsequently, with the development of anesthesia, they began the practice of performing biopsies for diagnosis as well as the use of techniques for more extensive laboratory testing. The high prevalence of TB and other bacterial diseases prompted the building of the first state laboratory at the Central Hospital for the Insane in Raleigh, North Carolina, in 1903 and a second at the State Sanatorium for Tuberculosis in Aberdeen, North Carolina, in 1909. The State Laboratory of Hygiene of the North Carolina State Health Department in Raleigh was without peer for agricultural and environmental testing by the mid-1920s, but laboratories associated with North Carolina hospitals before 1920 were generally inadequate. Standardization of laboratory supervision and quality control would come only later, and practitioners oversaw most laboratories. Pathologists had not yet identified clinical pathology as a discipline for which they would ultimately be responsible. In any case, there were few trained pathologists in the state.

Medical services for the general population were provided for a segregated population; in-patient services in Durham were at Watts Hospital for whites and at Lincoln Hospital for African Americans.[24, 25]

Watts Hospital, a gift from Mr. George Washington Watts, opened in February 1895 but was relocated in 1909 with expanded capacity to "minimize infections." Over succeeding years, Watts Hospital also expanded its mission and sought to create a medical school, but this plan met with some resistance in large measure because of the bequest in 1924 of millions by James B. Duke to create the Duke Endowment. To the extent that it is relevant to this book to mention the kinds of diseases treated at Watts, data are limited, but gonorrhea was said to be the most common, along with urinary tract infections and chronic appendicitis. Treatments prior to the availability of sulfa drugs included diet, topical antibacterial substances such as alcohol and carbolic acid, and vaccines such as typhoid. Whatever the excellence of the facility for the times, the contributions of the physician staff, and the benefits for the residents of Durham, it became clear that "voluntary hospitals" such as Watts were not a sustainable model, especially in this case, with Duke University and Duke Hospital located nearby. Funds for the construction and operation of Lincoln Hospital were provided by George Washington Watts[20] and by James B. and Benjamin Duke[17] in 1901 and again in 1921 to relocate the facility. Lincoln Hospital, though in some ways more vulnerable than Watts, has survived to this day.

There are excellent discussions of the development of these two facilities, their resources, patient populations, finances, and controversies in books cited above by P. Preston Reynolds, MD, PhD.[24,25] Although each facility would have cared for patients with the prevalent infectious diseases, it is not evident from the available literature if there were some diseases more prevalent than others. One can only surmise from the reports of communicable diseases to the county health department that they were more or less the usual diseases but likely had greater attendant morbidity and mortality among African Americans. In addition, a specialty hospital for ear, nose, and throat conditions (McPherson Hospital) was opened in 1926 in downtown Durham.

There was no recognized specialty of infectious diseases at that time, but there were persons uniquely associated with the management of TB. Because TB was prevalent worldwide and had no effective therapy, an organization called the National Association for the Study and Prevention of Tuberculosis was formed in 1904. In spite of the absence of a proven therapy, a variety of supportive measures were undertaken, including, most prominently, rest, fresh air, and sunlight. Those supportive measures prompted the development of numerous private and, later, public sanatoriums nationwide. In 1871, Dr. H. P. Gatchell opened the first sanatorium in Asheville (The Villa); in 1875, Dr. Joseph Gleitsmann opened a second one in Asheville (The Mountain Sanitarium for Pulmonary Diseases). At one point, in about 1930, Asheville claimed to have twenty tuberculosis specialists and twenty-five private sanatoriums.[26] Subsequently, Dr. Edward Livingston Trudeau opened the most prominent sanatorium in Saranac Lake, New York, in 1885. The state of North Carolina opened the first of four sanatoriums in 1909, in the Sandhills area. This ultimately accommodated seven hundred patients and was renamed the McCain Sanitorium [sic] after Dr. Paul McCain, a longtime superintendent of the facility. The others, at Black Mountain, Wilson, and Chapel Hill (Gravely), were opened in 1937, 1943, and 1953, respectively. Additional private sanatoriums were established in the Durham area as well. It is not clear what, if any, connection these facilities had with Watts or Lincoln Hospitals or with the physicians at these facilities.

In addition to the enormous disease burden of tuberculosis and, to a lesser extent, malaria, smallpox, typhoid, and diphtheria, in 1917–1918 another disease emerged as the most major calamity to beset the world, namely pandemic influenza, or the Spanish flu as it became known, and this at the same time as WWI.[5] Barry makes the case that each condition served to compound the problems of the other at a time when the infrastructure, the population, and the medical facilities were ill prepared, not to mention the rudimentary scientific understanding of infectious diseases. War at any time is disruptive, but siphoning off health-care providers from the private sphere to tend to battle casualties left the civilians with limited resources. The declaration of war by President Wilson in April of 1917 and the attendant overcrowding in barracks and movement of troops in various stages of mobilization both in the United States and abroad set the stage for spread of the disease by, and compromise of, the military. Ironically, it was this younger generation that was more at risk of acquiring the flu and experiencing the more serious complications. North Carolina was not spared, with thousands of Tar Heels killed in battle or killed from influenza in the interval from 1917 to 1920. Somewhat paradoxically, this convergence of disasters contributed to important scientific discoveries and established an initial connection among the Rockefeller University and Foundation, Dr. William H. Welch at Hopkins, and what later became known as Duke University.

CHAPTER 2

CREATION OF DUKE UNIVERSITY, ITS MEDICAL SCHOOL AND HOSPITAL

If anything augurs well for the success of this great undertaking in medical education, it is, in my opinion, the research spirit of the originating faculty and its willingness to start on an educational experiment as expressed in its curriculum, not following blindly the tested arrangements of other schools, but sailing out, in part at least, on the unknown sea of educational procedure.
—Dr. Lewis H. Weed, April 30, 1930[1]

Much of what came to be known as Duke University was, according to Duke historian James F. Gifford, due to four factors: (1) poverty and, especially, ill-prepared medical providers and inadequate facilities; (2) the influence of *The Flexner Report* and the accomplishments of the Johns Hopkins Medical Center; (3) the Depression; and (4) the James B. Duke indenture.[1] The bequest made by James B. Duke in December of 1924 came at an opportune time because it set the example for influential individuals in Durham to initiate selected projects that the growing population needed. Fueled by the success of the tobacco industry after the Civil War, Durham became the hub of a new business, and with it came not only more people but also the problems associated with the tremendous population growth (the population in Durham had increased from five thousand in 1885 to forty-eight thousand in 1927).[2] A primary problem was the inadequate infrastructure to support the growth, the consequences of which were the unsafe water supply, a nonexistent sanitation plan, and overcrowding. There were frequent epidemics of typhoid ("Durham fever"), malaria, dysentery, and TB in the city and region, which exposed the desperate need for more and better medical facilities and providers.[1,2] Watts and Lincoln Hospitals, although very good for the times, were simply inadequate to deal with the rising demands, and medical education in the state lagged well behind that in other states.

"Buck," as James B. Duke was known to friends, grew up in this environment. He was born in 1858 to a poor, conservative, and staunchly Methodist family who lived in northwest Orange County. Both his mother and his stepmother died of typhoid fever when Buck and his older brother, Ben, were quite young. Not long afterward, his father was captured and imprisoned in 1863 by the Union Army and not released until the conclusion of the Civil War. In the absence of either parent, their stepmother's parents raised both boys. Ben had some formal education, but Buck had little. The boys' father, Washington Duke, although a farmer, capitalized on the rising popularity of Golden Leaf tobacco and ultimately formed the Washington Duke, Sons and Company. The company initially marketed roll-your-own cigarettes but in 1885 was the first to produce machine-made cigarettes. By his early thirties, Buck had assumed leadership of the company. His business acumen was extraordinary, and he surrounded himself with talented, hardworking, and productive people.

Between 1890 and 1904, he built the American Tobacco Company through a merger of his own company and four other manufacturers. The American Tobacco Company, though later forced to dismantle owing to an antitrust suit, was the major source of the enormous fortune that Buck accumulated. Buck was minimally involved in the financial support of Trinity College, which later became Duke University, and other charities, those being left to Ben, but it is believed that as early as 1914, he was giving thought to what eventually would become his indenture. Over the years before committing to the indenture, he expanded his business interests to include textiles, railways, and electric power. His involvement in electric power, as stated by Robert Durden, a Duke family historian, was the "most creative economic endeavor of his life."[3]

In national, regional, and local forums, there were discussions regarding the need for increased medical facilities and well-trained physicians. *The Flexner Report* had placed the inadequate medical education system front and center, but implementation of its recommendations would require a coordinated and financially demanding effort on the part of government, industry, and private citizens.[4] Early in the century, The Rockefeller Foundation had embarked on studies of hookworm in the South and had begun to influence the direction of an evolving set of strategies through interactions with the political, financial, and academic leaders in Durham. Those leaders included the president of Trinity College (later president of Duke University), William Preston Few; James B. and Benjamin Duke; George Watts and John Sprunt Hill (son-in-law of George Watts); Dr. John A. Ferrell, the state director of the Rockefeller Sanitary Commission for the Control of Hookworm, who represented the Rockefeller Foundation; and William Henry Welch of Johns Hopkins.[5-7] They considered a variety of scenarios before deciding to establish a four-year medical school.

Earlier, there were attempts to establish a medical school at Trinity College, which Washington, Benjamin Duke, and two Methodist ministers purchased and moved in about 1891 from Randolph County to Durham. Dr. J. F. Crowell, then president of Trinity made the first attempt in 1891, enlisting the financial support of Washington Duke and Julian Carr, another successful tobacco industrialist. Crowell's plan was resisted from many quarters and ultimately failed. William Preston Few, the successor of Crowell as president of Trinity, made the second attempt. He proposed in 1920–1921 that a four-year medical school be established—in which the first two years of medical school would be pursued at the University of North Carolina in Chapel Hill for nonclinical training and the second two years pursued at Trinity for clinical training. Although there was general enthusiasm for this idea from the governor (Cameron Morrison), the UNC president (Harry W. Chase), the president of the board of Watts Hospital (John Sprunt Hill), and Abraham Flexner of the Rockefeller Institute (the primary author of *The Flexner Report*, which revolutionized the requirements for a degree in medicine), it also met with great resistance from competing cities and boards of trustees. Ultimately, that idea also failed because of a lack of financial resources.[2] When James B. Duke created the Duke Endowment in December of 1924 with a bequest of $40 million (by more recent estimates, the equivalent of nearly $450 million), this resource trumped all other actual or nascent ideas for the first four-year medical school in North Carolina. In fact, other four-year medical schools in the state did not open until 1941 (Wake Forest University), 1953 (the University of North Carolina), and 1977 (East Carolina University).

Mr. Duke's indenture was signed on December 30, 1924, and funded the creation of the Duke Endowment, the mission of which was to distribute funds "annually to specified educational, religious and medical institutions of both races in the two Carolinas."[8] Specifically, the endowment provided funds for the creation of Duke University, it's Medical School and Hospital; Davidson, Furman, and Johnson C. Smith Colleges; rural hospitals; orphanages for children of white and non-white races; and the Methodist church. A stipulation in the indenture was that revenue generated for the original bequest be deposited in the endowment in the first ten years after funding. Less than a year after signing the indenture, Mr. Duke unexpectedly died of pneumonia on October 10, 1925.

In the next five years, before Duke Hospital opened in 1930, there was extensive new construction and recruitment of the persons who would implement the various requirements outlined in the indenture. Of particular importance was the recruitment of the founding dean of the new medical school. Beginning in 1926, President Few of Duke University began that search. In 1927, acting upon the advice of Dr. William Henry Welch of Johns Hopkins, Few offered the position to Dr. Wilburt C. Davison, who was then a faculty member at Hopkins. Davison had been a Phi Beta Kappa graduate of Princeton and a Rhodes Scholar destined for study at Oxford in 1913. His mentor at Oxford had been Sir William Osler, then the Regius Professor of Medicine and Dean of the Medical School. Davison had attended the first three years of medical school there, had worked in various capacities in the European conflict with Germany, and had joined the US forces when war was declared in 1917.

After the war, he completed his fourth year of medical school at Hopkins, trained in pediatrics, obtained a faculty appointment, and pursued clinical work and laboratory research. In 1927, Davison, at

16

thirty-four years of age, accepted the position as dean of the Duke Medical School.[1] Davison then set about developing plans to assemble what was required to start a new medical school, including the buildings, library, faculty, students, and curriculum. Architects from Philadelphia and Baltimore provided plans for the construction of the university, medical school, and hospital buildings, and actual construction began in September 1927 on land previously used as the fairgrounds. Initially, the campus housing for the university, medical school, and hospital was for male students; the original Trinity College site was for housing female students.[1]

The priorities established by the curriculum emphasized preparation for the practice of medicine in rural communities, with the highest priority given to teaching, followed by patient care, training of technicians, and research—all to be based on state-of-the-art science, as was the case at Johns Hopkins and institutions abroad. Davison and others followed the construction with interest, even offering suggestions on specific issues such as the width of the windows. Provision was made for the separation of facilities for white and black patients. One further detail of interest is that the hospital wards were named for eminent physicians, initially fourteen and later twenty-two. Eight of the physicians had contributed to the field of infectious diseases or public health (italics denotes the names of the wards): James L. *Cabell* (public health), Henry F. *Campbell* (prevention of yellow fever and dengue), William S. *Halstead* (rubber surgical gloves), Oliver Wendell *Holmes* (the contagiousness of puerperal fever), Josiah C. *Nott* (yellow fever being of "probable insect or animalcular origin"), Watson S. *Rankin* (malaria eradication in Panama), Walter *Reed* (proved yellow fever was transmitted by mosquitoes), and William H. *Welch* (pathology of infectious diseases).[2]

Davison himself obtained lists of medical books from the libraries in Baltimore, Boston, Nashville, Rochester, and select medical schools in Europe. In addition to arranging to purchase the books common to several of the lists, he later accepted medical books donated by Mary Duke Biddle Trent Semans (granddaughter of Buck), among others. Davison resided in rooms at the site where Trinity College (now called *East Campus*) was located and is said to have continued his laboratory work on "enzymes" while construction was underway. As specified in Mr.

Duke's indenture, the total cost for construction of the medical school and hospital could not exceed $4 million (actual cost, $3,922,933).[6,7]

New faculty and staff recruitment was an obvious priority for Davison, given the projected opening of the hospital and medical school to be in 1930. His first appointment was for the chair of the Department of Medicine. For this position, he selected Harold L. Amoss, a Harvard graduate with MD and DPhil degrees who had worked with Simon Flexner at the Rockefeller Institute from 1912 to 1922, followed by an appointment as associate professor in the biological division at Hopkins from 1922 to 1929. Amoss had published fifty or more papers on his research on polio, typhoid, pneumococcus, and the herpes virus, all of which were published in 1956 in the *Journal of Experimental Medicine*, a publication of the Rockefeller Institute. He was a member of many prestigious societies and appeared to be a good choice for the first chair of the Department of Medicine at Duke. The Duke archives, however, contain discrepant views of his collegiality, ranging from "a kindly, thoughtful physician" by Dr. John R. Paul (an expert on polio) in a memorial piece written after Amoss's death in 1956[9] to a rather more "contentious disposition" noted in correspondence between other faculty members and the dean (Duke University Medical Center Library and Archives, circa 1933). Whatever the truth, Dr. Amoss was asked to resign in 1933, after only four years as department chair. For the remainder of his career, he was in private practice in Greenwich, Connecticut. His successor was Frederick M. Hanes, who served as chair from 1934 to 1946.

The second recruitment to the new medical school was Dr. Deryl Hart, a Hopkins-trained surgeon in 1930. At Duke, he served as the chair of surgery from 1929 to 1960. Hart initiated what became the Private Diagnostic Clinic, which generated revenues to support the attending physician, the hospital, and, later, research. In 1960, he was elected president of Duke University. Other key appointees were also Hopkins products: Dr. Wiley Forbus as chair of pathology from 1929 to 1964, Dr. William A. Perlzweig as chair of biochemistry from 1929 to 1949, and Bessie Baker, RN, as dean of the School of Nursing from 1929 to 1949. William Welch of Hopkins, for whom a hospital ward was named, was very influential in the makeup of early faculty and staff.[1,6,7]

Another influential appointee was Dr. Watson S. Rankin, a North

Carolina native who trained in obstetrics and pathology at the University of Maryland but developed an interest in public health. He was appointed as dean at Wake Forest at the age of twenty-six and worked in the Rockefeller Hookworm Eradication Program before his appointment to the Duke faculty as professor of public health. Rankin was appointed the first full-time health officer as secretary of the North Carolina State Board of Health from 1909 to 1925. Later he was appointed head of the Hospital and Orphans section of the Duke Endowment, a time when he articulated the conversion of private hospitals to community hospitals in a monograph titled *The Small General Hospital*.[10] In that later role, he facilitated collaborations with the School of Medicine and was most influential as a member of various committees that sought solutions to the cost of health care, including the AMA's Cost of Medical Care Committee. He was a signatory of an extremely controversial committee report that was perceived by opponents as socialized medicine.[11] From 1919 to 1920, he was president of the American Public Health Association.

David T. Smith, another recruit with Hopkins connections, was appointed in 1930 as professor of bacteriology and as associate professor of the Department of Medicine. Initially, the hospital lost money, as patient-derived revenues were insufficient to cover the costs, even though the charges were very high. Only one in four white patients and one in twenty-five black patients were able to pay. This dilemma was understandable given the crash of the stock market in 1929 and the subsequent Great Depression, which disproportionately affected blacks adversely. To address the financial problems, the newly created Hospital Care Association offered a health insurance plan. In addition, North Carolina counties were obliged to contribute to the support of the hospital, the rationale being that residents of those counties availed themselves of the resources offered by the new facility.

In 1930, a grant from the Rockefeller Foundation allowed for the admission of thirty-eight first-year students and eighteen third-year students. Admission to the medical school was initially based more on a candidate's personality and character, but this was to change later. Generally, the third-year students transferred from other two-year schools and graduated from Duke in 1932.

Each year had four terms, and students could elect to take three or four terms a year, with twelve terms required for graduation. The initial curriculum was designed for future family practitioners in the rural South, but over time, an increased emphasis was given to research and specialization. This might have been predicted given the challenges expressed by Dr. William Welch and Dr. Lewis Weed, both of Johns Hopkins (see the block quote introducing the chapter). Nonetheless, it was strongly encouraged for students to have externships with private practitioners, and two years of an internship was required for graduation.[6,7]

As demanding as those activities were, another addition to the demands was the collapse of the stock market and the ongoing fallout. In some respects, Durham was not as badly hurt as larger cities, especially those with financial institutions that were said to have made imprudent loans. The infusion of funds for creating a new university, medical school, hospital, and other concerns dictated by the endowment brought some level of protection to the city and state. Nonetheless, the Great Depression hit hardest those populations already shown to be vulnerable financially, socially, and medically. This would not pass for many years, leaving many victims in its wake. After some hard years in the 1930s domestically, the specter of serious trouble abroad began to raise its ugly head, ultimately setting off WWII. Scientifically, meanwhile, continuous progress was the rule. Over the course of the remainder of the twentieth century, unbelievable progress was made in the provision of scientifically based medical care and in understanding the basis and breadth of human and animal diseases—progress that no one would have believed in one's wildest dreams when Duke began its own contributions.

Photograph 2.1 shows individual pictures of two of the major figures in the evolution of Duke, Mr. James B. Duke and Dr. Watson S. Rankin, as well as a group picture of early faculty.

Early Faculty

Row 1—C. Johnston, Francis Swett, Harold Amoss, Oscar Hansen-Preuss, R. Baker, Eloise Smith, Frederick Bernheim, James Ruegsegger

Row 2—Harry Hudnall, Julian Ruffin, Duncan Hethorination, Francis Porro, David Smith

Row 3—Elbert Persons, Raymond Rigdon, Max Oates, Jerome Syverton, Edwin Alyea, Watt Eagle, Bayard Carter, Alfred Shands

Row 4—James Hicks, Walter Mayer, Frederick Reese, Rowland Bellows, Anne Lawton, Royal Calder, William Hollinshead, Robert Ross, Deryl Hart

Row 5—Wilburt Davison, W. Baker, Paul Preu, Sarah Thompson, Paul Sanger, Thomas Walker, Robert Reeves, Elbert Apple, George Eadie

J.B. Duke

W. S. Rankin

21

CHAPTER 3

TUBERCULOSIS

> Why, when one comes near consumptives ... does one
> contract their disease, while one does not contract dropsy,
> apoplexy, fever, or many other ills? With the phthisic the
> reason is that the breath is bad and heavy... In approach-
> ing the consumptive, one breathes the pernicious air. One
> takes the disease.
>
> —Aristotle[1]

Although the condition called *consumption* was recognized as early
as 1000 BC, it is hypothesized that the bacterium *Mycobacterium tuberculosis* (MTb) evolved from *Mycobacterium bovis*, a bacterium that
predominately infects cattle and likely existed many years earlier than
MTb. Clinical descriptions of the human disease (referred to here as TB,
to distinguish it from the organism MTb) began to appear in medical
texts in Greece and India in about the mid-600s BC. For example: "The
patient coughs frequently; his sputum is thick and sometimes contains
blood. His breathing is like a flute. His skin is cold, but his feet are hot.
He sweats greatly, and his heart is much disturbed. When the disease
is extremely grave, he suffers from diarrhea."[1] By the first millennium,
TB had been detected in Europe, Asia, and the New World. A major
advance for the diagnosis of pulmonary diseases was Rene Laennec's
invention of the stethoscope in the late 1790s. More important, however, was his postulation from postmortem examinations that there
are two distinct types of TB: cavitation and calcification.[1] Tragically,
Laennec contracted TB himself and died in 1826 at the age of forty-five.

It is somewhat surprising that he did not recognize the disease as being communicable.

By the end of the nineteenth century, TB had reached pandemic proportions worldwide. There were one thousand deaths per one hundred thousand persons, likely because of increased urbanization, crowding, and a poor understanding of the route of transmission. Fortunately, a concerted effort was made to correct those reasons for TB's spread, and by the early part of the twentieth century, the death rate had fallen to around two hundred per one hundred thousand persons. Jean Antoine Villemin performed further animal inoculation studies to assess its transmissibility in animals, but it was Robert Koch who, in 1882, discovered the organism MTb through differential staining of the tubercle bacillus and the surrounding cells.[1] Koch's discovery, however, was not widely accepted as the cause of TB because the germ theory of disease had not been embraced, so TB continued to exact a high price in chronic morbidity and mortality worldwide.[2]

After the discovery of the cause of TB and after the discovery of X-rays by Wilhelm Roentgen in 1895, which provided further proof of active disease, health departments developed and distributed educational materials aimed at preventing it. Also, national and state associations dedicated to the recognition of TB began to focus on its prevention. The associations included, in 1904, the National Association for the Study and Prevention of Tuberculosis (later called the American Lung Association), and its first president and vice president were Dr. Edward Livingston Trudeau and Dr. William Osler, respectively. Both men were giants in the field of medicine, reflecting the importance of the disease in those times. A similar organization, the North Carolina Association for TB Prevention, was founded in 1906. Dr. Watson Rankin, the first full-time state health director, was a member, as were Duke faculty members Drs. David T. Smith, William B. Tucker, and A. Derwin Cooper.

The epidemic spawned what came to be called the Sanatorium Movement, which began in the mid-1800s and lasted for about a hundred years. In that movement, patients with TB were voluntarily sequestered in institutions called "sanatoriums" to receive the particular putative treatment favored by the staff of the facility.[3] The American Sanatorium Association was created in 1905 and served as the principal professional organization (later called the American Thoracic Society). The range of treatments was broad and included the physical location of the sanatorium (sea level versus altitude), the climate (cold versus warm), and the activity level (exercise versus rest). All, or at least most, shared the view that exposure to fresh air and sunlight were critical to a successful outcome. Bacillus Calmette-Guerin (BCG), a strain of MTb isolated from a cow in the early part of the twentieth century, and tuberculin, a protein extract of *Mycobacterium tuberculosis*, were discovered and, for a time, were thought to provide both a protection against infection and a treatment for active disease. Although both proved ineffective, BCG continued to be used in certain circumstances in the United States and is still used abroad (http://www.lungusa.org/site/January 18, 2006).

Dr. H. P. Gatchell established the first American mountain sanatorium in Asheville, North Carolina, in 1871, and Dr. Joseph Gleitsman established a second in 1875.[4] Both men departed after several years, but Dr. S. W. Battle arrived in Asheville in 1885 and provided care for some prominent citizens with TB, including Mr. E. W. Grove, the founder of the Grove Park Inn, and Mrs. W. H. Vanderbilt, the daughter-in-law of Mr. Cornelius Vanderbilt, the industrial magnate and developer of the Biltmore Estate in Asheville. Battle also founded the first North Carolina chapter of the American Red Cross. Besides Drs. Gatchell and Gleitsman, another immigrant, who was named Karl von Ruck, arrived from Germany and pursued the development of a TB vaccine. Indeed, in a paper published in *The Journal of the American Medical Association,* he claimed to have discovered such a vaccine, a claim that was never confirmed.[5]

Although numerous sanatoriums were created around the world, at one point there were twenty-five in Asheville alone, which had become a mecca for the training and the treatment of TB. Ultimately, North Carolina supported four state sanatoriums: one each near the cities of Raeford, Wilson, Black Mountain, and Chapel Hill. Several sanatoriums existed in Durham—including one in northeast Durham on Nancy Rhodes Drive and one on Broad Street, in what became the WTVD television station.

Dr. Edward Livingston Trudeau founded one of the most famous sanatoriums in the United States, located at Saranac Lake, New York, in the late 1885s. As was true for many physicians of the time, Trudeau himself had acquired TB, probably from the care of his brother. After recovering his health, he continued the practice of medicine there and established a laboratory for the identification of the tubercle bacillus, which had been identified by Koch not long before. As described by Anne Davis, the overall value of the sanatorium movement depended on the perspective with which it was viewed.[3] From the perspective of public health, the benefit of limiting communicability was thought to be minimal because the number of patients who could be accommodated in sanatoriums would have been a small fraction of those with active infection. From a physician's point of view, however, confining large numbers of patients in sanatoriums afforded the opportunity to document the natural history of untreated TB, which became an invaluable reference base and provided a captive audience on whom to try diagnostic tests and to test new treatments.

For patients, there were both positives and negatives. On the positive side, a sanatorium provided a sympathetic environment, one that was free from the stigma attached to the disease, also allowing time for self-reflection, learning patience and tolerance, and giving the opportunity to see firsthand successful resolution of the active disease process. On the negative side, confinement for long periods required separation from family and friends and not being gainfully employed. Also, for many others, the structured discipline was challenging. A comparison of the morbidity and mortality of TB during hospitalization with that after discharge is fraught with confounding variables. Certainly there were significant relapses and readmissions,[6] some as high as 50 percent of discharges, but some were less than that discouraging figure.[7,8]

Parallel to the Sanatorium Movement, there was the Public Health Movement, which emphasized better housing and working conditions and likely complemented whatever good the sanatoriums were contributing. The decline in case rates in the general population over the first four decades of the twentieth century were more likely related to improvements in the factors that increased the likelihood of transmission, such as overcrowding, poor housing, better nutrition, and a better understanding of the mode of transmission.

Given this background, what case can be made for Duke's prospective investment in the expertise and resources to address this serious disease? The evidence can be seen as both circumstantial and direct—or merely happenstance. It is difficult to tell whether a specific person or circumstance was the critical factor that prompted another person to embark on TB management or research in the early years. The coexistence in time, space, and field of interest of two or more individuals is, however, at least circumstantial evidence of cross-pollination, if not evidence of direct mentor-to-trainee investment. Additional evidence that the connections were the result of a proactive effort might be found in the transfer of committee chairmanship from one to another, co-authorship of a senior member and a junior member on publications or grants, and active participation in specific projects. I do not claim that a master plan was put in place at the beginning of the century and was pursued systematically to its desired end. There is, however, every reason to think that an organization would want to rally persons to solve a problem, whatever the problem, and to provide an environment in which that could happen. In the case of TB, I believe that happened because of both circumstantial proximity and direct mentorship, not happenstance. A review of the careers of those identified with research on MTb or with the management of patients who had TB will illustrate that improvements were not by happenstance.

The first Duke connection with the disease seems to have been Dr. Watson S. Rankin, who was the first full-time health director of the North Carolina State Board of Health, from 1909 to 1925, and later a professor in the Department of Public Health at Duke University. He was responsible for the state's TB programs[8] and was a member of the organizational group who formed the National Association for the Study and Prevention of Tuberculosis (NASPT) at the time it was founded in 1906.[3] Official accounts of Dr. Rankin's career and accomplishments are sparse, especially so given his numerous and important contributions to medicine, public health, and the Duke Endowment. Born in Cabarrus County in 1879, Rankin received his medical degree from the University of Maryland. He then did postgraduate work at Johns Hopkins under Drs. Welch and Osler and residencies in obstetrics and pathology at the University of Maryland before becoming dean of the two-year Wake

Forest Medical School at the age of twenty-six. In 1909, he joined the State Board of Health as its first full-time health director and, during his tenure, was recognized for three major achievements: first, he organized public health agencies across the state; second, he recognized the importance of vital statistics for public health; and third, he recognized the importance of rural community hospitals in improving public health. He was the recipient of many honorary degrees, a member of many prominent professional societies, including the North Carolina TB Association, and president of the American Public Health Association.

Perhaps his greatest achievement, at least for Duke University, was in his advisory capacity to James B. Duke. Mr. Duke especially relied on him when contemplating and later completing his indenture, in which he committed a fortune for the creation of the Duke Endowment. Appointed as an originating trustee of the Duke Endowment and as head of the Hospital and Orphans Section, Dr. Rankin wrote a monograph titled *The Small General Hospital*, which articulated the vision of this section of the Duke Endowment to build a private hospital or to revise the administrative structure of private hospitals to become community-controlled hospitals.[9,10] Two statements in his obituary in 1970 reflect the esteem in which Dr. Rankin was held. From the North Carolina Board of Health came this: "His administration saw the establishment and growth of county health departments in North Carolina (among the first in the nation). Dr. Rankin was instrumental in placing prevention and control of communicable disease on a more scientific basis." From the Duke Endowment came this: "Dr. Rankin has contributed more to the health of North Carolina than any man in his generation."[11] Also, Dr. Rankin was considered to be of such sufficient importance to the Duke Endowment that a hospital ward in Duke Hospital was named for him. It also seems likely that Dr. Rankin had a strong influence on Dr. David T. Smith, who also had been trained at Johns Hopkins and was also among the first faculty selected for the Duke Medical School and Hospital.

After the identification of the tubercle bacillus, it would not be evident for many years that improvements in the management and treatment of patients with TB occurred because of the contributions of Duke faculty, including Dr. David T. Smith, Dr. William Tucker, Dr. Derwin

Cooper, Dr. Frederick Bernheim, Dr. Carol Dukes Hamilton, and Dr. Jason Stout. The link between those improvements and the early stages of Duke Hospital and Medical School, however, can be explored through the prism of these and other individuals' careers.

The first four of those listed above were born at roughly the same time, at the turn of the last century, and therefore were exposed to the same important historical events such as WWI, the Influenza Pandemic, and the Great Depression. An important incentive for their eventual careers was that they were exposed to an environment in which TB was highly prevalent. Smith and Cooper became infected with the tubercle bacillus, requiring their hospitalization in the Saranac Lake and Black Mountain sanatoriums, respectively, for extended periods before the availability of effective drug treatments. Although all men arrived together at the critical juncture when treatments did emerge shortly before or after WWII, they were minimally, if at all, previously acquainted with one another. They all came to this era by different paths, but all became involved intimately in one way or another in the treatment of patients with TB, and all had definite affiliations with Duke.

Before I discuss their individual connections with events unfolding in the decade from 1940 to 1950, I shall give a chronology of events that will help to demonstrate the contributions of these men. Despite the finding by Paul Ehrlich in 1910 that a chemical could treat the spirochete that causes syphilis and the further discovery by Gerhard Domagk in 1935 that sulfanilamide exhibited an inhibitory influence on the tubercle bacillus, there remained skepticism that an effective treatment for TB could be developed. Not all shared this skepticism, however, and Amberson and others conducted the first clinical trial in 1931, comparing a compound called sanocrysin (sodium-gold-thiosulfate) with bed rest. Regrettably, both interventions had the same outcome, with no real improvement.[12]

A decade or more later, in 1944, two graduate students working on soil microbiology in the laboratory of Dr. Selman Waksman at Rutgers discovered streptomycin, a compound derived from *Streptomyces griseus*, which had inhibitory activity against the tubercle bacillus in vitro.[13] Shortly thereafter, animal studies by William Feldman and Corwin Hinshaw confirmed its inhibitory activity.[14] In November of 1944, at the

Mayo Clinic, its first use in humans demonstrated a cure of a woman from Cannon Falls, Minnesota, with advanced pulmonary TB. The initial dosage of 0.4 grams per day was ineffective in clearing the sputum of the tubercle bacillus, but an increase to 1.2 grams per day resulted in a favorable microbiologic and clinical response.[15] Clinically successful small case studies followed, with the drug provided by Merck, but two problems emerged: an adverse effect on the eighth cranial nerve affecting the vestibular system (vertigo), and the emergence of strains of the tubercle bacillus resistant to streptomycin. These isolates were from patients whose treatment with streptomycin was failing.

Based on the positive clinical benefits, initiatives were set in motion to test this compound in larger clinical trials, first in the United States and later in the United Kingdom. The first trial was accomplished by the joint Armed Forces and Veterans Administration (VA) Cooperative Studies Program, in which seven VA and two armed forces hospitals were identified for participation initially. One VA was in North Carolina (the Oteen VA, now called the Asheville VA). A daily dosage of 1.2 grams per day for 120 days was chosen. Of note, concurrent controls were not used, but the patients' pretreatment course and historical controls of untreated patients were available for comparison. Subsequent studies by the British Medical Research Council and the United States Public Health Service (USPHS) did include controls, but as the first network to address both the safety and efficacy of this new drug, the VA Armed Forces group did not consider the use of concurrent controls justifiable.

Published in 1947, the results in 223 patients were generally positive, with clearance of TB from the sputum in 43 percent of the patients, but of those who remained TB positive, 65 percent had resistant organisms. In addition, toxicity was observed.[16-18] Subsequent studies by the British Medical Research Council[19] and the USPHS[20] more or less confirmed these findings and gave not only a hope of cure to many but also the recognition that monotherapy with this drug had some fairly serious limitations.

Fortunately, progress was being made on other fronts. One discovery was based on a report in *Science* in 1940 by another Duke faculty member, Dr. Frederick Bernheim—that para-amino-salicylic acid (PAS) could competitively inhibit salicylic acid and suppress the tubercle bacillus in vitro and, later, was shown to be active in humans.[21,22] A third and more impressive drug, isoniazid (INH), was discovered and developed independently by Hoffman-LaRoche, the Squibb Institute, and the Bayer Company in 1951–1952.[18] Clinical trials were conducted by all the clinical trial networks just mentioned, using various combinations, dosages, and durations. The ultimate finding was that the two-drug combination of either INH and streptomycin or INH and PAS was equally efficacious. This is not the end of the story about TB treatment; however, it should be sufficiently descriptive to understand better the importance of these events in the lives of the Duke faculty.

Dr. David T. Smith, like most of the other early Duke Medical School faculty, received training at Johns Hopkins (medical school in 1922 and internship in pediatrics in 1922–1923) and at Rockefeller Hospital (1923–1924).[23] Smith married Susan Gower in 1923, and in 1924 he developed hemoptysis (blood in the sputum) and was diagnosed with pulmonary TB. He was initially admitted to the Rockefeller Hospital and subsequently transferred to the sanatorium at Saranac Lake, NY. There, after a period of convalescence, he was allowed to work in the microbiology laboratory, thus beginning his career in TB and MTb research. With that experience and with his TB "cured," he was offered and took a job at the New York State Hospital for Incipient Tuberculosis in Ray Brook, New York, where he worked between 1925 and 1930, supervising the laboratory technicians. His research there focused on non-mycobacterial pulmonary infections that often accompanied, and confused, the diagnosis of TB. His initial focus was on fungal and anaerobic bacterial pulmonary infections.

Having known Wilburt Davison (the first dean of the Duke Medical School) and Harold Amoss (the first chair of the Duke Department of Medicine) at Johns Hopkins, Smith was a logical recruit to come to Duke, which he did in 1930, and he was named professor of the Department of Bacteriology (later Microbiology) and associate professor of the Department of Medicine. There is no indication that he had met James B. Duke or anyone in the Duke family, but he was named a James B. Duke professor later in his career, in 1954. Dr. Smith certainly would have interacted with Watson Rankin, who by this time (1930) would be in the employ of the Duke Endowment overseeing the

Hospital and Orphans Section. In addition to teaching medical students and technicians and serving as a pulmonary and infectious diseases consultant on the wards, he was the acting chair of the Department of Medicine when Dr. Amoss was away. He served in a similar capacity under Drs. Frederick Hanes and Eugene Stead, both chairs of the Department of Medicine, until he stepped down as chair of the Department of Microbiology in 1958. Dr. Smith briefly served as the first chair of the Department of Preventive Medicine from 1963 to 1968 and collaborated with numerous other Duke faculty in the course of his career, including especially Dr. Norman Conant, who later took over his position as chair of the Department of Microbiology. He collaborated more peripherally and on specific clinical issues with Duke faculty Drs. William Johnston, Herbert Saltzman, Herbert Sieker, Hilda Pope Willett, Stuart Bondurant, Robert Abernathy, Raymond Postlethwaite, Albert Spock, and Will Sealy.

In addition to his administrative accomplishments, Dr. Smith published over 150 papers in a wide range of journals, reviews, editorials and books, including multiple editions of the *Manual of Clinical Mycology* and *Zinsser's Textbook of Bacteriology*.[24] In some of these papers and certainly in the textbooks, Dr. Smith had coauthors in related fields, some at Duke. The themes of his papers on TB were diverse. They included aspects of epidemiology related to TB, such as variables inherent in the tuberculin test, called purified protein derivative (PPD), changes in the positivity of the PPD over time, changes between 1930 and 1960 in medical students' tuberculin reactions, the influence of atypical mycobacteria on the skin test, the "booster" effect of repeated tests, and the more common presentation of primary TB in adults as exposures to active cases occurred. In related papers, Dr. Smith proposed intensified public health efforts with yearly PPDs and X-ray examinations for active cases. This seemed like good advice, for deaths owing to TB declined steadily from two hundred per one hundred thousand persons in 1900 to twenty-five per one hundred thousand persons in 1949, before drugs became widely available. Dr. Smith was fortunate that his career spanned the period when Streptomycin, PAS, and INH became available. It is less clear, however, what his involvement was in the clinical trials sponsored jointly by the Armed Forces and VA Cooperative

Studies Program,[16-18] the British Medical Research Council[19], and the United States Public Health Service[20] that resulted in the adoption of specific regimens. He was named as a participant in the annual meetings at the Oteen (Asheville) VA, where progress in these studies was discussed and where his counterpart at the Durham VA, Dr. William Tucker, also would have been a participant.

A separate line of inquiry pursued by Smith and others at Duke led to what was called the "stream flow" theory of blood flow to explain the apical location of reactivated TB in the lungs.[25,26] That line of inquiry was based on the theory that blood from the superior vena cava (into which the thoracic lymphatic duct drained) and inferior vena cava remained in a largely layered conformation upon exit from the right ventricle and therefore preferentially perfused the lung apices with tubercle bacilli coming from a draining infected lymph node. Some experimental evidence was provided to demonstrate that emboli initiated in the lower extremities and entering the circulation through the inferior vena cava became deposited in the lower lobes of the lung.[27,28] With the availability of BCG and new effective drugs, he published papers on their utility in adults and children as preventatives and on their utility in animal models challenged with bovine and human MTb and treated with or without adrenocorticotropic hormone/cortisone and TB drugs.[24]

Smith served as president of the North Carolina Tuberculosis Association in 1944 and president of the National Association for the Study and Prevention of Tuberculosis in 1950, as well as serving on several other local societies. In these capacities alone, he would have worked with Dr. A. Derwin Cooper, who carried on the clinical work at Duke and the VA Hospital after Dr. Smith retired. Dr. Smith was the recipient of many awards, the most prestigious of which were the Trudeau Medal in 1957 (the highest honor conferred by the American Thoracic Society), the James D. Bruce Memorial Medal conferred by the American College of Physicians in 1966 for his work in public health, and his election to the Association of American Physicians. Dr. Smith died in 1981, and he was described by colleagues as a calm, kind, and skilled clinician and teacher. A former chair of the Department of Pediatrics said these words about Dr. Smith: "He "contributed enormously to the foundation and meteoric rise of the Duke University Medical Center."[29]

Bridging Dr. Smith's contributions to the Duke portfolio of accomplishments related to TB was Dr. William Tucker. Born in China to missionary physicians, Dr. Tucker received most of his schooling there before matriculating at Oberlin College. He received his MD degree and house staff training at the University of Chicago, after which he remained as a faculty member, with responsibilities for teaching and caring for patients with TB. He did not acquire TB himself, but at some point, he acquired polio and required a number of operations to allow him to walk with a cane. He was actively engaged in all the early trials of streptomycin, PAS, and INH, especially the trials conducted by the armed forces jointly with the VA Cooperative Studies Program at the Oteen VA as well as the USPHS-sponsored trials at the Durham VA.[16,30-37]

Streptomycin was the first truly promising drug that could be used to treat TB, and Dr. William Tucker played a pivotal role with all the TB drugs licensed for use in humans. Those discoveries bordered on the miraculous, and as a Durham VA-Duke faculty member, he was at the forefront of science as the chair of the Plans Committee of the VA Cooperative Studies Program for eleven years, from 1947 to 1956. Dr. Tucker was also active regionally and in the annual meetings of the TB symposia held at the Oteen VA to discuss new developments and progress. Virtually all his publications dealt with some aspect of TB management or treatment, many of them official publications of the results of research done as part of the VA Cooperative Studies Program, including analyses of the treatments of various non-pulmonary TB; one was an extensive review of the evolution of the Cooperative Studies Program.[16,38] Moreover, Dr. Tucker was chief of the VA Medical Service at the Durham VA Medical Center from 1955 to 1956.[39]

Dr. Tucker certainly depended on Dr. Derwin Cooper as a valuable resource for the management of patients on the TB ward at the VA. After one year as chief of the Medical Service at the Durham VA, he was recruited by VA Central Office (VACO) to first direct the Pulmonary Service at VACO and later, in 1961, the Medical Service at VACO.[38] He served as the first chair of the Cooperative Studies Executive Committee at VACO, appointed on March 11, 1966. Like Dr. Smith, he too was awarded the Trudeau Medal by the American Thoracic Society. He retired from the VA and assumed a faculty position in Gainesville, Florida, as professor of medicine. He died in 1979.

Complementing Dr. Tucker, for they were very much contemporaries, was Dr. A. Derwin Cooper. Dr. Cooper was described as a gentle soft-spoken self-effacing man. After graduating from George Washington Medical School, he did his house staff training in the departments of pathology and medicine at Duke from 1932 to 1934. Then, for a brief period, he went into general practice. Like many other physicians in the late 1930s, Dr. Cooper contracted TB and was hospitalized at the Black Mountain Sanatorium for two years, from 1939 to 1941. Effective drug therapy, as noted earlier, had not yet become available, indeed had not been discovered, but he recovered.

Following his recovery, he assumed a number of positions, including medical director of the Durham County Tuberculosis Sanatorium, assistant director of the Durham County Health Department, and director of the chest clinic there. He joined the faculty in 1953 as an attending physician at the Durham VA Hospital, where he served as a consultant for the inpatient management of TB.[39] In addition to his appointment as a clinical assistant professor of the Department of Medicine at Duke, he was appointed as a clinical associate professor of the Department of Medicine at the University of North Carolina and was an attending physician at the Gravely Sanatorium in Chapel Hill. During his appointment to the VA, Dr. Cooper supervised the Duke interns, residents, and fellows who were serving rotations on the TB ward, which consisted mostly of patients with advanced pulmonary TB but also with TB involving other body parts, including the kidney, pericardium, pleura, bone, and central nervous system. No one in North Carolina was more knowledgeable about the clinical management of patients with TB, and he was an invaluable resource for the Medical Service in the oversight of care provided on the TB ward.

Dr. Cooper was a member of the national and state TB societies. He was president of the North Carolina Tuberculosis Association and a member of the American Thoracic Society, American College of Chest Physicians, and American Society of Internal Medicine.[10] In those various professional roles, Dr. Cooper would have interacted extensively with Drs. D. T. Smith and Tucker and greatly influenced the training of future phthisiologists at Duke, such as Drs. Alexander Spock in pediatrics and Harry Gallis and me in internal medicine. His election to numerous leadership positions, including president of the Durham Rotary and YMCA and chairman of the Durham Chapter of the National Foundation for Infantile Paralysis, and awards for his civic and professional activities for the prevention and treatment of TB are testament to his extraordinary contributions to Duke and to the population of the surrounding area. To honor him, in 1983 the Adult Division of Infectious Diseases and the Pediatric Division of Pulmonary and Allergy established the Derwin Cooper Lecture Series, which has been held thirteen times since his retirement in 1981. Lecturers have included some of the most prominent clinicians and investigators in the field. The names of the lecturers and the years of their lectures are noted as follows:

Dr. Emanuel Wolinsky	1983
Dr. Edward Kendig	1984
Dr. David A. Smith	1986
Dr. Rosalind Abernathy	1989
Dr. Thomas Cate	1991
Dr. Fred Gordin	1993
Dr. Richard O'Brien	1997
Dr. Marvin Pomerantz	1999
Dr. William Burman	2005
Dr. Jeffrey Starke	2006
Dr. Peter Cegielski	2007
Dr. David Ashkin	2008
Dr. Carol Hamilton	2010

Dr. Cooper retired from full-time employment in 1975 but continued as a consultant to the Tuberculosis Program at the Durham County Health Department until shortly before his death in 1984.

Prior to Dr. Cooper's retirement, other Duke staff and faculty became involved in some aspect of TB management and treatment. One of those was Dr. Ron Karpik, who himself contracted TB as a result of his contact with TB patients who were routinely hospitalized on ward

8B at the VA in the late 1960s and early 1970s or who passed through the Duke Private Diagnostic Clinic on their way to one of the Durham or Chapel Hill sanatoriums (on Broad Street or Nancy Rhodes Street) or at Gravely Sanatorium at the University of North Carolina. Of course, with increased appreciation of the transmissibility of the tubercle bacillus in the hospital setting, policies and procedures were put in place to minimize this possibility, but by the mid-1960s, these procedures were not likely to be foolproof. No formal reports of evaluations exist to reflect the risk to hospital staff, but certainly members of the staff were exposed and some infected, whether they became symptomatic or not.

Dr. Karpik's experience is best reflected in his own rendition of the events.[17,40] In summary, Dr. Karpik was a junior resident in the Department of Medicine at Duke in 1967. At the time of his diagnosis with pulmonary TB, his only symptoms were a chronic cough and fatigue, the latter of which he attributed to the demands of his residency. His diagnosis was discovered after a chest X-ray revealed a soft infiltrate in the upper posterior lobe of the left lung and a stained slide of sputum revealed acid-fast bacilli. He was promptly relieved of his clinical responsibilities and referred to the Gravely Sanatorium in Chapel Hill, where he began his twenty-four months of treatment under the supervision of Dr. Derwin Cooper. His treatment included INH, PAS, and, initially, injections of streptomycin. After two to three months, his sputum became culture negative and he returned to the Duke training program, albeit not as a resident but now as a fellow in pulmonary medicine, a subject he then, and even later as a private physician, found engaging. Dr. Karpik's story has been repeated many times in other places, and I am aware of two additional cases of TB in employees of the VA Hospital in the 1970s.

I arrived first as a fellow in 1970 and then as a new faculty member at Duke in 1971. I had trained in internal medicine and infectious diseases in Cleveland with Dr. Charles Rammelkamp, chair of the Department of Medicine, and Dr. Emanuel Wolinsky, chief of the Division of Infectious Diseases, a world expert on mycobacterial disease, and, like Drs. Smith and Tucker, a recipient of the Trudeau Medal from the American Thoracic Society. As chief of the Infectious Diseases Section at the Durham VA Medical Center, I was responsible for the inpatient and outpatient clinical infectious diseases services and was greatly assisted by Dr. Cooper in the oversight of the TB ward, which consisted of forty to fifty patients who had all manner of mycobacterial infections.

By 1973, Dr. Harry Gallis shared the clinical responsibilities through rotations on the inpatient Infectious Diseases Service at the VA, although his primary responsibilities were at Duke. In addition, as the only full-time faculty member in the Infectious Diseases Service at the VA, I chaired the Hospital Infection Control Committee and the Pharmacy and Therapeutics Committee. Responsibilities for these committees included updating and, in some cases, establishing policies and procedures for the management of active TB cases and employee safety from hospital-acquired TB. For the latter, I was charged with evaluation and adoption of new anti-TB drugs, of which rifampin was the newest. I served in similar capacities at Duke, although not as chair. The burden of TB cases at Duke was substantially less because cases of suspected or active disease were referred more routinely to one of the county or state sanatoriums. Supporting these activities were the exceptional clinical microbiology laboratories at Duke and the Durham VA. Notable leaders and staff at these facilities included Drs. Dolph Klein, Gail Hill, and, later, Dr. Barth Reller and others at Duke; and Dr. Peter Zwadyk, Betty Crews, and others at the VA.

TB activities were not confined solely to the adult medical services at Duke and the VA. In fact, Dr. Laura Gutman was responsible for these issues in the Department of Pediatrics and as a representative to the Infection Control Program at Duke. Dr. Gutman came to Duke in 1971 from Seattle, where she'd trained in pediatrics and infectious diseases, with some emphasis on TB through the Department of Epidemiology and the Firland Sanatorium in Seattle. As one of four initial faculty in pediatric infectious diseases, Dr. Gutman emphasized clinical and operational research, particularly that related to TB. Although there were relatively few children with TB, acute disease was managed on Howland Ward in reverse isolation rooms (air pressure in a patient's room being less than on the ward so the air flows into the patient's room, not out of it). Dr. Gutman quickly came to realize that the disease in children is different than it is in adults. It tends to be more rapidly progressive in children, with what seems like a blood-borne illness complicated by TB meningitis or other extrapulmonary disease and almost invariably results in an unfavorable outcome for neurological sequelae. Recognizing this, she advised the relevant county and state health departments and pediatricians to move quickly to identify the children who were in contact with adults newly reported as having TB.

Moreover, a diagnosis of TB in children poses different challenges. Their immature immune systems cause tuberculin tests to be unreliable, and their less vigorous cellular immune responses results in minimal respiratory secretions for examination and a modulated cellular response in the spinal fluid. In addition, she recognized that infected or exposed children who require treatment or prophylaxis are wholly dependent on parents to oversee medication compliance. This is often not seen as a priority by parents once the child becomes less symptomatic. In the early 1990s, she led a study of 14,038 children who were born to mothers infected with the human immunodeficiency virus (HIV).[41] Of these children, 75 were proven to have an active coinfection with MTb, another 40 had asymptomatic disease, and another 71 were thought to merit prophylaxis treatment for TB.

Coinfection with TB and HIV became increasingly common in the early 1990s, when the case rates of TB were unexpectedly increasing secondary to the enlarging epidemic of HIV and the somewhat earlier decline in the quality of the public health infrastructure. Surprisingly, 20 percent of patient isolates and 15 percent of adult-source isolates were resistant to INH and rifampin. These and other issues remained the focus of Dr. Gutman's attention, as reflected in her resume through the mid-1990s, when her career began to focus more on the maltreatment of children. Her successor as the leader of the pediatric TB area is less clear, but according to Dr. Thomas Murphy, a pediatric pulmonologist, the number of cases no longer merits the attention of full-time faculty. As he points out, however, this has left a significant gap in the medical education of future physicians, and as has happened in the past, this could negatively affect the ability to respond to future increases in TB.

Dr. Harry Gallis preceded both Dr. Gutman and me at Duke. He was a Duke Medical School graduate in 1967 and an intern in medicine in 1967–1968, which was followed by an appointment in the Laboratory of

Microbiology at the National Institute of Allergy and Infectious Diseases (NIAID), with Dr. Roger Cole as director. There, from 1968 to 1970, he worked on streptococcal antigens and their role in the pathogenesis of glomerulonephritis. He returned to Duke in 1970 and served in various training capacities, including as a fellow in infectious diseases, until appointed to the faculty in 1973. Although he pursued basic laboratory investigations that were extensions of his work at the National Institutes of Health (NIH), his career evolved as the preeminent consultant for pulmonary conditions for which TB was a possible differential diagnosis. Other conditions included the myriad of fungal infections that were becoming increasingly frequent with the emergence of the fields of solid organ and bone marrow transplantation.[42-44] Recognized as the Duke expert on pulmonary infectious diseases, Dr. Gallis was named as the recipient of many teaching awards and was recruited by the State Board of Health to chair the Advisory Committee on TB from 1983 to 1985. This appointment was likely prompted by his role at Duke, overseeing employee health as it related to TB[45] and his paper on Miliary TB.[46]

Curiously, in 1959, at age fifteen, he was found to have a spot on his lung, and his father, a physician in Athens, Georgia, was sufficiently concerned about the possibility of TB that he had his son, then a high school student, undergo a full workup for MTb, including skin tests and gastric aspirates for culture. Luckily, all the tests for MTb were negative, but his skin test for Histoplasma was, and still is, 4+ positive, making histoplasmosis the most likely cause of the spot on his lung. Eventually, Dr. Gallis assumed responsibility for Continuing Medical Education at Duke and later at the Carolinas Medical Center in Charlotte, North Carolina, where he continued his interest in pulmonary infections. Possibly even greater interests of his are opera and orchids, both nearly lifelong passions. One of the most memorable Medical Grand Rounds at Duke was Dr. Gallis's presentation on "Tuberculosis in Opera."

Dr. Barbara Seaworth's entry into the realm of TB was somewhat of a surprise because of her earlier career interests in infection control, general clinical consultative activities in infectious diseases, and research on viral diseases. Her research on viral diseases included hepatitis C, cytomegalovirus, and HIV. Dr. Seaworth obtained her medical degree at Washington University in St. Louis in 1977, was a house officer at

Wilford Hall, San Antonio, Texas, while in the US Air Force, and was an infectious diseases fellow at Duke from 1980 to 1982. Later she distinguished herself as the director of the TB Education Center for the Texas Department of State Health Services from 1993 to 2005 while serving as a TB consultant for the same government agency. She was appointed the state of Texas's TB Controller in 2000 and as the medical director of the Heartland National TB Center located in San Antonio, Texas, in 2005. Dr. Seaworth served on numerous expert and review panels for various government agencies and in 2010 was awarded the William Stead Award for the Physician of the Year by the National TB Controllers Association. The emergence of multidrug-resistant (MDR) MTb and extensively drug-resistant (XDR) MTb, about which she is considered an expert, occupied substantial amounts of her time and energy in dealing with an outbreak in Texas.[47-49]

The latest TB aficionados to arrive as Duke faculty in the twentieth century were Drs. Carol Dukes-Hamilton, Peter Cegielski, Mark Perkins, Richard Frothingham, Elizabeth Talbot, and Jason Stout. None came to Duke specifically because of a connection with one of the previously discussed faculty, but more likely because of all the prior accomplishments in the field of infectious diseases. Some entered through the Infectious Diseases Training Program and others through the Clinical Microbiology Training Program. Their contributions to the field are perhaps most efficiently discussed in that order, for it reflects the continuum of interest and accomplishment in the field of TB at Duke.

Dr. Carol Dukes-Hamilton obtained her MD degree from the University of Utah, was house staff and an infectious diseases fellow at Duke in the Department of Medicine, and joined the faculty at Duke in 1991. She remained a Duke faculty member for the next seventeen years. During that time, she was an active clinician and a clinical and basic science investigator, progressing to an academic status of full professor. She began her research career by working on the molecular pathogenesis of HIV and published a number of important papers with her mentor, Dr. J. Brice Weinberg.[50-53] Her major research interest, however, was in the field of public health, in particular in the study of TB and TB coinfection with HIV/AIDS, in which she has been active at the local, national, and international levels since the late 1980s. A number of publications resulted from her clinical research work on TB and TB/HIV.[54-56] Two publications resulted from her membership on the Scientific Advisory Committee of the TB Trials Consortium, which was funded by the Centers for Disease Control and Prevention (CDC). She participated in the design, conduct, and analysis of Study 22, a study to assess the benefits of rifapentine used once weekly with INH. The study demonstrated that this regimen was not as effective as a standard twice-weekly regimen using rifampin.[57] In a secondary analysis of the data from Study 22, she showed that the end-of-treatment X-ray findings were highly predictive of disease relapse.[58] She served as the director of TB Control for North Carolina from 1998 to 2009 and chair of its Advisory Committee from 1995 to 2001. In those roles, she consulted on cases statewide, set TB public health policy, and oversaw program review and quality. She is an internationally recognized expert in the management of TB, including TB-HIV/AIDS and multidrug-resistant (MDR) MTb, and has continued to publish the results of her ongoing clinical research on MTb. In 2008, she moved her professional activities to Family Health International, located in the Research Triangle area.

Dr. Peter Cegielski was in the same Duke internal medicine residency class as Dr. Carol Hamilton. He came to Duke after a Harvard BA degree, a University of California at San Diego MD degree, and a one-year primary care internship at the University of Vermont in 1988. He expressed an early interest in infectious diseases (ID) and enrolled in the Duke infectious diseases fellowship program. During that time, he lived and worked in Dar es Salaam at our partner institution, Muhimbili Medical Center, for two years, from 1988 to 1990. While there, he engaged in the clinical management of inpatients, many of whom had HIV or TB, or both. His research at that time reflected the prevalent diseases as follows: he published on the clinical presentations of TB in patients with and without HIV, clinical hypersensitivity reactions in patients with HIV being treated for TB, skin testing for MTb in HIV-infected patients, diagnostic testing for MTb, epidemiology of HIV in Africa, developing criteria for clinical staging of HIV, coinfection with HIV and TB, and non-pulmonary TB.[54,55,59-62]

Subsequently, he worked as an associate professor in the Duke

Department of Medicine until 1994 and then as an assistant professor in the Department of Medicine at the University of Texas in Tyler, Texas. In Texas, he focused on TB, serving as the TB consultant for thirty-five counties in northeast Texas until 1996.[63,64] He then joined the Department of Epidemiology at Johns Hopkins as an assistant professor and devoted his efforts to the support of the Hopkins HIV NET site in Chiang Mai, Thailand, where HIV was the main object of study.

Following that, he embarked on what has become his primary focus on TB for the remainder of his career thus far by joining the Division of TB Elimination at the CDC in 1998. Rising to supervising medical epidemiologist in 2001, his attention has been occupied by multiply resistant MTb (MDR MTb) and extensively resistant MTb (XDR MTb) outbreaks around the world, including in Russia[65] and South Africa. He has published nearly a hundred papers on these topics from a range of perspectives, including epidemiology, policy, prevention, and coinfection with HIV. Because of his publications, he has served on numerous policy work groups and discussion panels, and his prominence in the field is testament to his contributions.

Following a somewhat different path, Dr. Mark Perkins first joined the Duke faculty in 1993 after house staff training (in medicine and pediatrics) and fellowship training at Vanderbilt (1984 to 1988 and 1991 to 1992). In between, he had a two-year stint in the Laboratory of Infectious Diseases/Respiratory Virus Section (Dr. Robert Chanock, director) between 1988 and 1990 and a year as a clinical microbiology fellow at Duke from 1990 to 1991. He was hired as faculty (1993) specifically to codirect a newly created infectious diseases unit in Vitoria, Brazil, at the Federal University of Espirito Santo, in collaboration with Dr. Reynaldo Dietze. Credit for this unique arrangement goes to several parties, but there is no doubt that it was Dr. Perkins's primary inspiration and desire to embark on this new adventure that made it possible. The emphasis of the unit, at least initially, was to do TB research with Dr. Perkins as director of the Mycobacteria Reference Laboratory, which was part of the NIH's TB Research Unit. It was overseen by Dr. J. J. Ellner, then of Case Western University. These activities were pursued productively over the ensuing five years, with numerous publications on clinical laboratory diagnosis[66,67] and molecular techniques both to identify the organism and to follow the course of the disease as a surrogate marker for clinical outcome.[68-72] Setting up a new unit in a foreign country required overcoming countless obstacles, not the least of which was learning Portuguese.

In 1998, Dr. Perkins accepted a position as medical officer for research in the Global TB Program of the World Health Organization (WHO) in Geneva. This was followed by an appointment as manager of the Diagnostics, Communicable Diseases Research Program within WHO for another four years (1999 to 2003). Subsequently, he became the cofounder and chief scientific officer for the Gates-funded Foundation for Innovative New Diagnostics (FIND), where his responsibilities revolved around the identification of novel diagnostics for TB. In this role, he was a member of numerous highly visible and prestigious TB-related advisory committees and published widely in this area. Most of his publications at FIND were understandably concerned with TB diagnostics[73], but seven were concerned with malaria diagnostics. He continued in that role through 2011.

Dr. Richard Frothingham graduated from Duke Medical School in 1982, and in preparation for what he thought would be a career as a medical missionary, he completed a medicine/pediatrics residency at Strong Memorial in Rochester, New York. Immediately after this, he began a three-year commitment in Haiti as a missionary, accompanied by his wife and three children. In this capacity, he saw many cases of TB, some with and some without coinfection with HIV. After two to three years in Haiti, he concluded that he wanted to pursue a more academic career and entered the Infectious Diseases Fellowship Program at Duke in 1990, during which he worked in the Molecular Ecology Laboratory, overseen by Dr. Ken Wilson at the VA and supported by the NIH Interdisciplinary Research Training Program on AIDS.

Using molecular techniques, he attempted to identify unculturable microorganisms from patients with suspected infections, techniques that later were applied to mycobacteria and led to the development of high-throughput sequence-based differentiation of typical and atypical mycobacteria.[74-79] He was appointed to the faculty as an associate in the Department of Medicine in 1993, and over the next fifteen years, he published over thirty papers on basic or clinical research aspects of mycobacteria. While working on these laboratory projects, he has served as a member of the North Carolina State Board of Health's Medical Advisory Committee on TB for the past ten years. He also served as the laboratory mentor of an infectious diseases trainee (Dr. Elizabeth Talbot) and has progressed academically, achieving the rank of associate professor with tenure.

Dr. Elizabeth Talbot was an intern and resident in the Department of Medicine and a fellow in Infectious Diseases at Duke from 1993 to 1998. As a fellow in the Mycobacterial Genetics Laboratory with Dr. Frothingham, she began her career with a specific interest in TB and related subjects. This was a period when she also worked with Dr. Perkins in Brazil for four months. She was productive while working in the laboratory on clinical diagnostics,[80] on molecular studies of vaccines for *M. avium*, on the use of PCR to identify *M. bovis* BCG, on studies of the pyrazinamidase gene,[81-82,83] and on studies of multidrug resistance domestically and internationally.[84-86] With her background, it was not a surprise that she was accepted as an officer of the Epidemiology Intelligence Service at the CDC and assigned to its International Activities Division of the Program on TB Elimination from 1998 to 2000. At the end of her two-year commitment, she was appointed associate director of the TB/HIV Research Program in Botswana, where she remained for another three years (2000 to 2003).

CONCLUSIONS

Duke's contributions to the prevention and treatment of TB are considerable in the twentieth century and are due to the recognition that TB is a serious problem, adversely affecting the health of the residents of the state. Its contributions are also due to the increasing support of the health care and public health infrastructure of the state. This has happened neither by chance nor by a grand strategy that was put in place at the outset. It is the result of sequential investments on the part of individuals and programs to confront what was, and still is, a serious problem in the world. Sometimes a direct link can be identified between succeeding generations of professionals, but just as commonly, opportunity attracts further engagement, as happened in the response to this disease in this country and abroad. Whatever the explanation, Duke students, staff, and faculty have stepped up to the challenges that TB poses. The challenges persist, however, in the form of unpredicted resurgences of new cases that occurred in the mid-1980s and early 1990s, due in large part to the emergence of epidemic HIV infection; the enormous burden of morbidity and mortality especially in the developing world; and the limitation of sufficient resources, diagnostic tests, drugs, and vaccines. In addition, there is the emergence of serious drug resistance and the predilection for TB to find sanctuary in hard-to-reach populations. Fortunately, a new cadre of persons interested in responding to these new challenges has emerged since the end of the twentieth century. When this chapter is written at the end of the twenty-first century, the names of Drs. John Crump, Jason Stout, David Holland, Elizabeth Reddy, and others will likely figure prominently.

Photograph 3.2 is a composite of pictures of six prominent figures in the field of tuberculosis in the twentieth century.

David T. Smith

William B. Tucker

A. Derwin Cooper

Carol D. Hamilton

Mark D. Perkins

Barbara J. Seaworth

CHAPTER 4

HUMAN IMMUNODEFICIENCY VIRUS

From first symptoms to death in six years at age twenty-eight, the case of MW was not unusual in the early years of the HIV epidemic, nor was his migration to the big city unique for an admitted or closeted gay man for whom sexual encounters were both frequent and unconventional. Those encounters often led to an explosion of new and nearly always fatal infections. That, of course, was not the only way in which the then-unknown virus could be acquired, but MW's response was typical: denial of sexual orientation or risk group; identification with an accepting peer group; indiscriminant gay or heterosexual sex or drug use; insidious onset of symptoms and denial of their potential implications; late presentation to a health-care provider because of stigma and fear; extremely limited therapeutic options; often rejection by loved ones and, if fortunate, reconciliation with time; onset of opportunistic infections; progressive onset of wasting; and, in short order, premature death. Although such a scenario no longer prevails in most of the developed world, cases like MW's tragically persist in many, if not most, resource-limited countries.

—John Hamilton, 1995

This chapter will describe Duke's contribution to the understanding and management of HIV/AIDS—a rich history indeed. At some level, all the clinical department chairs at Duke from 1930 to the mid-1950s had some role in the evolution of what became the response to HIV. The department chairs were Deryl Hart and Clarence Gardner (surgery); Harold Amoss and Frederick Hanes (medicine), Wilburt Davison (pediatrics); Robert Ross, E. Bayard Carter, and Roy Parker (obstetrics-gynecology); and Richard Lyman, Ewald Busse, and Keith Brodie (psychiatry). Increasingly, however, subsequent chairs played more direct roles in the response to HIV: James Wyngaarden, Joseph Greenfield, and Barton Haynes (medicine); Jerome Harris, Samuel Katz, and Michael Frank (pediatrics); David Sabiston and Robert Anderson (surgery); Charles Hammond and Haywood Brown (obstetrics-gynecology); and Allen Frances and Ranga Krishnan (psychiatry). Some of them played roles through policy decisions, some through faculty appointments, and some through financial support. Several basic science chairs also played important roles in the fourth quarter of the century: Wolfgang Joklik and Jack Keene (microbiology) and D. Bernard Amos and Thomas Tedder (immunology).*

EARLY CONTRIBUTIONS

RECRUITMENTS

Among the more important contributions was the recruitment of Dr. Joseph Beard. Beard, frequently described as "difficult," trained in surgery at Vanderbilt and, soon after that, worked at the Rockefeller Institute under the supervision of Dr. Peyton Rous, known for his work on tumor-associated viruses (the Rous sarcoma virus) and for being among the first to recognize the connection between viruses and cancer in animals and man.[1,2] Beard also became committed to the study of virus-associated tumors. In the early 1930s, Beard discovered the avian myeloblastosis virus (AMV), a cause of cancer in chickens.

---

* Definitions:
  HIV denotes infection by the human immunodeficiency virus. AIDS denotes the acquired immunodeficiency syndrome, a diagnosis applied when the HIV infection progresses to a state in which CD_4 cells in the immune system fall below two hundred, resulting in immunodeficiency, or when an opportunistic infection or an unusual cancer occurs.

Experience at the Rockefeller Institute stood him in good stead for future work on viral vaccines, in particular on a vaccine for a virus that was posing a serious threat to the thoroughbred horse population in the United States. This virus, the eastern equine encephalitis virus (EEE), caused a fatal infection in horses (commonly called the "blind staggers") and seemed epidemic in the United States. The virus was transmitted to horses by the mosquito, with some intermediate hosts. Beard had developed a technique to purify the virus and to inactivate it so that it could not replicate but still stimulate an immune response when injected into animals (and eventually into humans). This inactivated virus would not treat, but it would prevent a new infection with EEE. Somewhat surprising, there was a significant market for such a product because it was an early version of an effective, protective, killed vaccine. The pharmaceutical firm Lederle, a subsidiary of the American Cyanamid Company, was interested in its application and initiated a collaboration with Beard that supported much of his research for many years. The vaccine prepared according to his specifications proved highly effective and halted the epidemic in horses somewhere in the mid-1930s.

Riding a wave of quick successes, Dr. Beard and his wife, Dorothy, who was also his laboratory partner, were recruited to Duke by the then-chairman of surgery, Dr. Deryl Hart. Beard was made a professor of experimental surgery. How this recruitment proceeded is not altogether clear but likely involved referrals from the Rockefeller Institute and the prospects of the infusion of commercial dollars. Whatever happened, Beard was not one to sit by idly. With funds from Lederle, he and his wife established the Dorothy Beard Research Foundation "to develop and systematically apply methods for the physical, chemical, morphological, and biological characterization of viruses causing disease in animals, most recently tumor viruses." Beard's work with avian tumor viruses identified unique strains that were able to cause distinctive pathologies for which later specific oncogenes with implications in human cancers were identified—e.g., avian myeloblastosis (myb), avian erythroblastosis (erb), and avian myelocytomatosis (myc). Notably, the laboratory produced and marketed AMV-derived reverse transcriptase, an enzyme with enormous value at the time for basic research and even more valuable later when AIDS emerged. Beard was, it seems, the default source for this enzyme. Whatever the revenue from it and its ultimate patent and largesse from Lederle, it seems all for the good of science, not for Beard's own personal worth. It is not surprising, then, that at that time in the century, he was one of very few virologists who was a basic science investigator, especially one with a molecular focus, and certainly one of the few concerned with tumor-associated virology. He was, therefore, in my opinion, a *giant* in the field. In words from a Duke icon, Dr. Eugene Stead, Jr., Beard was "one of the great men of Duke."[2]

Naturally, Beard was a member of many scholarly societies and national committees and was the recipient of many awards. He was closely connected with the National Cancer Institute, the National Institutes of Health (NIH), and the American Cancer Society. Dr. Beard accepted Dr. Dani Bolognesi as a postdoctoral fellow in 1965, who later succeeded Beard as head of experimental surgery when Beard left Duke in 1973. After leaving Duke, Beard formed and led the Life Sciences Company in St. Petersburg, Florida, to make and distribute reverse transcriptase until his death in 1983. Because he was the initial champion of viral diseases and vaccines at Duke, there could hardly have been a better choice for an early faculty member.

THE RESEARCH TRAINING PROGRAM

In the 1950s and 1960s, Duke Medical Center began to emerge as an interesting and, in some ways, a unique training site for aspiring PhDs and MDs in the basic and clinical sciences. In the early post–WWII period, scientific inquiry was a growth industry nationally, and Duke capitalized on this wave of opportunity. Among other advances, the Research Training Program (RTP) was initiated in 1960, largely because of the vision of Dr. Philip Handler, the chair of the Department of Biochemistry. With a focus on the advanced training of physicians (and some postdoctoral students) in the basic sciences, Dr. Eugene Stead, the chair of the Department of Medicine, appointed Dr. James B. Wyngaarden to head this program, and he appointed six other basic science-oriented associate professors to serve as core faculty. They were Walter Guild, PhD (biophysics), Samson Gross, PhD (genetics), Kenneth McCarty, PhD (biochemistry), Montrose Moses, PhD (anatomy), Daniel Tosteson, MD (physiology), and Salih Wakil, PhD (biochemistry). Dr. Wyngaarden, who had trained with Dr. Dewitt Stetten at the NIH, had come to Duke three to four years earlier with an appointment in biochemistry. Drs. Handler and Wyngaarden described the program more fully in a publication in the *Journal of Medical Education,* which confirmed their intention to nurture the development of skills in medical students and physicians and selected postdoctoral trainees so that they might become independent clinical investigators.[3] Success of this "experiment" was more apparent for the postdoctoral trainees than it was for the medical students and junior physicians, seemingly because of the internal pressures for physicians to engage in clinical work. Nonetheless, as an experiment in medical education, the program was deemed a success. It ultimately contributed to the change in the Duke curriculum that led to a full year (the third year) of research in the four-year curriculum and possibly led to the creation of the Medical Science Training Program (MSTP) for MD and PhD candidates. Obviously, HIV/AIDS was not a topic of research within the RTP, but the fundamentals of science were to become the tools for such investigations later.

When he became chair of the Department of Medicine in 1967, Dr. Wyngaarden directed a research-driven academic department that nurtured would-be basic science-oriented MDs to conduct fundamental research in the course of their academic careers. The basic science emphasis changed over time, with recognition that clinical research had become a rigorous discipline, one that was a necessary complement to basic science. That point became obvious when the HIV/AIDS epidemic began. Dr. Wyngaarden was visionary in that regard when he published a paper in which he sounded the alarm that the physician investigator was becoming an "endangered species," a trend that would handicap the response to the epidemic that was to come.[4] After Dr. Wyngaarden became director of the NIH in 1982, he oversaw and personally initiated some major activities with important implications for HIV/AIDS.

FACULTY MEMBERS IN THE ADULT INFECTIOUS DISEASES SECTION

In the 1960s, the Department of Medicine had only a modest commitment to most subspecialties, including infectious diseases, but had supported

Dr. Suydam Osterhout, a recent chief resident (1955–1956) and nephew of Wilburt Davison, to train for several years at the Rockefeller Institute and then to join the Duke faculty in the Department of Medicine. With his training in microbiology, Osterhout was also appointed as the director of the Clinical Microbiology Laboratory at Duke. Infectious diseases was assuming an increasing clinical role in the aftermath of the Hong Kong flu epidemic in 1968 and the rise of other subspecialties, including the pulmonary, renal, and gastrointestinal divisions (Wyngaarden, personal communication). Although infectious diseases had not yet become a division at Duke and was not yet a board-certifiable discipline by the American Board of Internal Medicine, an Infectious Diseases Service was created in 1970. Until the Infectious Diseases Division was created in 1977–1978, the Infectious Diseases Section resided administratively within the Pulmonary Division.

Dr. Wyngaarden recruited Dr. Thomas Cate in 1968, bolstering the Infectious Diseases Service. Dr. Cate came to Duke with a background in influenza research at the NIH, where he worked in the laboratory with Drs. Vernon Knight, Robert Chanock, Paul Gleeson, and Robert Couch. He continued this research, which was funded by the NIH, during the eight years he remained at Duke while serving, along with Dr. Osterhout, as an infectious diseases consultant at both Duke and the VA hospital. Dr. Cate left Duke in 1978 to join the Influenza Research Center located at Baylor in Houston.

I was appointed to the faculty in 1971 to head the Infectious Diseases Section at the Durham VA Hospital. Previously, I had completed a one-year fellowship (1967–1968) in infectious diseases in Cleveland, Ohio, under Dr. Emanuel Wolinsky, a world authority on mycobacteria; served two years as an epidemiology intelligence service officer of the Centers for Disease Control (CDC) in Atlanta, Georgia; was assigned to the State Board of Health in Raleigh, North Carolina (1968–1970); and completed one further year as a fellow at Duke (1970–1971). Between 1970 and 1977, I attended on general medicine and provided clinical consultations on the tuberculosis and general medical wards. I increasingly spent time on research related to viral diseases; the NIH and the VA funded that research. In particular, an evolving outbreak of hepatitis B in dialysis patients, staff, faculty, and students at both Duke and the VA occupied much of my time in the 1970s, until an effective vaccine became available in the early 1980s. In addition, I spent a one-year sabbatical in the Netherlands, working in the immunogenetics laboratory of Dr. Jon J. van Rood (1977–1978). Because of the development of an expanding kidney transplantation program at Duke and the VA, cytomegalovirus emerged as a serious complication. The necessary immunosuppression by drugs that was then required to prevent loss of the transplant allowed cytomegalovirus to activate and to cause life-threating infection as well as rejection of the transplanted kidney. This virus also became the primary focus of much of my later research career and served as an entrée to the AIDS epidemic.

Additional specialists joined the infectious diseases faculty from the Duke Department of Medicine house staff, including Drs. G. Ralph Corey and Harry Gallis. Both men became exemplary clinicians, and the former became director of the house staff training program in the Department of Medicine, a position he held for twenty years.

FACULTY IN PEDIATRIC INFECTIOUS DISEASES

The new chair of the Department of Pediatrics, Samuel Katz, was appointed in 1968 and came with substantial personal expertise in virology, having trained in the laboratory of Dr. John Enders at the Boston Children's Hospital. Dr. Enders was a corecipient of the Nobel Prize for his work on the successful in vitro cultivation of the poliovirus. Dr. Katz did research on measles and helped to develop a measles vaccine. Later he was a member of the Advisory Committee on Immunization Practices, which provides guidelines for the immunization of adults and children.

In addition to these contributions, Dr. Katz brought three additional pediatric faculty from Boston, all of whom had an interest and accomplishments in virology and also had trained in Dr. Enders's laboratory. One was Dr. David Lang, who became chief of pediatric infectious diseases. Dr. Lang also had a laboratory interest in cytomegalovirus, which as previously stated was found later to be a serious complicating infection in patients with HIV/AIDS. The second pediatric faculty recruited by Dr. Katz was Dr. Catherine Wilfert, whose virologic expertise served

Duke well in her role as head of the Viral Diagnostic Laboratory. Prior to the HIV/AIDS epidemic, she published over thirty papers, primarily on viral and rickettsial diseases. Much later, in 1996, after retiring from Duke, she was recruited to be the scientific director of the Elizabeth Glaser Pediatric AIDS Foundation (EGPAF). The third recruit from Boston was Dr. John Griffiths, a pediatric neurologist with expertise in the herpes viruses. A fourth recruit, Dr. Laura Gutman, with experience in tuberculosis, came from Seattle. Each brought skills that later became relevant for patients with HIV/AIDS.

BASIC SCIENCE FACULTY

Subsequent recruits to Duke with indirect implications for infectious diseases and HIV were two basic scientists. They were D. Bernard Amos, MD, a scientist in the evolving field of immunogenetics and transplantation, and Wolfgang (Bill) Joklik, PhD, a classical basic science virologist. They came to Duke in 1960 and 1968, respectively. Amos, a native of England, came to the United States after graduating from medical school in 1951 and working with Peter Gorer in the laboratories of Peter Medawar and Peter Snell from 1952 to 1955. Initially working on tumor antigens in mice at the Roswell Park Cancer Institute, Amos became increasingly interested in human transplantation and the immunologic response to it. Drs. Phillip Handler and Barnes Woodall recruited him to Duke in 1960 to set up the Immunogenetics Research Group, which ultimately was integrated into the Department of Microbiology, then headed by Dr. Norman (Bill) Conant. Dr. Amos was inclined to come to Duke because of the "air of tremendous potential,"[5] which he came to realize as true as evidenced by his subsequent experiences. Amos was one of perhaps ten pivotal scientists to discover the human leukocyte antigen (HLA) system, which had such profound influences on solid organ transplantations and, later, on bone marrow transplantations and infectious diseases. He was elected for membership in the National Academy of Sciences.

Dr. Delford Stickel performed the first human kidney transplant at Duke in 1965, and the scientific basis for this revolutionary treatment has made ever more dramatic interventions possible. So too have come dire infectious complications, many of which have been partially ameliorated over time but have posed serious challenges to the adult and pediatric infectious diseases faculty. The discovery of the HLA system by Amos and others opened vast new fields of inquiry that reflect upon and affect the human response to animate and inanimate objects.

Dr. Joklik, an Austrian by birth who grew up in Australia, was appointed as the chair of the Department of Microbiology and Immunology after working in many prestigious departments around the world.[6] He was prolific with seminal publications on reoviruses[7-14] and poxviruses[15-17] and notably for his publications on the avian sarcoma virus, a retrovirus.[18-23] Although Dr. Joklik was not involved subsequently on studies of HIV, the Department of Microbiology expanded dramatically under his leadership as chair of the department and provided expertise to many in the field. He also was the leading force in the successful application for a new regional cancer center, which was later to assume considerable importance in the response to the AIDS epidemic. He served on many prestigious committees and study sections and was elected as the first president of the American Society for Virology in 1982.

The infusion of infectious diseases talent more than doubled the faculty in this field at Duke. The generous support by the more senior basic scientists—Joklik, Amos, Katz, and Wyngaarden—to the newer faculty members created a formidable team to confront an ever-increasing array of epidemic and endemic viral, bacterial, and, later, fungal diseases. Indeed, in a few short years, Duke went from a faculty with almost no expertise in a discipline that dominated other prominent medical schools to one that was on a par with many of these schools. Clinical collaborations between the adult and pediatric teams also flourished, with pediatricians attending on the adult Duke and VA consult teams and vice versa. Research collaborations increased later when HIV emerged, but even in the 1960s and 1970s, the Department of Surgery fostered creative research in basic science laboratories by surgery residents who often worked in the field of immunology with Dr. Amos and also with Dr. Joseph Beard, in the experimental surgery section, on animal and human tumor viruses. The chair of the Department of Surgery, Dr. David Sabiston, also recruited additional PhD faculty to lead major programs related to infectious diseases.

THE CLINICAL MICROBIOLOGY LABORATORY AND THE VIRAL DIAGNOSTIC LABORATORY

The late 1960s and 1970s were times when the Clinical Microbiology Laboratory and the Viral Diagnostic Laboratory began to respond to a sense of heightened expectations on the part of clinicians. Newly available tests for infections not heretofore recognized demanded an expansion of the technologic capacity of these independent laboratories to support the expanding programs for solid organ transplantations (kidney initially), cancer chemotherapy, and immune defects. Standard bacteriology was reasonably prepared for these new demands, but anaerobic microbiology was a newfound discipline, at least in its application to human disease. It found a champion in Dr. Gail Hill, a trained anaerobic bacteriologist in the Clinical Microbiology Laboratory who worked alongside the original director, Dr. Suydam Osterhout, and his successor, Dr. Dolph Klein. The erstwhile arcane discipline of mycology was brought to the fore because of the interests and expertise of Dr. David T. Smith and, even more so, by that of Dr. Norman Conant, a world-renowned basic and clinical mycologist who had been the chair of the Department of Microbiology for a number of years. Dr. Conant's expertise became increasingly useful as new and strange fungal diseases were uncovered with the growth of solid organ and bone marrow transplantations, aggressive cancer chemotherapy, and ultimately HIV/AIDS.

When Dr. Barth Reller became the director of the Clinical Microbiology Laboratory at Duke in 1988, the laboratory received a boost in technical rigor and a newfound role in training physicians destined for careers in clinical microbiology. Before coming to Duke, Dr. Reller trained at the University of Washington in Seattle under Dr. John Sherris and then served as the laboratory director at the University of Colorado for fifteen years. To his credit, a board-eligible and board-certified training program was established at Duke in medical microbiology and served as a training program for fellows specializing in adult and pediatric infectious diseases.

The following individuals played further critical roles in the laboratory: Drs. John Perfect, Barbara Alexander, Aimee Zaas, and Kimberly Hanson in the Department of Medicine; William Steinbach and

Harmony Garges in the Department of Pediatrics; and Thomas Mitchell, Joseph Heitman and John MCusker in the Department of Microbiology. Although anaerobic organisms and fungi had been recognized as pathogens previously, the capacity to detect and identify these organisms in patient specimens was dramatically increased owing to the expertise of these individuals.

The same can be said of the Viral Diagnostic Laboratory, which was overseen by Dr. Catherine Wilfert and later joined by Dr. Laura Gutman. Technicians and trainees named Carolyn, Clementine, Nancy, Ed, and Mary supported the laboratory work. It may be obvious, but the capacity to actually isolate and identify a virus in a patient with a particular illness made it likely that more patients infected by the same virus would be identified and, when appropriate, interventions initiated or at least sought. In addition, many of the same conditions that prompted the need for more sophisticated bacteriology and mycology also prompted the need for additional sophisticated viral diagnostic tools and expertise. Over subsequent decades, that need has dramatically increased. Among the earliest molecular tools was immunofluorescence antibody staining of clinical specimens for specific identification of viruses—in real time. Others are even more astounding, and those will be discussed as we move into the later part of the twentieth century. Also, there were some rough roads ahead with changes in personnel, authority, and budget issues, but there remained the need for this incredibly important technology for the clinical management of patients.

In parallel, the traditional career pathway in science flourished at Duke, particularly in the field of infectious diseases, with the recruitment of Dani Bolognesi, PhD, an individual who would play a major role in the response to the HIV virus ten to fifteen years later. Dr. Bolognesi's entry into the retrovirus field began with his PhD work under Dr. Joseph W. Beard at Duke University (Bolognesi, personal communication). Dr. Beard's interests centered on the pathogenesis of avian RNA tumor viruses (parenthetically the viruses isolated by Beard were later found to contain cognates of human oncogenes such as myc, myb, erb, and so forth). His aim was to understand better the composition of that class of agents. Dr. Bolognesi's work focused on the molecular nature of the RNA species contained in virions, and he was one of the first to identify the high molecular weight of parental RNA in these particles.[24] He did his postdoctoral work at the Max Planck Institut in Tubingen, Germany, with Professor Heinz Bauer, a noted avian retrovirologist. His work there elucidated the nature of various morphological structures of the virus, including the proteins associated with the core as well as the spikelike structures containing the viral glycoproteins.[25-27] On his return to Duke to succeed Dr. Beard, Dr. Bolognesi turned his attention to mammalian RNA tumor viruses. His work continued on virion structure and composition, which led to a model of an RNA tumor virus published in *Science*.[28] Most of his attention, however, turned to evaluating the potential of virion proteins and glycoproteins to serve as vaccine immunogens to protect against infection and tumorigenesis.[29-31] This work was carried out in close collaboration with Professor Werner Schafer of the Max Planck Institut. This work led to the next stage of Dr. Bolognesi's career, when he entered the AIDS field and established a major effort at Duke to develop its HIV vaccine program, which continues today under the able leadership of Dr. Barton Haynes.

Events apart from Duke independently served to set the stage for later developments, including the important discovery of the human T-cell leukemia virus (HTLV-1), the first recognized human retrovirus by Robert Gallo and others at the National Cancer Institute in 1980.[32] Dr. Barton Haynes returned to Duke from the NIH in 1980, and in the course of evaluating a patient from Japan with a T-cell lymphoma, he collaborated with Dr. Bolognesi and Dr. Richard Metzger and made use of their laboratory facilities. Not long thereafter, he successfully isolated the causative virus from that patient and, in so doing, documented only the second case of human T-cell leukemia caused by HTLV-1.[33]

Dr. Bolognesi had recruited other basic scientists to his laboratory, including Drs. Kent Weinhold, Thomas Matthews, and Alphonse Langlois, in the early 1980s, but after the recognition of AIDS and the isolation of the human T-cell leukemia virus-III (HTLV-III) by Robert Gallo and the lymphocyte-associated virus (LAV) by Luc Montagnier, which were done at the National Cancer Institute and the Pasteur Institute, respectively, contact became more frequent among all the Duke and NIH retrovirus investigators. Although not a member of the Infectious Diseases Section, Dr. Barton F. Haynes trained in medicine at Duke and then joined the Laboratory of Clinical Investigation, which was under the direction of Dr. Sheldon Wolff at the NIH in 1975. Another member of that laboratory and Haynes's mentor was Dr. Anthony Fauci, who later was to become head of that laboratory and, even later, the director of the National Institute of Allergy and Infectious Diseases (NIAID).

While at the NIH, Dr. Haynes became board certified in both infectious diseases and allergy and in clinical immunology, with his research focus on immunology, particularly on B cell biology. This experience led him to studies of Wegener's granulomatosis, then in the provenance of the rheumatology division at most major medical centers, including Duke. For that reason, he was recruited back to Duke in 1980 as a rheumatology faculty member under the division chief, Dr. Ralph Snyderman. Haynes became chief of the Division of Rheumatology in 1985. When Haynes arrived at Duke, he began his work—with Dr. Louise Markert, who led the effort—on human T-cell ontogeny and thymus development, which culminated in the development of a cure for the DiGeorge syndrome by transplanting human thymus.[34,35] Together, Haynes's work on T-cell biology and T-cell lymphomas coupled with his work on HTLV-1 primed him for the work to come on the etiology of AIDS and later work on vaccine development for HIV-1.

David Barry, MD, later a pivotal player in the drama to unfold after the recognition of the gay-related immune deficiency (GRID) syndrome, relocated to the Burroughs Wellcome (BW) Company in the Research Triangle Park, North Carolina, from the Bureau of Biologics at the Food and Drug Administration in 1977, where he had been since 1972.[36-39] Dr. Barry had been acting deputy director of the Division of Virology and director of the Influenza Task Force in the period when a swine flu epidemic was predicted in 1976. At BW, he was the head of the Anti-Infectives Section, where he oversaw the development of acyclovir, an antiviral agent active against some of the herpes viruses. This agent was arguably the first effective antiviral agent in an era when viruses were only beginning to be considered treatable. Soon after his arrival at BW, Dr. Barry was appointed to the Division of Infectious Diseases at Duke as an adjunct assistant professor of medicine. His principal role was to attend the weekly infectious diseases clinic at the Durham VA Medical

Center and the Duke Infectious Diseases Case Conference, which he did for five to six years. His later involvement revolved around the identification, by in vitro and in vivo testing, and the licensure of the first effective antiretroviral drug.

CREATION OF THE ADULT DIVISION OF INFECTIOUS DISEASES

In 1978, Dr. David T. Durack was recruited as the first chief of the newly created Division of Infectious Diseases at Duke. Dr. Durack, a native of Australia, came with impressive credentials. He had been a Rhodes Scholar at Oxford, where he worked on endocarditis with Dr. Paul Beeson and, later, with Dr. Robert Petersdorf, both legends in infectious diseases. Instead of returning to Australia as originally planned, he accepted an appointment as the chief resident in medicine at the University of Washington in Seattle, where Dr. Petersdorf was then chairman. There, Dr. Durack continued his investigations on endocarditis in animal models as a junior faculty member from 1974 to 1977. In 1977, Dr. Wyngaarden recruited him to head the new Division of Infectious Diseases at Duke, which he led until 1994. During his seventeen-year tenure, he was academically productive, with over one hundred publications, mostly on endocarditis and meningitis. His publications included the clinical and epidemiologic studies on endocarditis that further refined the diagnostic criteria for that infection. Those criteria led to the establishment of the "Duke criteria," a reference used even today.[40-46]

With the acknowledgment by the American Board of Internal Medicine in 1974 that infectious diseases was a board-certifiable subspecialty and with Dr. Durack's arrival, the Duke Infectious Diseases Division formally became a training venue for would-be infectious diseases subspecialists. Dr. John Perfect became the first official fellow of the division, but before the division was established, Drs. Jack McCloskey, Harry Gallis, Ralph Corey, and I completed requirements in the Duke Infectious Diseases Section for what would become the requirements for board eligibility.

Other faculty in the departments of medicine and pediatrics were already at Duke and engaged in research that would later be relevant to address the AIDS epidemic, including Drs. Ross McKinney and Brice Weinberg.

THE HIV EPIDEMIC

The first reports in 1981 of a unique syndrome,[47,48] known shortly thereafter as the gay-related immune deficiency (GRID) syndrome, and subsequent publications[47,49-51] ushered in an era dominated by HIV/AIDS, which extended well beyond the end of the twentieth century worldwide. The complexity of the disease and the attendant societal issues proved to be extremely challenging. Current and former Duke faculty and personnel were involved extensively in new developments, whether medical, social, or political. What made this period even more challenging was that many of those developments were occurring simultaneously. A discussion of the developments may give the impression that they were occurring in isolation, but they were not.

From the clinical perspective, patients with the symptoms associated with GRID began to appear in the Duke and VA infectious diseases clinics in the months following the initial reports and somewhat later in the Duke pediatric clinic, perhaps in 1985 or 1986. Symptoms and signs typical of GRID included unexplained fever, weight loss, lymphadenopathy, Kaposi's sarcoma lesions on the skin and elsewhere, other unusual cancers and opportunistic infections. In children, the usual presentation was *Pneumocystis carinii* pneumonia (PCP), which in a child less than one year old was rapidly fatal. Other presentations in children included oral thrush, developmental delays, and wasting.

The Duke clinic's attending staff and fellows before 1985 included Dr. David Durack, Harry Gallis, Donald Granger, Charles Ellenbogen, Mimi Cameron, and Gunther Lallinger. The VA clinic included Duke faculty Drs. David Barry, Mary Klotman, and me, as well as infectious diseases fellows and medicine residents. The Duke pediatric clinic included Drs. Samuel Katz, Catherine Wilfert, Cornelia Dekker, Sandra Lehrman, John Frank, Ross McKinney, and Robert Drucker. The recognition that we were experiencing the first wave of a frightening epidemic, the cause of which was unknown, raised major concerns among many constituencies. As the face of the disease became better known, it seemed that those infected came from a distinct population, namely sexually promiscuous gay men. Immediately, those men were stigmatized, and overt hostility or disguised disapproval was applied. As the epidemic progressed, it became apparent that other at-risk groups were becoming infected, and the stigma that had been uniquely applied to gay men was now extended to them. The hostility and disapproval came from many sources, including politicians, health-care providers at many levels, religious leaders, and others. One illustration was a quote by columnist James J. Kilpatrick in the *Washington Post* in 1987: "One can weep for the children, but it is hard to work up much sympathy for the sodomists and addicts who have brought this on themselves."[52] To be sure, there were diverse opinions about how best to respond to the threats posed by the epidemic and to individual patient management, as well as how to abort the epidemic. Suffice it to say, it was a very stressful time for patients, their families, health-care personnel, loved ones, and public health officials. The unknowns accompanying the epidemic were perhaps the major source of anxiety.

The clinics for adults at Duke and at the VA were similar in many respects, with both clinics providing care for patients who had a complex new disease for which there were no specific treatments. The VA, being a government facility, had the advantage of financial resources and of a commitment to eligible veterans to provide them with state-of-the-art health care that had no, or at least minimal, financial burdens, whereas the Duke clinic did not possess that potent financial and medical infrastructure. Of course, Duke patients were, at some level, eligible for health insurance, but it was not automatic. Indeed, the clinic patient populations increasingly consisted of persons with insufficient or no means. The first patients at Duke and the VA with GRID were recognized in 1982, and neither institution was fully prepared to deal with this new disease, although the presenting opportunistic infections were to some extent treatable. Nevertheless, it is no overstatement to say that the patients were sick, sick, sick. Initially, for patients at Duke, few faculty members were involved in their inpatient and outpatient management, whereas patients at the VA were incorporated in the regular infectious diseases clinic managed by the VA faculty and fellows.

In 1987, Dr. Durack, then the chief of the division, asked Dr. John

A. Bartlett, a fellow in infectious diseases at Duke, to set up a clinic in the Medical Outpatient Clinic to manage the increasing number of patients. Dr. Bartlett describes the uncertainty about how to proceed as being physically and emotionally draining. New faculty members were recruited to assist him, but burnout of faculty and staff was the norm because of the perception that a diagnosis of AIDS was a virtual death sentence.

By this time, there was increased emphasis on clinical trials to test promising new treatment strategies as part of the newly formed Duke AIDS Treatment Evaluation Unit (ATEU) of the NIH. Clinical trials included studies on azidothymidine (AZT), which had been discovered less than a year earlier. Dani Bolognesi was the principal investigator of the ATEU, and Michael Cairns, a newly minted faculty member in the Infectious Diseases Division, was the lead physician for the Duke ATEU. The staff of both clinics operated creatively to be as certain as possible that needed physical or emotional care was provided in an atmosphere that felt more like a family. New models of care were established in which outpatient treatments were provided. The professional staffs attracted those with special commitments to care for patients often described as marginalized populations. Dedicated staff members at Duke, such as Tony, Ken, Gordon, Les, Scott, Sandy, Trish, Martha, Barlett, Carl, Robert, and at the VA, such as Kay and Pam, were stalwarts in the days before and after effective treatments became available. One former staff member describes the personnel as "cowboys" and "renegades" committed to overcoming whatever obstacles were encountered. Whatever tragedies drained the zest for life among the staff, they tried to maintain a semblance of levity. One family member describes her first visit to the clinic with her infected husband as follows: "I walked in the door and encountered the social worker, a woman, dressed as a man, complete with fedora and penciled-on mustache. Nearby stood a male nurse dressed as a female nurse, in cap, tights, and dress. Vampires lurked in the background. It was Halloween. It was also an altogether fitting introduction to the ID [infectious diseases] clinic, where the craziness of HIV and AIDS were met, and to some small degree neutralized, by an intentional counter-craziness borne of a recognition that this was, in fact, no normal disease, as modern Western medicine liked to think of normal."[53]

Vocal and often aggressive activism by those already infected and those at risk of being infected occupied the national and international stage. It is understandable that the activists had questions and wanted answers. What was the cause of GRID? How was it transmitted? Who was at risk? What was the prognosis? How could it be prevented? How could it be treated? Though answers to these questions eventually became apparent, the activists, and everyone else for that matter, wanted action *now*. In several different venues, the frustrated activists daily confronted Duke-connected faculty members. Two prominent figures included Dr. James B. Wyngaarden, the former chairman of the Department of Medicine at Duke and then, in 1982, the director of the National Institutes of Health, and Dr. David Barry, a consulting Duke faculty member and attending at the infectious diseases clinic at the Durham VA, also a director of the Virology Section at the Burroughs Wellcome company (BW). At one time or another, both had activists literally sitting on their doorsteps and demanding action. Activism took a somewhat different approach locally, with the creation of a community-based support network in 1982 consisting of infected and at-risk individuals, both homosexual and heterosexual men and women. This network was called the Lesbian and Gay Health Project (LGHP).[54] This group of volunteers established Health Line, which was probably the most notable source of accurate information on GRID in the region, and developed the buddy system, in which patients were paired with one or more healthy but sympathetic and helpful volunteers for physical and emotional support. The LGHP obtained financial support from industry and state government, the most visible use of which was to purchase a house for infected individuals who had no other place to stay. The network was not able to address all the problems confronting patients, but it was an enormous help to many.

Clinical demands were enormous but so too were the demands on basic scientists to discover the cause of the disease. Major laboratories around the world redirected their research efforts, and many possible causes of GRID were advanced. It was, however, Francois Barre-Sinoussi and Luc Montagnier who were ultimately given credit by the Nobel Prize Committee for the first isolation of the virus that caused GRID; they called it lymphocyte-associated virus (LAV).[55,56] Their publication in 1983 was followed by the isolation of HTLV-III by Robert Gallo (including Haynes and Palker of Duke)[57] and Popovic,[58] described in publications in 1984. Dr. Barton Haynes contributed specimens to Dr. Gallo from patients with hemophilia who were being followed by Dr. Gil White at the University of North Carolina.[59,60] Isolates from those specimens were among the first isolates from the Gallo laboratory and, along with other work, established patients with hemophilia as an additional population of at-risk persons, along with IV drug users, recipients of blood transfusions, and Haitians. Not long afterward, heterosexual contact among men or women with AIDS was added to the list of at-risk populations.

Drs. Haynes and Bolognesi had established a collaboration with the Gallo lab, Haynes initially through the process of identifying the second isolate of HTLV-I in 1983[33] and Bolognesi because of his decision to shift the focus of his laboratory from animal tumor viruses to human retroviruses (Dr. Beard had died in 1983, and Dr. Bolognesi had become director of the Division of Experimental Surgery). Dr. Gallo invited Haynes and Bolognesi to join the Acquired Immunodeficiency Syndrome (AIDS) Working Group at the National Cancer Institute of the NIH. Relocation to the NIH was not possible for a number of reasons, but Haynes and Bolognesi agreed to work with the Gallo group and attended the monthly meeting of the Gallo Lab at the NIH for the next two years. Haynes's task was to organize the work on investigation of hemophiliac patients in the United States, hence the partnership with Dr. Gil White at UNC. Haynes and Bolognesi set up a BioSafetyLaboratory-4 (BSL-4) unit in the Duke Animal Laboratory Isolation Facility (ALIF) for the culture of HIV. The laboratory had a functioning Blickman Line (a specimen and concentrated virus stock processing unit that prevented laboratory personnel exposure to HIV) in which all the initial work was performed for screening hemophilia patients' plasma for retrovirus antibodies. In ensuing years, the same group of original investigators collaborated on many projects and a number of basic science faculty were recruited.

There remained some skeptics that LAV or HTLV-III was the cause of AIDS, including Dr. Peter Duesberg of the University of California at Berkeley[61] and Prime Minister Mbeki of South Africa (Reported by PE Sox at the IAS meeting in Durban, SA in 2000), but their skepticism was eventually refuted. Still, isolation of a virus or not, a treatment or cure had not yet been identified, and all concerned became more and more vocal and assertive.

Filling a crucial need, a blood test for antibodies to HTLV-III became available in 1985, but as an illustration of how little was understood about the disease, an article in *Frontpage*, the gay and lesbian paper, wrote this about the test: "A positive result means that at some point you were infected with the HTLV-III virus. It does not mean you have or will have AIDS" (*Frontpage* 1985). Of course, we now know that without treatment, very few infected persons will escape progression of the disease.

CLINICAL TRIALS OF AZIDOTHYMIDINE

Once many of the early questions were answered, the question of treatment became the focus. Antiviral drugs were few in number in the mid-1980s. The BW Company in the Research Triangle Park was the company most prominent in the field of antiviral therapy but had no drug active against HIV. With the support, however, of Drs. George Hitchings and Trudy Elion, both later recipients of the Nobel Prize for their work on purine-derived drugs, Dr. David Barry, head of the Anti-Infectives Section since 1977, led a team at BW dedicated to the discovery of a drug active against HIV. He had been recruited from the Bureau of Biologics, where he was the director of the task force on influenza. The BW Anti-Infectives Section had worked earlier on and licensed acyclovir, a unique antiviral with activity against herpes virus I and II and the varicella zoster virus, all members of the *Herpesvirus* family. With the onset of AIDS and the recognition that its cause was a human retrovirus, however, Dr. Barry pursued a very active discovery program and, in the course of that, unearthed a compound that previously had been tested as a cancer drug and failed. Under Dr. Barry's leadership, Marty St Clair, a laboratory technician at BW discovered that "Compound S," as it was initially named, was found to have potent inhibitory activity against the reverse transcriptase of a mammalian retrovirus, a critical enzyme for the replication of HIV. Then, in blinded studies at the National Cancer Institute and at the Duke Surgical Oncology Research Facility, the inhibitory activity of Compound S against live HIV in culture was confirmed independently in both laboratories.[62] Compound S was then identified as azidothymidine, or AZT. Claims to have discovered AZT came from several laboratories, but BW was ultimately awarded the patent. AZT, however, was found to be difficult to synthesize, requiring exogenous thymidine derived initially only from whale sperm. That characteristic posed a major obstacle to the production of amounts large enough to conduct the necessary phase I, II, and III clinical trials, but such trials, if successful, could lead to licensure and marketing to an increasingly demanding population who were experiencing enormous morbidity and mortality.

ADULT TRIALS

Within months of the laboratory demonstration of inhibitory activity, a phase I study of drug safety was conducted at the National Cancer Institute (ten patients), Duke University Medical Center (eight patients), and the Durham VA Medical Center (one patient,) for a total of nineteen patients in 1985 and 1986.[63] The results of the study demonstrated that the drug, when administered either intravenously or orally, was sufficiently safe to proceed with a phase II trial. In addition to demonstrating safety, the drug resulted in some positive clinical benefits, but the number of patients was insufficient to conclude much about the efficacy of the drug. With accelerated approval from the FDA to proceed to a multicenter phase II clinical trial, 282 participants were enrolled in an efficacy study designed to detect a clinical benefit in a placebo-controlled trial in symptomatic patients with CD4 cell counts less than two hundred per cubic milliliter, chosen because this low level of CD4 cells indicated that the infection was progressing.[64]

Thirteen patients were enrolled at Duke. The study was terminated early by the Data Safety and Monitoring Board after the board concluded that there was a reduced mortality (nineteen in the placebo arm and one in the drug treatment arm) and a delay in the onset of AIDS-defining events in the drug treatment arm compared to the placebo arm. The study design using a placebo was extremely unpopular, and widespread demonstrations by activists across the country were common. The FDA deemed apparent drug efficacy and drug safety measures sufficient, and plans were set in motion to proceed with phase III trials. The AIDS Clinical Trials Units (ACTU) of the NIH, the Department of Veterans Affairs Cooperative Studies Program (VA CSP), and a European/UK consortium all wanted to sponsor this trial. The supply of AZT, however, was insufficient to support all the proposed studies at once, and Dr. Barry concluded that the Cooperative Studies Program of the VA was best positioned to embark on a rigorous test of safety and efficacy. He therefore committed the available supply of AZT for that phase III trial to the VA CSP. His decision was not without controversy, and every effort was made to redirect the initial supply to the NIH studies. That was not done, but after a relatively brief time, sufficient amounts of the drug were available for all planned studies to proceed. The VA and the NIH initially proposed randomized, placebo-controlled trials. The NIH proceeded with that design in patients with lower (less than two hundred) CD 4 cell counts per cubic milliliter (ACTG 016) and with higher (less than five hundred CD 4) cell counts per cubic milliliter (ACTG 019). In response, however, to an unprecedented action by the FDA in which the drug received accelerated approval for use in symptomatic patients with CD 4 counts less than two hundred in 1987, the VA altered its study design from a placebo-controlled trial to one in symptomatic HIV infected patients that compared early treatment (CD 4 count greater than two hundred) with delayed treatment (after the CD 4 count fell below two hundred) to accommodate the FDA's mandate that patients with CD 4 counts less than two hundred were now eligible to receive AZT.[65] This action by the FDA and the subsequent action taken by the VA were pivotal in the interpretation of future studies and set a standard that had ramifications beyond the phase II trial that were perhaps not anticipated. The European/UK consortium likewise converted their study design to a comparison of early versus later treatment.

Both NIH trials enrolled substantial numbers of patients, and both studies were terminated early by the Data Safety and Monitoring Boards because of highly statistically significant reductions in morbidity in those patients receiving AZT compared to the placebo controls after a mean follow-up of forty-six weeks in ACTG 016 and 55 weeks in ACTG

019. Too few deaths occurred to conclude anything about the efficacy of AZT in decreasing mortality. The results of both studies were published in the *New England Journal of Medicine* in 1990.[66,67] Moreover, ACTG 019 demonstrated equivalent efficacy at five-hundred- and fifteen-hundred-milligram dosages of AZT, allowing future studies to reduce the dosage of AZT and thus minimize the predictable side effect of AZT, namely a reduction in levels of hemoglobin at the higher dose. The Duke Infectious Diseases Clinic participated in ACTG 019 by the enrollment of 156 patients but was not a participant in ACTG 016. The results of these studies further solidified the strategy for use of antiviral chemotherapy and were received with great enthusiasm by the infected patients, at-risk personnel, health-care providers, public health officials, and by some segments of the pharmaceutical industry. Access to treatment resulted in complicated treatment regimens and annoying side effects. Most important, the cost remained a problem, but on balance, this first true antiviral drug for HIV brought a ray of hope for the future. The FDA acted upon the results and broadened the indications for use of AZT in March 1990 to one that included asymptomatic patients with CD 4 cell counts of five hundred per cubic milliliter or less.[68]

A somewhat different perspective emerged as the results of the VA Cooperative Study Program 298 (CSP 298) became known. Codirected by me at the Durham VA Medical Center and Dr. Michael Simberkoff at the New York VA Medical Center, the study compared early with delayed monotherapy with AZT in 238 patients in seven VA Medical Centers and Walter Reed Army Hospital with symptomatic HIV infection and CD 4 cell counts greater than two hundred per cubic milliliter. After a follow-up of a mean of 27.2 and 28.2 months for early and later therapy, respectively, no significant difference in survival was noted between those treated early versus those treated later, and the benefit of early treatment in terms of progression to AIDS was limited to three to six months.[69] This study allowed a direct comparison of strategies for early versus later therapy and demonstrated that there was no survival benefit and that the benefits in delayed progression of disease were time limited, both of which exposed the limitations of monotherapy. This study was published in the *New England Journal of Medicine* in 1992. It was therefore not surprising that there was considerable skepticism about these results. The study was conducted, however, under rigorous guidelines, a characteristic for which the VA Cooperative Studies Program is known. Moreover, the European/UK trial, named the Concorde Trial, which had larger numbers of patients and a similar study design, basically came to the same conclusion as the VA study did.[70]

Many, including Dr. Barry of the BW Company, predicted that combination therapy would be required to achieve a sustained benefit, in part because of the development of antiviral drug resistance by the virus. CSP 298, in a collaborative study with the virology laboratory at BW, was, in fact, the first to demonstrate this in a correlation of in vitro resistance with decreased activity of the drug in vivo.[71] In addition, in collaboration with Chiron, the developer of the assay to detect viral load, VA CSP 298 was the first to demonstrate that plasma viral load served as a surrogate marker for drug efficacy.[72,73]

PEDIATRIC TRIALS

Another important risk group for acquiring HIV infection included infants and children born to infected mothers. Most of the risk occurred at the time of birth, but breast-fed neonates were also at risk. Prior to the development of antiretroviral drugs, the prognosis was dismal, with death occurring within a few years of birth. With the discovery of the effectiveness of AZT in adults, Drs. Wilfert, Katz, McKinney, and others collaborated with Dr. Phillip Pizzo at the University of Maryland and Dr. Gwen Scott at the University of Miami to conduct some of the earliest antiretroviral trials in children.[74-81] In fact, under the supervision of Dr. Catherine Wilfert, the first dose of AZT in children at Duke was administered in 1986.

A pediatric section of the ACTG (PACTG) was established, and Dr. Catherine Wilfert, a Duke professor of pediatrics, became its first chair. The PACTG conceived of and sponsored the landmark study that established the feasibility of interrupting the transmission of HIV from mother to child by the administration of AZT to the mothers in the few weeks before delivery and to the newborn child at delivery and for several days after delivery (PACTG 076). This study was published in the *New England Journal of Medicine* in 1994 and heralded the decline in the transmission of HIV from mother to child.[82] Additional studies amplified on the results of PACTG 076.[83,84] Because of PACTG 076, new cases of HIV in newborns in the developed world are now almost nonexistent; the caseload of pediatric HIV is therefore dwindling. Dr. Wilfert was named the scientific director of the Elizabeth Glaser Pediatric AIDS Foundation (EGPAF) in 1996 and president of the Infectious Diseases Society of America in 1999. She continued her demanding work for EGPAF well into the twenty-first century.

Dr. Sandra Lehrman came to Duke as a pediatric infectious diseases fellow in 1979 after working with Dr. Robert Chanock at the NIH (1969 to 1972), obtaining an MD from Brown (1972 to 1976), and doing a pediatric internship and residency at the Massachusetts General Hospital (1976 to 1979). Duke was a logical choice because of her interests in virology and because Dr. Katz (measles) and his early recruits of Drs. Wilfert (diagnostic virology), Lang (cytomegalovirus), and John Griffith (herpes viruses) would be her mentors. Later, as a faculty member from 1979 to 1983 (and then part time from 1983 to 1995), she would have them as her colleagues. In 1995, Dr. David Barry recruited her to join his group at the BW Company as a senior research scientist, where she focused her attention initially on fundamental and clinical aspects of antiviral drugs such as acyclovir for herpes viruses and later AZT for HIV. Over succeeding years, she assumed additional responsibilities, and until BW was taken over by Glaxo, she rose through the ranks, achieving the positions of vice president and general manager of BW Manufacturing, Inc., and BW's international director of biotechnology. Dr. Lehrman was very involved in the development of AZT through all the clinical trials. After BW's merger with Glaxo, she became increasingly involved with the biotechnology industry and joined Triangle Pharmaceuticals, which Dr. David Barry had created and led. She participated in drug development until Dr. Barry's sudden death in 2002, and then she worked for several biotechnology and pharmaceutical firms. Over the course of her career, she published widely, mostly on antivirals for herpes viruses and clinical trials of AZT in children.[62,63,78,85,86]

Dr. Ross McKinney came to Duke as a pediatric resident in 1979 and initially engaged in the management of children born to women later found to be infected with HIV. After the discovery of AZT, he was increasingly involved in clinical trials of antiretroviral drugs, including phase I and multicenter trials through the PACTG, which resulted in high-profile publications in *The New England Journal of Medicine*, the *Journal of Pediatrics*, and the *Journal of the American Medical Association*.[74-76,78,87-89] In addition, he was a coauthor of an article about AZT resistance in children, which was published in *The Lancet*.[79] Dr. McKinney served as a member or chair of numerous PACTG committees and as chief of the Division of Pediatric Infectious Diseases at Duke from 1998 to 2003.

BASIC SCIENCE INVESTIGATIONS AND THE CENTER FOR AIDS RESEARCH

Meanwhile, basic science investigators at Duke were discovering many important facts about HIV and the immune response to it. With Dr. Bolognesi as principal investigator, Duke investigators successfully applied for an NIH-funded Center for AIDS Research (CFAR) in 1989, and except for a lapse of five years, between 2000 and 2005, the Duke CFAR has been funded continuously, now through 2015. Numerous Duke faculty in the departments of medicine, pediatrics, surgery, and microbiology (later known as molecular genetics and microbiology) and from the VA were either directly or indirectly affiliated with the CFAR. These faculty included Drs. Haynes, Weinhold, Matthews, Langlois, Palker, Lyerly, Montefiori, Bartlett, Hicks, Ferrari, Barry, Lehrman, Cullen, Greene, Wild, Oas, Markert, Buckley, Hale, Liao, Staats, Bohjanen, Garcia Blanco, Dukes Hamilton, Weinberg, Klotman, Frothingham, Wilson, and me. There was much collaboration between these faculty members and faculty members from other institutions. What was to become an extraordinary series of major contributions to the field began when the second isolate of HTLV-1 from a patient with adult T-cell leukemia and cutaneous lymphomatous vasculitis was accomplished at Duke in 1983.[33,90-93]

The following examples represent a small minority of the scientific output of these CFAR investigators but are some of the more prominent discoveries. In 1985, CFAR investigators participated in the initial blinded in vitro and in vivo testing of AZT .[62] Moreover, they demonstrated the following:

1) development of neutralizing antibodies to an envelope glycoprotein of HIV and detected antibodies to the V3 loop structure of the virus;[94,95]

2) determined the mechanism of viral neutralization;[95]

3) showed that the virus mutated from its founding strain to a diversity of strains or quasi-species;[96]

4) evaluated a synthetic peptide immunogen containing the V3 neutralizing domain and an immunodominant T-helper region of gp120, which was done in preclinical trials[93,97] and early stage clinical trials;[98]

5) performed proof of concept studies that demonstrated the usefulness of mucosal immunity to selected immunogens and adjuvants;[99-101]

6) detected a cluster of native North Carolinians infected with HTLV-1;[102]

7) discovered HIV circulating as recombinant forms

8) determined the origin of HIV from chimpanzee simian immunodeficiency virus;[103,104] in 1992

9) discovered a novel peptide that prevented fusion of HIV with the target cell;[105]

10) in 1998, performed in vivo tests with that peptide in humans;[106]

11) discovered that the thymus shrinks in patients with HIV infection but with highly active antiretroviral therapy (HAART) and thymus transplantation, their immune systems can be restored to make T cells.[107,108]

With critical financial support from Dr. David Sabiston, the chair of the Department of Surgery, and from DuPont, a stand-alone research facility (the Surgical Oncology Research Facility) was constructed on the Duke campus so that state-of-the-art HIV research could be conducted. In the early 1990s, this laboratory, overseen by Dr. Bolognesi, was named the Central Immunology Laboratory of the Human Vaccine Trials Network. The Central Immunology Laboratory later became the nucleus of what was to be called the Human Vaccine Institute.

Another basic scientist, Mariano Garcia-Blanco, MD, PhD, was recruited to Duke in 1990 as an assistant professor of microbiology and of molecular cancer biology from MIT, where he had been a postdoctoral fellow with Dr. Phillip Sharp. At MIT, he had become interested in the biochemical activity of the HIV-1 protein Tat and its RNA binding site, the TAR element, which is found at the 5' end of HIV-1 transcripts. Working with a pre-doctoral student, they developed an in vitro transcription system that recapitulated Tat-dependent transactivation. Using biochemical structure probing[109] and two-dimensional nuclear magnetic resonance spectroscopy, they determined the structure of the terminal loop of the RNA element[110,111]. Again using in vitro transcription, they focused on the identification of cellular proteins that interacted with Tat and TAR, and with affinity chromatography and mass spectrometry they discovered a protein originally termed CA 150 (now TCERG1), which is involved in transcript elongation and processing.[112-115] Further, they explored the use of RNA decoys as potential anti-HIV tools and used these tools to explore the mechanism of Tat trans-activation.[116,117] During that time, Dr. Garcia-Blanco served as co-scientific director of the VA Research Center on AIDS and HIV Infection (RCAHI), where he nurtured relationships between faculty, the adult Division of Infectious Diseases, and the Department of Microbiology.[118]

David Montefiori, PhD, came to Duke as an associate research professor in the Department of Surgery in 1993 from Vanderbilt, where he had been pursuing immunologic research relevant to HIV. His major research interests have been in the areas of viral immunology and AIDS vaccine development, with a special emphasis on neutralizing antibodies. One of his highest priorities has been to identify immunogens that generate broadly cross-reactive neutralizing antibodies for inclusion in HIV vaccines. Over the years, he has explored multiple types of assays for neutralizing antibodies and other potential antiviral antibodies, focusing on assay optimization, standardization, validation, and high throughput. The scope of his research covers HIV-infected humans,[119-122] nonhuman primate models of simian immunodeficiency virus (SIV), and simian-human immunodeficiency virus (SHIV) infection.[119,123-125] He directs a large AIDS vaccine immune-monitoring program that serves as a national and international resource for standardized assessments of neutralizing antibody responses in preclinical and clinical trials of candidate AIDS vaccines. He chaired the Antibody Laboratory Standardization Working Group, which assisted in the design of the Global HIV Vaccine Enterprise Strategic Plan for HIV Vaccines. Somewhat later, he assumed positions of leadership with the Human Vaccine Trials Network (HVTN) and the NIH preclinical HIV-1 vaccine development program as well as directed a Comprehensive Antibody Vaccine Immune Monitoring Consortium as part of the Bill and Melinda Gates Foundation's Collaboration for AIDS Vaccine Discovery.

Herman Staats, PhD, too, came to Duke in 1993 and pursued his research in the field of mucosal immunology. The crucial antibodies for effective mucosal immunity are predominantly of the IgA isotype and protect against infection at mucosal surfaces. Some of his early work demonstrated that parenterally administered HIV vaccines rarely induce IgA isotype antibodies comparable to those induced by mucosally administered vaccines, whereas effective mucosal vaccines are able to induce both systemic IgG and mucosal IgA.[100,101] Notably, in addition to stimulating serum antibody responses, nasal immunization induces significant levels of IgA in secretions of the female reproductive tract. Because HIV is commonly contracted through sexual contact, Staats's observation is at least of theoretical importance. Subsequently, the Staats Laboratory began to evaluate the contribution of adjuvants administered with nasally administered vaccines.[99] The results indicated that cytokines were effective adjuvants for nasally administered vaccine. Later work expanded on this theme in the twenty-first century.[126-128]

Drs. Warner Greene and Bryan Cullen in the Duke Rheumatology and Immunology Division in the Department of Medicine made important contributions to understanding the regulatory pathways that influenced the replication of HIV, which they summarized in a 1989 publication.[129] Dr. Greene performed work at Duke "dealing with the biology of the NF-KB transcription factor, accessory proteins of HIV and HTLV-1, and a surprising functional interplay between the HTLV-1 Rex and HIV Rev proteins" (Warner Greene, personal communication).[130-136]

Similarly, Dr. Cullen cites his most important publications in the twentieth century while at Duke. In a personal communication, he describes the following contributions:

1) The demonstration that the HIV-1 Rev protein functions as a sequence-specific nuclear export factor[137] and the identification of the RNA target site for Rev, the Rev Response Element.
2) The functional dissection of the Rev protein, including the identification of the RNA binding and multimerization domains and, most importantly, the leucine-rich activation domain. This led to the identification of dominant negative mutants of Rev.[138,139]
3) The identification of the envelope V3 loop as the determinant of tissue tropism and coreceptor specificity in HIV-1.[140]
4) The first demonstration that HIV-1 could infect non-dividing cells.[141]
5) Analysis of the host cofactors required for HIV-1 Tat function.[142-144]

Dr. Guido Ferrari joined the Duke HIV laboratory in 1993 as a research associate and for the next six to seven years focused on the cellular immune response to HIV, with particular attention to the implications for an effective HIV vaccine. Dr. Ferrari led or collaborated with others from the laboratory on a number of high-profile and high-impact publications, a number of which are cited here.[96,145-149]

In 1988, the Durham VA Medical Center was named as a Research Center on AIDS and HIV Infection (RCAHI), one of five nationally, and I was the director. The center had three components:

1) clinical programs (AIDS Clinic and Inpatient Service, AIDS Patient Database, AZT Cost Effectiveness Analysis, and Hospice Education Demonstration Project);
2) clinical investigations (VA Cooperative Study 298, AZT Resistance, HIV Viral Load as a Surrogate for Disease Progression, HIV Vaccine Preparedness, and Quality of Care);
3) basic science programs (Retrovirus Immunity/Immunization, led by Dr. Thomas Palker, a co-scientific director; Viral Latency and Tropism, led by Dr. Mary Klotman and me; HIV Mononuclear Cell Interactions, led by Drs. Brice Weinberg and Carol Dukes; Molecular Analysis of Opportunistic Infections,

led by Drs. Richard Frothingham, Kenneth Wilson, and John Perfect; Transcription Factors, led by Drs. Mary Mudryj and Kenneth Schmader; and Molecular Therapeutics, led by Dr. Mariano Garcia-Blanco, a co-scientific director. (References reflecting the activity of these programs are listed under the individual investigators.)

The center provided laboratory space, including a designated isolation facility level three. Four RCAHI-affiliated faculty members were VA Career Development awardees, and the center was responsible for the successful application for a T32 training grant on AIDS (the Interdisciplinary Research Training Program on AIDS) from the NIH and the Minority International Research Training Program from the Fogarty International Center of the NIH. The RCAHI was active until 2000, when the program was discontinued nationally.

My background before the onset of AIDS was in the study of the molecular control of viral latency of murine cytomegalovirus,[150-155] in clinical trials of the hepatitis B virus vaccine in dialysis patients and staff,[156,157] and the use of hyperimmune Hepatitis B Immune Globulin (HBIG) after needlestick.[158,159] After the onset of the AIDS epidemic, the majority of my AIDS work stemmed from my involvement in VA Cooperative Study 298, discussed earlier in this chapter.[69,71,160-162]

Dr. Klotman collaborated on much of the murine cytomegalovirus work but left the Durham VA RCAHI in 1988 to join the Laboratory of Tumor Cell Biology at the National Cancer Institute under the direction of Dr. Robert Gallo. She remained there until 1994 and made major contributions to the understanding of HIV pathogenesis.[163-165] She identified alternatively spliced mRNAs of HTLV-1, which were found to code for unique, previously unrecognized viral proteins. She was one of the first to study the kinetics of HIV-1 mRNA expression in the infection of primary T cells and macrophages. After moving to Mount Sinai as chief of the Infectious Diseases Division in 1994, she continued with productive research in HIV, HIV-associated nephropathy (HIVAN), and cellular and innate immunity.[166-172] As stated by her chair at Mount Sinai, Dr. Barry Coller:

Her work in defining the tissue-specific patterns of HIV mRNA expression in HIV transgenic renal epithelium provided the first evidence that renal cells could support the HIV-1 life cycle. She was responsible for the virology in collaborative studies that identified the renal epithelium as a unique compartment for HIV replication. Her lab then defined the kidney as a long-term reservoir for HIV replication. She continued to focus on the interactions between HIV-1 and primary cells through the identification of important signaling pathways required for replication. Her lab demonstrated the role of STAT-1 activation in the soluble HIV-1 inhibitory activity of CD_{8+} T-cells. She defined a novel mechanism by which the antimicrobial peptide, alpha-defensin, inhibits HIV by altering host cell signaling.

Dr. Brice Weinberg, a hematologist by training, joined the staff at the Durham VA medical center in the 1970s and initially pursued an understanding of the fundamental biology of human mononuclear cells and macrophages. However, in 1991 he began the publication of a number of studies on the effects of HIV infection on the functional activity of human blood monocytes and peritoneal macrophages, including anti-cryptococcal activity cellular tropism and inhibition by specific antibodies, by synthetic peptide analogs of HIV-1 protease and by the drug class cobalamins.[173-180] Whereas earlier understanding held that HIV-1 required cells to be proliferating to allow integration of viral DNA into host DNA, Dr. Weinberg and colleagues showed for the first time that HIV-1 could infect and integrate into nondividing cells.[141]

THE INFECTIOUS DISEASES TRAINING PROGRAM

The Infectious Diseases Training Program proved to be a valuable pool of potential recruits for HIV-related clinical and basic science research. Among these recruits, Drs. Jerome Kim and Paul Bohjanen stood out as basic science investigators in the field of HIV.

Dr. Jerome Kim joined the Infectious Diseases Training Program as a fellow in 1987, and after twelve months of clinical training, he joined the laboratory of Dr. Warner Greene at Duke for two years as a fellow and a further year as an associate in medicine at Duke. With the assistance of Dr. Greene, Dr. Kim successfully competed for a Pfizer postdoctoral fellowship award and collaborated with Dr. Sarah Hanly on the Rex protein of HTLV-II, which shares functions with the Rex protein of HTLV-I and with the REV proteins of HIV and SIV. Drs. Kim and Hanly showed that Rex-II, acting on an RNA-responsive element (the Rex-responsive element) regulated the nuclear export of unspliced and partially spliced HTLV-II RNA that encoded the structural proteins Gag, Env, and Pol.[181,182] In collaboration with another investigator in the same laboratory, they looked at the proliferative phenotypes of different cell lines from HTLV-I-infected persons from Japan.[183] Last, Dr. Kim engaged in studies using Rev-response element decoys to suppress HIV replication through gene therapy.[184] The remainder of his career was primarily research done in military facilities after he was called-up at the onset of the first Gulf War in 1991. At the Walter Reed Army Institute for Research (WRAIR), he continued his work on various aspects of retroviral research, including immune responses to HIV and HIV vaccine prevention of HIV infection in Thailand. Selected publications before 2000 are noted in the bibliography.[185-188] Subsequently, he has played a major role in HIV vaccine development supported by the Department of Defense.

Dr. Paul Bohjanen came to the Duke Medicine House staff Training Program as an MD, PhD, in 1993 with a serious interest in laboratory research. He therefore entered the clinical investigator path in 1993, which involved two years of clinical work as a resident followed by two years of research. He began his research years in 1995 with funding from the Howard Hughes Medical Institute Postdoctoral Fellowship for Physicians. He worked in the laboratory of Dr. Mariano Garcia-Blanco, where he studied HIV transcriptional regulation, in particular the biochemistry of the HIV Tat protein and the mechanisms by which it usurps the human cellular gene expression machinery to promote transcription of the HIV RNA. From that work, he published two papers in *Nucleic Acids Research*.[116,117] He then joined the Infectious Diseases Training Program and continued his research, this time with the support of a K08 career development award from the NIH. With this support, and under the mentorship of Dr. Jack Keene, he focused on understanding how mRNA degradation regulates the expression of cytokine genes and other genes involved in activation of the immune system. From 1999 to 2000, Dr. Bohjanen completed twelve months of clinical infectious diseases training, venturing into the field of clinical research. Under the mentorship of Dr. Charles Hicks, he studied the steady-state pharmacokinetics of lamivudine[189] and reviewed HIV as a target of gene therapy.[190] He was then recruited to the departments of medicine and microbiology as an assistant professor at the University of Minnesota, where he has continued as a highly successful NIH-funded basic scientist and clinical investigator on various aspects of HIV/AIDS, including HIV/AIDS in Africa.

OTHER CLINICAL TRIALS AND CLINICAL RESEARCH

Dr. John A. Bartlett: After working with Professor Wolfgang Joklik in the laboratory, Dr. John A. Bartlett shifted his research focus to clinical HIV research, including clinical trials. Most of his published work fell into three major categories prior to 2000. Selected examples include the following: (1) HIV clinical papers;[191] (2) clinical trials of antiretrovirals, antibiotics for opportunistic infections, and vaccine candidate trials;[67,192,193] and (3) collaborations with Drs. Bolognesi, Haynes, Weinhold, Matthews, and Markert on studies of immune function in patients with HIV infections.[145,194-197] He was a coauthor of the article in the *New England Journal of Medicine* that described the results of the ACTG 019 study, among others. Dr. Bartlett was very active in the AIDS Clinical Trials Group (ACTG), rising to levels of leadership in the Executive Committee and successfully competing for ACTG support. Arguably, his greatest contribution to the HIV/AIDS efforts at Duke in the twentieth century was the leadership he provided for the adult HIV clinic. He had a strong commitment to the many AIDS patients under his care and to the community organizations that supported the People Living with AIDS (PLWA) and the Lesbian Gay Bisexual and Transgender (LGBT) groups. Because of his commitment, he was the recipient of many civic and academic awards. After 2000, his contributions internationally increased significantly, with multiple grant awards from the NIH for work on clinical trials on AIDS, HIV, and malignancies and for medical education in Tanzania.

Dr. Charles Hicks: Another major figure in the clinical HIV arena at Duke was Dr. Charles Hicks. He was a West Point graduate and served in the military as a physician for twenty years before coming to Duke in 1994 with the goal of working in the HIV clinic and on the inpatient infectious diseases consultation services. At that time, the volume of both inpatient and outpatient services was increasing dramatically because no effective and sustained antiretroviral regimens had been established and the patient population was increasingly ill. As a resident in San Francisco from 1979 to 1982, he experienced the first wave of patients with GRID and later developed his career as an authority on sexually transmitted diseases, including HIV. Dr. Hicks concentrated on industry-sponsored trials of new and improved antiretrovirals for management of his patient population, and he published four papers prior to 2000.[98,198-200] With the arrival of effective combination therapy in the mid-1990s, he was able to expand his publication record and his portfolio of activities to include leadership positions in the ACTG (formerly ATEU), receipt of an award from the NIH for a senior mentoring program (a K24 grant) for two five-year cycles, taking over the direction of the Interdisciplinary Research Training Program on AIDS, and creating the Acute HIV Infection Network.

Dr. Nathan Thielman: Dr. Nathan Thielman's first connection with Duke was as a medical student from 1986 to 1990 and then as an intern and resident from 1990 to 1993. Because he was interested in international health and HIV, he participated in several clinical studies in Tanzania on diagnosing and screening and on delayed-type hypersensitivity in patients with HIV.[201,202] Subsequently, he has become increasingly engaged in international health as it relates to HIV and education. He has been the initiator and coinvestigator of numerous NIH grants, particularly collaborating with faculty members who have initiated studies using social and behavioral science methodologies. In addition, he plays a major role in the coordination of the Duke domestic research site for studies by the AIDS Clinical Trials Group.

Other contributors: Many individuals have played major and important roles in complementing the basic and clinical research, clinical trials, and programs. They have also played important roles in the day-to-day care of the many HIV-infected patients over the nearly twenty years since the first cases of AIDS were described. Their contributions to Duke's response to the epidemic are incalculable. Those individuals are listed here: *in medicine*, Barbara Alexander, June Almenoff, Andrew Alspaugh, Michael Cairns, Mimi Cameron, Gary Cox, David Durack, Charles Ellenbogen, John Engemann, Vance Fowler, Richard Frothingham, Harry Gallis, Donald Granger, Carol Dukes-Hamilton, Allison Heald, Chris Ingram, Souha Kanj, Kathy Kirkland, Gunther Lallinger, Diego Miralles, John Perfect, Jason Stout, Charles van der Horst, Hetty Waskin, and Ken Wilson; *in pediatrics*, Dennis Clements, Robert Drucker, Sam Katz, Ross McKinney, Cathy Wilfert, and others. To be sure, by 2000 the epidemic had evolved from a frightening untreatable and ultimately fatal acute infection to a treatable chronic infection. Still, there remain many unanswered questions, some of which have been addressed in subsequent years. Accounts of these achievements will remain the topic of future writings.

Photograph 4.3 is a composite of pictures of five prominent figures in the field of HIV/AIDS at Duke in the twentieth century.

Joseph W. Beard

Dani P. Bolognesi

Barton F. Haynes

Mary E. Klotman

John A. Bartlett

CHAPTER 5

HOSPITAL INFECTION CONTROL AND EMPLOYEE, STUDENT AND OCCUPATIONAL HEALTH

In 1874, an English surgeon said the following:

> I have unwillingly and almost tremblingly proceeded to operate in the hospital. I come to the conclusion in my own mind that pyemia if it does not find its birthplace, does find its natural home and resting place in hospitals; and although the hospital may not be the mother of pyemia it is its nurse.[1]

In an extensive review of the history of infection control, Sydney Selwyn cites the existence of records describing efforts at hospital infection control in Egypt, Palestine, and Greece as early as 500 BC.[2] Those records include descriptions of the hospitals as follows:

> The Abaton [ward] was a lofty and airy sleeping chamber, its southern side being an open colonnade ... This provision of abundance of pure fresh air for the sick by day and night, which is so beneficial now, was undoubtedly so then also, and probably brought much credit to the God and his shrine ... The precinct was as beautiful as the noblest works of Greek art could make it.[3]

After the fall of the Roman Empire, however, attention was paid less to cleanliness of the body and more to cleanliness of the soul, particularly in the early rise of the Christian church.

Early studies of nosocomial infections began in Scotland in the first half of the eighteenth century, and Selwyn's article names Scottish pioneers, including J. Pringle, who were active in their attempts to manage hospital infections.[2] In 1764, Pringle endorsed what he called *animate contagion* as the cause of various diseases in the army and said:

> In camp, the contagion (dysentery) passes from one, who is ill, to his companions in the same tent; and from thence perhaps to the next. The foul straw becomes very infectious … But of what nature is this infection? In the former editions of this work I consider the spreading of the distemper as owing to putrid exhalations from humors of those who first fall ill of it; and that when this miasma is received into the blood, I conceived it to act upon the whole mass as a ferment, disposing it to putrefaction …But, having since perused the curious dissertation, published by Linnaeus, in favor of Kircher's system of contagion by animalcula, it seems reasonable to suspend all hypotheses till that matter is further inquired into.[4-6]

According to Selwyn, however, it was Dr. Charles White who in 1773 first described puerperal fever[7] and that Dr. Alexander Gordon in 1795 postulated that it was caused by an infection.[8] He also notes that two Edinburgh professors, Francis Home and Thomas Young, in 1819, described puerperal fever as being caused by an infection.[9]

No mention is made of the two physicians who are most often cited as being among the first to describe puerperal fever, a known cause of serious morbidity and mortality in women after childbirth. Those two physicians are Oliver Wendell Holmes[10] and Ignaz Semmelweiss.[11]

Whatever the cause of these conditions, there was no doubt in Richard Brocklesby's mind that hospitals were dangerous places. He wrote:

Infirmaries, or hospitals, in all countries, are for the most part unclean infectious places, and tho' every precaution is taken to purify them, such as washing with vinegar, burning brimstone, gunpowder, or resinous substances, scouring the boards and such like; yet a perfectly safe purification, in some cases, can never be fully effective ... The seeds of infection, once sown, continue, in some instances, to spread contagious diseases, and to contaminate the house, as much as ever the walls of the Israelites were infected with the filthy leprosy, which is said to have germinated from the walls of their tents or huts.[12]

Just as obstetrical wards were plagued by the morbidity and mortality that accompanied puerperal fever, so too were surgical wards plagued by postoperative infections. James Simpson and Florence Nightingale recognized the problem and advocated more modern preventive measures to avoid these infections.[13-18]

Subsequent developments in infection control occurred through the partnership of Florence Nightingale and William Farr, who together demonstrated the influence of crowding on contagious diseases and proposed the collection of standardized data through surveillance by attending nurses. The concept of *antisepsis* was proposed by Joseph Lister, who performed experiments that demonstrated a dramatic reduction in infected wounds when the wounds were packed with lint soaked in carbolic acid; those experiments were published in 1867.[19,20] Lister's work improved acceptance of the *germ theory* of disease, and growing acceptance gave rise to the development of laboratories to support clinician's activities.[21] The initial focus on obstetrical and surgical patients expanded to medical patients, and the routes of transmission became defined as being airborne and by hand. Increasing attention was given to patients' risk factors, the carrier state of bacteria, and known animal and human reservoirs of bacteria. Eventually, the concept of antisepsis (actively killing germs) was joined by the concept of *asepsis* (the prevention of introducing germs or being free of germs). These concepts spread to Europe, where German investigators embraced them. For many years in the early part of the century, *Streptococcus pyogenes* was the dominant organism with nosocomial potential. After WWII, *Staphylococcus aureus* became the dominant organism. With the discovery of antibiotics, some felt that postoperative infections could be eliminated. It was found, however, that antibiotics not only did not eliminate infections but also posed additional challenges by creating resistant organisms.

By the late 1950s and early 1960s, the era of modern hospital infection control and epidemiology had gained credibility. This was accomplished primarily by statements from the Centers for Disease Control (CDC) at a meeting on *Staphylococcus aureus* in 1958, by publication of the book entitled *Hospital Infections* in 1960,[22] and by the introduction of the concept of *infection control practitioners* in 1962.[23] Developments thereafter included studies of the magnitude, costs, and implications of hospital-acquired infections[24-27] as well as an organizational structure to deal with those infections.[23,28-30] A final piece that confirmed the cost effectiveness of a formal infection control program was the completion of "The Study of the Efficacy of Nosocomial Infection Control," known as the SENIC study.[31]

At about the same time, it became increasingly apparent that health-care providers were themselves subject to acquiring infections by virtue of their work responsibilities. An example of such an infection was hospital-acquired tuberculosis (TB) in employees. Partly because of that risk, Dr. Harvey Estes at Duke led the push for a new department, the Department of Family and Community Medicine at Duke. Its creation[32] had been met with resistance, especially by the clinical department chairs, who objected to having another new department. Nevertheless, Estes and others gradually overcame their resistance, and the department became the home of a new program on employee health. From the outset, political wrangling continued, ultimately leading to the departure of Dr. Estes.

The establishment of a more stable model of employee health began with Duke's recruitment of additional faculty. Dr. George Jackson was recruited to head up a new department that would encompass the disciplines of employee, occupational, and student health in 1982; it was called the Department of Employee Health.[33] Dr. Leonard Goldwater came from Columbia University in occupational health and Dr. William

Christmas from the University of Vermont College of Medicine in student health.

Prior to that, in the mid-to-late 1970s, a program in hospital infection control led by Dr. Suydam Osterhout was established at Duke (called the Hospital Infection Control Committee), and one led by me was established at the Durham VA Medical Center. By the 1970s, guidelines were few, and personnel who chose to work in this field were often inexperienced. Leaders of those programs were typically trained in infectious diseases, but infection control practitioners consisted mostly of nurses who were often inexperienced. Specific activities embraced by the Hospital Infection Control Committee and the Department of Employee Health typically consisted of establishing and implementing infection control policies and investigating outbreaks of infections due to common Gram-positive and Gram-negative bacteria such as *Staphylococcus aureus*, *Serratia marcescens*, *Pseudomonas aeruginosa*, and Acinetobacter. In addition, outbreaks of influenza arose in patients and staff, including the pandemic Hong Kong flu in the late 1960s, and the threat of a swine flu outbreak in the mid-1970s required responses from the Department of Employee Health and the Hospital Infection Control Committee. Patients were still admitted to Duke and the VA with untreated active TB, and new policies and procedures were required. Efforts by the infection control team focused often on improved compliance with hand hygiene, on the implementation of a rigorous sterile technique when invasive procedures were to done, and on the adoption of "reverse isolation" in caring for patients with TB. The Department of Employee Health was proactive in proposing the administration of effective vaccines, first for influenza and later for many of the then-common childhood illnesses such as pertussis, measles, mumps, and rubella and for travelers going out of the country. The Department of Employee Health also began to focus on more preemptive and preventive strategies than heretofore used to improve the health of employees, including blood and body fluid precautions, stop-smoking campaigns, weight loss, and exercise.

A specific outbreak that involved both the Hospital Infection Control Committee and the Department of Employee Health was an epidemic of acute hepatitis B virus infection, which was recognized in employees and students and continued through the 1970s. Over the decade, 108 employees became infected, and the epidemic was of sufficient magnitude to justify the creation of a hepatitis subcommittee of the Hospital Infection Control Committee. Because of my interest in virology and experience with epidemic investigations as an EIS (Epidemiology Intelligence Service) officer at the CDC, I was asked to lead the investigation of this epidemic as chairperson of the subcommittee, with critical input from Dr. Robert Gutman, director of the VA Dialysis Unit, and from George Jackson, Wayne Thomann, and Debra Hunt of the Department of Employee Health.

Our findings are summarized as follows: Investigations of this epidemic were facilitated by the recognition in the mid-1960s of the so-called *Australia antigen* by Baruch Blumberg,[34] which was later shown to be the surface antigen of the hepatitis B virus. This antigen and accompanying antibodies proved very useful in the detection and management of persons infected with the hepatitis B virus. The acutely infected employees were physicians, nurses, medical students, and laboratory personnel. Many of them had had needlestick exposures to needles that had been used previously, but others had cutaneous or mucous membrane exposure to the blood or body fluids. Generally, the illness persisted for two to four weeks, with significant signs (fever, jaundice, and tender liver) and symptoms (fatigue, anorexia, and nausea), followed by complete recovery. That was true for 100 of the 108 employees ultimately diagnosed. Of the other 8 employees, 7 became chronic carriers of the virus, and 1 employee, a clinical chemistry worker, died. The infected employees shared common types of exposures as well as shared contact with distinct patient populations or their body fluids. The population that seemed initially to be the focus, or reservoir, of the hepatitis B virus consisted of persons with chronic renal failure who were on hemodialysis.

In a seroepidemiologic study done in the 1978–1979 academic year, dialysis staff had what turned out to be a very different serologic profile than that of the dialysis patients. Whereas otherwise healthy dialysis staff typically did not have circulating hepatitis B surface antigen, an extraordinarily high percentage of them were antibody positive for the virus (81 percent of them were antibody positive at Duke and 56 percent at the VA), suggesting that they had been infected in the course of their work but had had a subclinical or asymptomatic infection. In contrast, otherwise well dialysis patients typically had an extraordinarily high prevalence of circulating hepatitis B surface antigen and little or no antibody. We compared our data with that derived from the New York Blood Center, where hepatitis B surface antibody had been measured in transfusion recipients. Of those recipients, 29 percent were antibody positive. That comparison lent further credence to our tentative conclusion that Duke and VA dialysis staff had been exposed to hepatitis B in the course of their work and that the source of their exposure was the dialysis patients. It also supported the hypothesis that non-dialysis staff who acquired hepatitis B infection did so through contact with the blood or body fluids of dialysis patients. That proved to be the case.

Further, focusing now on those staff who had been antibody negative prior to the 1978-1979 investigation, at Duke 17 percent had sero-converted over that year, whereas at the VA and in the New York Blood Center, only 6 percent and 8 percent had converted, respectively. These data suggested that at least the Duke staff were still being exposed to the hepatitis B virus frequently, as judged by the high percentage of those who seroconverted in that year alone in spite of infection control efforts to minimize exposures to the blood and body fluids of the dialysis patients. Inasmuch as 34 percent of all Duke dialysis patients were hepatitis B surface antigen positive, inadvertent transmission to dialysis staff is perhaps not surprising. In addition, in those patients who did not have surface antigen, substantial numbers of them had prior exposure to the virus as reflected in the 45 percent and 33 percent detection of surface antibody in dialysis patients at Duke and the VA, respectively. Notably, unlike dialysis patients, infected staff cleared the circulation of the hepatitis B surface antigen (as the surrogate of the virus itself) upon recovery, whereas the dialysis patients remained chronic carriers. There were also substantial numbers of patients who converted their antibody status in the course of that year: 34 percent at Duke, 7 percent in the VA, and 14 percent in the New York Blood Center population. Collectively, the data suggested the existence of a reservoir of potentially contagious HBV infected dialysis patients to whom employees, trainees and students were inadvertently exposed in the course of their routine work owing to their contact with patients' blood and/or body fluids. This was true for dialysis staff, medical and surgical providers, house staff and medical students, ward staff, and laboratory technicians. The data further suggested that there was some cross infection within the dialysis unit. This suspicion was confirmed in studies published by Drs. Robert Gutman, director of the Duke and VA Dialysis Units, Milford Hatch (a CDC virology laboratory director), and me. In those studies, we demonstrated that in both the Duke and the VA dialysis populations and staff members, the subtype of hepatitis B virus (ay) was different than that which was prevalent in the community at large (ad).[35] This further suggested that cross-contamination in the respective dialysis units had occurred. Some aspects of the data discussed earlier suggested that the epidemic in the VA began earlier than it had in the Duke unit. If true, was that a possible explanation for how both units had the same subtype? Although uncommon, some patients received dialysis at both the VA and Duke.

This outbreak obviously captured the attention of the Hospital Infection Control Committee, the Department of Employee Health, and the hospital administration. Hospital policies changed, with an emphasis on minimizing and eliminating the possibility of cross-contamination, including the use of specific policies and practices for hepatitis infection control. Duke and the VA were participants in a clinical trial testing the value of hyperimmune hepatitis B immune globulin (HBIG) in the prevention of active acute hepatitis B after a high-risk needle-stick exposure.[36] The study demonstrated efficacy in the prevention of subsequent hepatitis B viral infection, and this practice was implemented at Duke and the VA.

Beginning in the late 1970s, Duke and the VA participated in the trial of the first version of the hepatitis B virus vaccine. Merck and Company had prepared the vaccine and required a complicated physical and chemical process to purify the hepatitis B surface antigen. The antibody to this antigen confers immunity to the virus. That vaccine prototype was administered to a number of different at-risk populations, including dialysis staff and dialysis patients. The vaccine, Heptavax B, was shown to be effective in dialysis staff[37] but was of no significant benefit for dialysis patients,[38] possibly reflecting the compromised immune systems in dialysis patients. Complicating the strategy to immunize staff with Heptavax B vaccine was that the initial preparation was derived from known chronic carriers of hepatitis B virus. The manufacturers of the vaccine were confident that no infectious hepatitis B viruses remained in the preparation, but what then became apparent was that the carriers of hepatitis B virus were often some of the same individuals who had been recognized as having the Gay Related Immune Deficiency Disease (or GRID), a disease that later was found to be caused by the HIV virus. Those facts made a wide-scale immunization program complicated because of the fears that accompanied the administration of the vaccine. Nonetheless, a hospital-wide vaccination program at Duke and the VA hospital was initiated in 1984, and the outbreak in employees halted. The Employee and Occupational Health and Safety Committee and the Hospital Infection Control Committee were heavily involved in organizing and implementing the program.

Several years later, the vaccine preparation process was altered and the vaccine was engineered by recombinant DNA technology, thus eliminating the possibility that live infectious hepatitis B or HIV were present in the vaccine preparations. Selected populations at Duke and the VA were chosen to receive the vaccine, with the recognition that not everyone was at risk. In the initial year of the vaccine program, the costs were significant (about three hundred dollars per employee) and were borne by the Duke and VA hospitals.[39] As would be expected, the Joint Commission on the Accreditation of Hospitals (JCAH) mandated federal regulatory policies and practices, which were initiated in hospitals nationwide. Those policies were further endorsed by the CDC's publication of its "Universal Precautions" in 1987[40] and updated in 1988 and mandated by the Occupational Safety and Health Administration's "Bloodborne Pathogens Standard" in 1991.[41] The hospital administrations at Duke and the VA enhanced the program for Hospital Infection Control and the program for Employee Health, with both programs responsible for dealing with those policies and future problems affecting employees and patients.

Apart from the hepatitis B epidemic, the more standard hospital infection control and employee health activities that were initiated in the late 1970s were continued through the 1980s, but the unexpected occurrence of the first cases of GRID in 1981 occupied most of the attention of the Hospital Infection Control Committee and the Department of Employee Health. Early efforts to respond to this new disease proceeded from the outset with multifaceted educational programs for patients and providers. When it became known, however, that GRID was caused by a retrovirus, HTLV-III (later called HIV-1), and a test was licensed to measure antibodies to the virus in 1985, a much more organized and complicated array of policies and procedures was put in place, led by the Ad Hoc Committee on AIDS, chaired by Dr. Osterhout. In addition to posing a new array of controversial societal issues, AIDS also posed incredibly challenging medical issues that affected both patients and caregivers. Official Duke Hospital policies were published on April 6, 1988, on the following topics: (1) HIV testing, (2) informed consent, (3) patient notification of test results, (4) patient pretest and posttest counseling, (5) test standards, (6) use of test results, and (7) confidentiality. In addition, in December of 1992 the final Duke policy and procedure recommendations were made for employees with HIV or hepatitis B.

One important activity with major implications for hospital infection control and employee health was the creation of what later became known as an antibiotic stewardship program. In the late 1970s and early 1980s, this program was referred to as the Antibiotic Evaluation Team and somewhat later as the Antibiotic Decision Support Team. Conceived in part because of the increasing costs of antibiotics at Duke Hospital, especially the new class of cephalosporins, the program had another impetus—the recognition that health-care providers were not well informed about the indications for antibiotic use. Working together on an informal basis, Drs. Catherine Wilfert and Harry Gallis concentrated on an informational program that included an evidence-based list of indications for using selected antibiotics and a list of recognized organism-susceptibility patterns. After a daily notification of the patients who were to receive a cephalosporin, the ward clerk simply placed the lists prepared by Drs. Wilfert and Gallis in the patients' charts, even though the attending physicians had not requested consultations. The intent was to optimize antibiotic therapy and, in so doing, potentially reduce costs by eliminating unnecessary antibiotic treatments and adverse side effects. The practitioners perceived this activity variously, some thankful for the helpful advice and some resentful of the unwanted intrusion into their practice. There was no personal financial incentive for either Dr. Wilfert or Dr. Gallis.

In the 1980s, it began to become clear to most providers that the adult and pediatric infectious diseases divisions should provide some oversight for the newly introduced and the increasingly expensive anti-infective agents. It was at this point that pharmacists began to play a more integral role in the oversight of anti-infective therapy. This was particularly true for the pharmacists who had worked side by side with the infectious diseases physicians in the Department of Medicine on clinical trials for new anti-infective drugs (Richard Drew, Pharm D, and William Pickard, MS Pharm). Chris Rudd, Pharm D, served in a similar role with the Pediatric Division of Infectious Diseases. In parallel, all served in a complementary role in the conduct of novel clinical trials of new agents. Drs. Gallis and Drew concentrated on the route of administration and dose optimization of a new class of antibiotics, namely fluoroquinolones, the intravenous preparation of which (Ciprofloxin) was commonly being used without clear-cut indications and at great cost. Typically, Dr. Drew reviewed the charts on all patients receiving those drugs and, for the patients without clear-cut indications, reviewed the clinical courses with Dr. Gallis. He then entered unsolicited comments in the patients' charts. In a sense, this activity increased the role of the infectious diseases and pharmacy departments in the oversight of antibiotic use and arguably improved the patients' clinical courses and certainly reduced the attendant costs. Also, optimizing antibiotic therapy in principle had the additional benefit of minimizing the emergence of bacterial resistance to the benefit of both hospital infection control and employee health. Those benefits captured the attention of the hospital administration and resulted in the establishment of a formal program to optimize the use of antibiotics and other anti-infective drugs, which were, in the late 1980s and early 1990s, dramatically increasing in number. The success of this program paradoxically led to the cancellation of financial support after one to two years owing to the mistaken notion that the problem had been solved and therefore needed no further support.

Meanwhile, Pickard was working closely with another Department of Medicine faculty member, Dr. James Hathorn, who was trained both in infectious diseases and in hematology/oncology, on clinical trials in patients with neutropenia, or low white blood cells. At the VA, I interacted directly with Michael Martz (Pharm D), who represented the pharmacy on the Pharmacy and Therapeutic Agents Committee, and Peter Zwadyk (PhD), who was head of the Clinical Microbiology Laboratory. To guide drug treatment decisions, we generated antibiograms that indicated susceptibility or resistance of more common bacteria to a range of antibiotics. Policies for the use of specific anti-infective drugs were established and dictated which drugs could be prescribed for which indications. This was a notable difference from subsequent practices at Duke and was only possible because proscriptive policies could be administratively introduced at the VA..

The complexity of patients with, or at risk for, infections only accelerated in the 1990s because of a number of factors, including expanding transplantation programs, ever more aggressive chemotherapy for patients with cancer, the recognition of serious immune defects in children, and the expanding burden of persons living with AIDS prior to the availability of effective therapies. The common theme among those populations was their compromised immune systems. If this were not enough, personnel changes, both additions and departures, resulted in realignments of responsibilities and expectations. Drs. Gallis and Hathorn left Duke, and Dr. Carol Dukes Hamilton assumed responsibility both at the VA and later at Duke under the newly named Antibiotic Decision Support Team (ADST), under the administrative umbrella of the expanded Pharmacy and Therapeutic Agents Committee, which was led by Dr. Peter Kussin. Additions to the program included Pharm Ds Elizabeth (Libby) Dodds and Melissa Johnson, both of whom joined the Adult Division of Infectious Disease. These developments increased the connections between the ADST and the School of Pharmacy at Campbell University, resulting in appointments both at Duke and at Campbell. Dr. Drew enhanced his reputation at both universities through the creation, initially, of a Duke-specific pocket guide for antibiotic use, available to all providers and, later, as a web-based Duke-specific set of guidelines and policies called "Custom ID." This product attracted international attention, including its purchase by The Netherlands to serve their nationwide health-care system.

By 1990, it had become clear that a more proactive infection control program and employee health program would be required to meet federal standards. Additions to the Infectious Diseases Division and to the Clinical Microbiology Laboratory provided the needed clinical expertise, as well as additional personnel and infrastructure. Dr. Daniel Sexton assumed leadership of the Hospital Infection Control Committee, and Dr. Osterhout turned his attention to his duties as director of admissions in the Duke Medical School. Drs. Wayne Thomann and Debra Hunt assumed positions dealing with a new program, the Employee and Occupational Health and Safety (EOHS) Program, where they assumed responsibility for policies and procedures for patient and employee safety. In addition, Dr. Kathryn Kirkland, a former fellow, became interested in infection control and after serving as an EIS officer assigned to the State Board of Health in Raleigh joined the faculty at Dartmouth and led the Hospital Infection Control Committee. In the latter part of the 1990s, she co-published papers with Dr. Sexton, later independently publishing her own papers as well.

Outreach activities were already under way, with infectious diseases faculty serving as consultants to community hospitals around the state, including the medical centers in Concord, Burlington, Henderson, Fayetteville, and the federal prison in Butner. In a special program with Saudi Arabia, nine faculty members each consulted on-site with the medical center in Riyadh for two-month intervals. Consultations consisted of guidance on clinical management in some instances, but more so on issues relevant to hospital infection control. Ten or more adult infectious diseases faculty provided these consultations. Prompted by a suggestion from Dr. Eugene Stead to Dr. Sexton that he should do his infection control research in the community instead of in only the large tertiary and quaternary care medical centers, the idea of an infection control network was considered.[42] In 1997, Dr. Sexton proposed the creation of such a network, first to me, as the division chief at Duke, and then to Mr. Paul Thacker, the business manager of Duke's Department of Medicine.

Two issues arose in the course of early discussions, and both revolved around money. First, it was clear that Dr. Sexton would require some financial and ultimately human resources and was unlikely to be sufficiently compensated by the hospitals that agreed to be part of this network—at least initially. Second, the infectious diseases faculty had already been performing outreach activities in the area of hospital infection control, and revenues from those activities were being used to support part of their salaries. To include all these consenting hospitals in the proposed network would possibly direct the generated revenue to the needs of the network and not necessarily to the support of the faculty.

To obtain the financial support needed to begin this effort, we held a series of conversations with Dr. Barton Haynes, then chair of the Department of Medicine, and with the leaders of the Private Diagnostic Clinic. It was generally agreed that this venture was worthwhile, and Dr. Haynes made funds available to begin the development of a network, including the hiring of two infection control practitioners, establishing a workable database, and marketing the concept to community hospitals. Marketing the concept of a network was based on a business plan prepared by Dr. Sexton. Almost from the outset, the network was to be called the Duke Infection Control Outreach Network (DICON).

To solve the second issue, we held conversations with individual faculty members, and somewhat surprisingly, most of them agreed to the establishment of a network. Because of Dr. Sexton's high clinical profile at many of the outreach hospitals and his extraordinary publication record,[43-47] most local physicians supported the establishment of contracts for DICON to provide support for the local infection control activities. Initially, six hospitals signed the contracts. The hospitals were in Person County, Oxford, Rocky Mount, South Boston, Henderson, and at Durham Regional. It was with these hospitals that models were developed to provide either full-service infection control support or a more limited service.

Although involved in infection control as a fellow to a limited extent, Dr. Kirkland subsequently joined the CDC as an EIS officer assigned to the State Board of Health in Raleigh, where she collaborated with the Duke HIC committee faculty on papers and projects. Later she was appointed to the faculty at Dartmouth, where she served as chair of the Hospital Infection Control committee for twelve years.

The concept of using community hospitals proved novel and did not compete with the CDC's National Nosocomial Infection Surveillance Program. The initial infection control practitioners (Melissa Bronstein and Mary Oden) were remarkably successful in achieving the stated goals, which boded well for the network's future and provided options to leverage resources from industry and from the federal government.

Photograph 5.4 is a composite of pictures of four prominent figures in the fields of infection control, employee and student health at Duke in the twentieth century.

Daniel J. Sexton

George Jackson

William A. Christmas

Kathryn B. Kirkland

CHAPTER 6

THE IMMUNOCOMPROMISED HOST

What brought me to Duke apart from the campus and general ambiance and the support of Handler, the most important thing was that air of tremendous potential down here ... It was the presence of a number of young, vital scientists, and the sense that Duke was going to go on in that direction. The administration tended to leave you very much alone. It was a very, very permissive and liberal campus at that time.

—D. Bernard Amos[1]

Patients with compromised immune systems are not a new phenomenon, but only in the last half of the twentieth century were the causes understood. One cause, ironically, was the use of certain medical interventions that were introduced during this period (for example, organ transplantations), interventions meant to prolong life. Conversely, these same interventions also led, in part, to a better understanding of the immune system. Because the consequences of a compromised immune system include increased susceptibility to new and unusual infections, the Division of Infectious Diseases at Duke has been involved both in the therapeutic approaches used to treat those infections and in the research studies done to describe how the immune system works. This chapter will describe the faculty and the scientific developments that contributed to the understanding and management of the immuno-compromised host. In addition, it will touch briefly on the disciplines of bacteriology, mycology, virology, and immunology, which are inextricably related to the immunocompromised host.

BACTERIOLOGY, MYCOLOGY, AND IMMUNOLOGY

Dr. David T. Smith: Dr. Smith was mentioned in chapter 3 because of his involvement with tuberculosis. His importance for the discussion of the immunocompromised host, however, resides primarily in his role at Duke as the chairman of the Department of Bacteriology (later Microbiology) and his collaboration with Dr. Norman Conant, an expert in mycology. Dr. Smith was recruited to the faculty at Duke in 1930, when the hospital first opened. His role on the faculty was to oversee the Clinical Bacteriology Laboratory, to teach the elements of bacteriology to the students, and to conduct research. Much of his research revolved around various aspects of mycobacteria, including the performance and interpretation of the tuberculin test, but his major focus in his first twenty years at Duke was a study of lung infections caused by anaerobic bacteria and fungi.[2-7,8]

Dr. Norman Conant: Not too long after Dr. Smith came to Duke, Dr. Norman Conant was recruited from Hans Zinsser's laboratory at Harvard, where he had been studying fungi and the diseases that they cause. In addition to obtaining his PhD under Dr. Zinsser, Conant did postdoctoral studies in South America, in Washington, DC, and in Paris; in Paris, he worked on studies led by Dr. Raimond Sabouraud, the founder of medical mycology.[9] Dr. Conant's connection with some of the major figures in mycology brought a certain cachet to the Duke Department of Bacteriology and Mycology. Conant had a productive academic career in which he contributed chapters to many prominent textbooks and published sixty-three peer-reviewed papers on various aspects of fungal diseases.[7,10-18]. He and coauthor J. G. Downing did a comprehensive review of mycotic infections, published in *The New England Journal of Medicine*.[19] Dr. P. H. Hiss authored the then-preferred textbook of bacteriology and mycology, with the first edition published in 1910. It was later titled *Zinsser's Microbiology*, and editorial responsibility was transferred for several editions to Drs. David T. Smith and Norman

Conant, among others. In addition, Smith and Conant collaborated on another most influential book, *The Manual of Clinical Mycology.*[3] Dr. Conant had many prominent collaborators, including Rhoda Benham, PhD, from Columbia University, Charles Smith, MD, from Stanford University, Chester Emmons from the National Institutes of Health (NIH). The collaborators at Duke included Drs. Roger D. Baker and J. Lamar Calloway, the first chief of the Division of Dermatology at Duke. Among Conant's major achievements was the development of the Duke summer mycology course, which he taught from 1948 to 1973.[9] The course was very popular and was attended by aspiring and accomplished mycologists from around the world. Because of his contributions to the field, Conant was the recipient of the coveted Rhoda Benham Award for outstanding contributions to the field by the Medical Mycological Society of the Americas. He was a member of a number of scholarly societies and the recipient of several Duke teaching awards during his career. He replaced Dr. Smith as the chair of the Department of Bacteriology and Mycology in 1958 and continued in that role until 1968.

The recognition of new and unusual infections had not occurred in the 1950s, but by the mid-1960s, new scientific discoveries paradoxically led to what came to be called *opportunistic infections.* Those discoveries included the following: first, the ability to perform tissue typing, which led to the possibility of performing solid organ and ultimately bone marrow and stem cell transplantations; second, the discovery of immunosuppressive drugs, which were critical for successful transplantations and other medical conditions and allowed more aggressive treatment of certain cancers that heretofore had been incurable; third, immunologic, molecular, and genetic testing to identify acquired and inherited defects in the immune system that led to deaths in newborns and neonates; and fourth, the recognition that the immune systems of the very young and the elderly were not sufficiently robust to ward off poorly controlled infections. Much later, patients with HIV/AIDS became the prototype of the immune-compromised host. Eventually, improved understanding of immunity gave rise to the possibility that the defect in the immune response could, in some cases, be corrected. Central to these discoveries, however, was a better understanding of the elements of a normal immune system.

Although the immune system was not fully recognized as a network of processes designed to prevent infection or eliminate foreign matter, a variety of scientists made observations touching on this concept as early as the fifth century AD. Their observations included the inoculation prospectively of the contents of smallpox pustules into susceptible recipients to induce a milder form of smallpox.[20] Somewhat later, Jenner used a variation of that technique in the latter part of the eighteenth century. He used cowpox rather than smallpox to induce an unknown protective factor that prevented or modified clinical smallpox. Protective vaccines were developed for, among other things, anthrax, rabies, and typhoid in the latter part of the nineteenth century. In the beginning of the twentieth century, serum-based therapy for tetanus and diphtheria was adopted without a full understanding of the mechanism by which serum provided resistance to infection. In the first half of the twentieth century, improved equipment and technology emerged and allowed for a better understanding of the components of the immune system. In the last half of the century, the importance of cellular immunology and innate immunity were added, increasing the understanding of the immune system. A detailed discussion of the evolution of immunology is beyond the scope of this book, but for further reading on this topic, the chapter on immunology in *Fundamental Immunology* by Mazumdar is extremely informative.[20]

Dr. D. Bernard Amos: Duke faculty members were uninvolved for the most part in the study of immunity, until the arrival of Dr. D. Bernard Amos in 1960. Amos was born and schooled in England, attended medical school there, and gained some of his earliest research experience working in the laboratory with Dr. Peter Gorer, who had worked in a laboratory in the United Kingdom, overseen by Drs. Peter Medawar and Peter Snell, arguably the founders of immunogenetics and transplant immunology.[21] In 1955, Amos moved to the Roswell Park Memorial Institute in the United States, where he worked with Theodore Hauschka on tumor immunology. In 1960, Amos was invited by Dr. Philip Handler, then the director of the National Institute of General Medical Sciences at the NIH, to give a report on what he knew about the genetics of tissue antigens. Subsequently, he was recruited by the then-dean of the Duke Medical School, Dr. Barnes Woodhall, to lead a genetics research group at Duke.

Amos joined Duke with appointments as professor of immunology and as professor of experimental surgery. Amos brought Drs. Eugene Day and Richard Metzger with him from Roswell Park, and the three of them formed the nucleus of the genetics research group. A further attraction for Amos was that at Duke there "was no master plan—there was only opportunity."[1] Although research space was so limited, his position was stable, having successfully competed for an NIH Research Career Award, and he could focus on the science of human transplantation.

To this end, he collaborated with other international scientists interested in immunogenetics as it applied to tissue typing. These scientists included Drs. Rose Payne, Paul Terasaki, Jon J. van Rood, Jean Dausset, Ruggero Ceppelini, and Fritz Bach. He also collaborated with a clinical team that was to be responsible for the first organ transplantations at Duke. The team included Drs. Delford Stickel, Caullie Gunnells, Roscoe Robinson, James Glenn, and, somewhat later, Hilliard Siegler, Fran Ward, and others. Amos concentrated initially on studies in mice[22] and later on family studies of the genetics of the immune system's recognition of foreign antigens in humans. He was recognized as a co-discoverer of the Major Histocompatibility Complex/ Human Leukocyte Antigen (MHC/HLA) system.[23-33]

At the national level, the interest in organ transplantation was increasing. The New York Academy of Sciences therefore sponsored the creation of the Transplantation Society, with Dr. Amos as chair of the organizing committee.[1,21] Dr. Amos organized the first "wet workshop" for standardizing a set of serum samples for use in tissue typing, a critical step for gauging HLA compatibility of donors and recipients in organ transplantation. That workshop was held at Duke in1964.[1,21] Because of the workshop's successful development of standardized tissue-typing reagents, kidney transplantations were performed in the United States and abroad, first in living related identical twins, next in living related family members, and later with cadaveric donors.

In 1965, Dr. Delford Stickel, the lead surgeon, accompanied by the aforementioned clinical transplant team, performed the first kidney transplantation using newly defined techniques to establish levels of compatibility between the donor and the recipient.[26,34-36]. Tissue-typing reagents became increasingly specific over time and were developed by the core investigators in the Transplantation Society. The new techniques included in vitro leukocyte agglutination tests, cytotoxicity tests, and skin grafts. The first kidney donor selected based on those in vitro tests was one of the recipient's seven siblings. The patient's outcome was favorable and better than earlier transplantations had been, in which those tests were not used. This one case, however, was not sufficient to conclude that the tests should be used as the selection criteria. Nonetheless, the favorable outcome in the restoration of normal renal function and in only a modest need for ongoing immunosuppressive therapy prompted intensified efforts in the United States and Europe to refine and to test those techniques in larger populations of patients with chronic renal failure.

Ultimately, the complexity of the tests became understood more completely and provided additional support for their adoption. With time, this expanded the pool of potentially eligible unrelated kidney donors. In addition, however, some practical issues arose because of the need to expand the opportunities for the use of living related donors and later for using cadaveric donors. One of the more important issues was the ability to identify correctly suitable donors and to match them with suitable recipients, sometimes at long distances. To address that issue, Amos cofounded with Dr. David Hume of the Virginia Commonwealth University an organ procurement program, the South Eastern Regional Organ Procurement Program.[21] The program was funded in part by the NIH and spawned additional research areas related to the requirements for preserving a resected kidney during a period of complete ischemia. It also spawned refinement of the in vitro donor-recipient compatibility test. Although the leukocyte agglutination test became standard for donor selection, it was almost impossible to identify donor-recipient pairs that were completely compatible by that technique unless the pair were identical twins.

How aggressive immunosuppressive therapy should be in individual patients postoperatively was not known. That question prompted research into identifying better-tolerated regimens that would pose less risk for developing the infectious complications of immunosuppression.

Significant success fortunately occurred in all those areas, leading to a realistic alternative to lifelong dialysis in patients with kidney failure.

Although the kidney transplant program expanded dramatically over the next fifteen to twenty years, it was not until 1984 that other transplant programs were established at Duke: liver and adult bone marrow in 1984, heart in 1985, kidney and pancreas in 1989, pediatric blood and bone marrow in 1990, outpatient autologous bone marrow, pediatric allogeneic unrelated cord blood in 1993, thymus in 1994, and lung in 1995.

As these transplantation programs became more routine and the HLA testing became more sensitive and specific,[37,38] additional problems were encountered. The problems included surgical technical difficulties and the infectious complications that arose from the need for more intensive immunosuppression. Subsequently, the ingenuity of investigators in this field, including pharmaceutical firms, has been very creative and has overcome many of these obstacles. Meanwhile, a substantial proportion of patients were immunosuppressed and experienced unusual opportunistic infections that required additional vigilance in their diagnosis and enhanced microbiologic diagnostic tests.

Dr. Rebecca Buckley: Another major figure at Duke in the early 1960s was Dr. Rebecca Buckley, a pediatric immunologist whose scientific interests were in the identification of the genetic basis of the severe defects in T and B cells that led to the clinical occurrence of the severe combined immunodeficiency syndrome (SCIDS) in newborns.[39-50] Initially trained in the laboratory of Dr. Amos and later in the laboratory of Dr. Richard Metzger, Dr. Buckley has, along with others, found twelve genes that, when mutated, result in severe combined immunodeficiency syndrome (SCIDS).[51] Treatments initially were limited because a compatible bone-marrow or stem-cell donor was not commonly available. Transplantation of incompatible cells not only did not provide the needed normal B and T cells but also did not reject the transplanted foreign cells, resulting in the occurrence of severe graft-versus-host disease. Nonetheless, Dr. Buckley pioneered a procedure whereby the T cells are removed from the donor bone marrow, allowing for a successful transplantation. In her over twenty-five years of research on babies with SCIDS who underwent transplantation by that method at Duke, 123 of

157 were still living in 2007.[51] That technique was developed initially in work done at the Sloan-Kettering Institute and its usefulness validated at Duke and other facilities.

In addition, Drs. Buckley and Schiff showed that when a transplantation is done before the infant is three and a half months old, the outcome is dramatically better than if it is done when the infant is older. When it was done before the age of three and a half months, the percent of survival was 96 percent, whereas for older infants the survival was 71 percent.[39] Delay exposes the infant to an acute lethal infection.

Dr. Buckley's laboratory research focused principally on T- and B-cell development, immunologic tolerance, and HLA restriction. Because of the importance of early transplantation, Dr. Buckley has advocated for neonatal screening to avoid predictable mortality from infection and, in fact, for cost-effectiveness. Dr. Buckley has tirelessly advocated for women in medicine and science and has received numerous awards, including election to the Institute of Medicine in 2003 and receipt of the Anlyan Award for Lifetime Achievement at Duke in 2006. In addition, she has served as an elected officer in professional societies, including president of the American Academy of Allergy and Immunology in 1980.

Dr. Ralph Snyderman: Dr. Snyderman returned to Duke in 1972 after spending five years working in the Dental Institute at the NIH on periodontal disease. He joined the Division of Rheumatology because of his background working in the general area of inflammation. Over succeeding years, Dr. Snyderman acquired expertise in clinical rheumatology and succeeded Dr. William Kelley as chief of the Division of Rheumatology in the early 1980s. He focused most of his attention on the molecular pathogenesis of inflammation, an important component in the response to infectious diseases. He maintained a loose connection with the Division of Infectious Diseases, studying unusual infections from an immunologic perspective. Dr. Snyderman was extraordinarily productive; his research focused on defining the mechanisms by which leukocytes accumulate at sites of inflammation.

In 1967, he developed the first reliable in vitro technology to quantify leukocyte chemotaxis. His work led to a definition of standard methodology for the study of this critical component of inflammation. Shortly thereafter, he identified C5a, a cleavage product of the fifth component of complement (C) as a major chemotactic factor, which was produced by complement activation or by proteolytic cleavage of C5. This seminal discovery advanced the field of inflammation as well as provided an important therapeutic target. Along with colleagues, he was the first to identify a chemotactic lymphokine (cytokine) produced after lymphocyte activation, and he identified the first chemotactic factor receptor on leukocytes. This and related work demonstrated coupling of the receptor to G proteins, leading to activation of phosphorylation protein C rather than adenylate cyclase, which at the time was thought to be the exclusive target of activated G proteins. Snyderman and colleagues further defined the detailed mechanisms of leukocyte activation and desensitization by chemoattractants and identified novel pathways for "receptor class" desensitization. The following citations provide the scientific details.[52-61] Because of his achievements, he was selected for high-level positions outside of Duke. Nevertheless, he ultimately returned to Duke in 1989 as its chancellor.

Drs Dani Bolognesi and Barton Haynes: Both Drs Bolognesi's and Haynes's contributions to the understanding and management of what came to be recognized as perhaps the consummate immunocompromised host, AIDS, have been discussed in some detail in chapter 4 and will not be further discussed here.

MICROBIOLOGY, VIROLOGY, AND CLINICAL INFECTIOUS DISEASES

Two additional iconic figures with enormous impact on the basic science and clinical care of infectious diseases were recruited to Duke in 1968: Wolfgang Joklik, PhD, and Samuel Katz, MD, respectively. Both were recruited to Duke as department chairs, Joklik as chair of the Department of Microbiology (succeeding Norman Conant, PhD) and Katz as chair of the Department of Pediatrics (succeeding Jerome Harris, MD).

Dr. Wolfgang Joklik: Dr. Joklik's scientific interests were in the areas of molecular biochemistry and the genetics of pox viruses.[62] When Joklik arrived at Duke, he joined the relatively small Department of

111

Microbiology, one that included the Division of Immunology, led by Dr. Amos. Dr. Joklik served in many leadership positions in professional societies, including the Association of American Virologists, and was elected to the National Academy of Sciences. Among his major achievements, Dr. Joklik cites the application for, and success in, obtaining one of the initial cancer centers supported by the NIH and the nurturing of young scientists with interests in virology.[63] As will be discussed in chapter 9 on vaccines, Dr. Joklik played a very important role in the policies adopted for the disposition of the smallpox stocks, but his importance for this chapter resides in his pivotal appointments to his Department of Microbiology.

Dr. Samuel Katz: Dr. Katz came to Duke from the Boston Children's Hospital and the laboratory of Dr. John Enders, a co-recipient of the Nobel Prize for his work on the tissue-culture isolation of the poliovirus. Dr. Katz, however, worked on the measles virus while in the Enders laboratory. He was nonetheless supported by the National Foundation for Infantile Paralysis, and he, in collaboration with Dr. David Feingold, obtained the first training grant for a joint training program in infectious diseases between the Children's Hospital and the Beth Israel Hospital in Boston.[63] At the time of his arrival at Duke, his department (Pediatrics) was dealing with the usual infectious diseases of childhood, including *Haemophilus influenzae* meningitis, empyema, epiglottitis, pneumococcal and staphylococcal diseases, and a variety of viral respiratory and enteric conditions. Dr. Katz brought with him from Boston three other pediatric infectious disease specialists: Dr. Catherine Wilfert, Dr. David Lang, and Dr. John Griffith. Somewhat later, he recruited Dr. Laura Gutman from the University of Washington. These recruits substantially enhanced the impact of the Pediatric Infectious Disease Service, and in the wake of the HIV epidemic, they managed an increasing number of cases of neonatal HIV.

As chair of the Department of Pediatrics, Dr. Katz's involvement in infectious diseases revolved principally around his expertise in the field of vaccinology and as a member of the Advisory Committee on Immunization Practices (ACIP) of the Centers for Disease Control (CDC). Dr. Katz encouraged collaborations between divisions in his department and counterparts at the University of North Carolina; in addition, he supported collaborations between the adult and the pediatric divisions of infectious diseases at Duke. He had a most enlightened view of the role of the infectious disease services in a complex quaternary care medical center and was a powerful advocate for both the faculty and the staff. He was also a prominent figure in the national infectious disease leadership and was a recipient of many awards. A popular teacher and a highly respected clinician, he contributed enormously to the Duke Department of Pediatrics and Division of Pediatric Infectious Diseases.

With the increasing frequency of kidney transplantation, the increasingly aggressive chemotherapy for cancer and other diseases, the progressive aging of the population, the improved survival of premature infants, and the recognition of AIDS, the demands for increased expertise in the broad field of infectious diseases prompted further recruitments to the basic and clinical sciences. Moreover, the nature of the so-called opportunistic infections posed many challenges, and new recruits were added to the faculty to address these challenges.

Dr. Thomas Mitchell: Prominent among those recruits in basic science was Dr. Thomas Mitchell. Mitchell obtained his doctorate in microbiology and immunology at Tulane University School of Medicine and was recruited to Duke by Dr. Joklik in 1974. As a basic science faculty member, Mitchell devoted most of his time to teaching graduate and medical students and to doing research in mycology. His contributions included basic studies of the dimorphism in *Candida albicans*, the structure and biological properties of the capsule of *C. neoformans,* and the characterization of the leukocyte chemotaxin produced by *Blastomyces dermatitis.*[64-73] For about a quarter of his time, he served as the director of the Clinical Mycology Laboratory and reinstated the popular Duke summer mycology course. He continued to oversee this course for the next seventeen years. Of note, visiting speakers consisted of a virtual Hall of Fame of clinical and basic mycologists.[9]

Dr. David Durack: Another recruit to Duke in the late 1970s was Dr. David T. Durack. Dr. Durack became the first chief of the Adult Division of Infectious Diseases and came following completion of a Rhodes

Scholarship at Oxford and a year as chief resident at the University of Washington. He had a background in experimental endocarditis and within several years applied similar techniques to a rabbit model of cryptococcal meningitis supported by a R01 grant from the NIH.[74-83] In addition, he forged important collaborations with others concerning fundamental discoveries of the eosinophil neurotoxin and the earliest clinical trials of AZT in the treatment of AIDS.

Dr. John Perfect: Working with Dr. Durack as a new fellow in 1978 in the Division of Infectious Diseases was Dr. John R. Perfect. Dr. Perfect continued this line of investigation by using animal models to study fungal pathogenesis and treatment. In 1987, however, he redirected his career to focus more specifically on the molecular pathogenesis of *Cryptococcus neoformans*. In preparation for this career shift, he took a sabbatical to work in the molecular mycology laboratory with Dr. Peter McGee at the University of Minnesota. By his own description:

> My primary model for these studies has been *Cryptococcus neoformans.* My laboratory has created an infrastructure of studies which have clearly defined the boundaries of cryptococcal virulence. With the merger of animal models of Cryptococcus in rabbits, mice, worms, grubs and zebrafish and the tools of molecular yeast manipulation, we have pushed forward to understand why and how Cryptococcus becomes a pathogen.[68,77,79,81,84,85]

In addition, his research group began to conduct translational research by applying basic science to clinical questions relevant to human disease. The knowledge base included studies of epidemiology, diagnostics, drug treatment, and therapeutic strategies for invasive fungal infections. Collaborations at Duke with Dr. Stephan Johnson in the Department of Botany and Dr. Joseph Heitman in the Department of Genetics greatly facilitated his entry into this new field and led ultimately to R01 funding from the NIH. Subsequently, Drs. Perfect, Heitman,

Mitchell et al. successfully competed for a Program Project Award, which established the Duke University Mycology Research Unit (DUMRU). Dr. Perfect's many ongoing accomplishments have led to his recognition as one of the most eminent mycologists in the United States. Moreover, he has remained closely connected with the clinical inpatient and outpatient services, with an emphasis on the immune-compromised host, and has continued to serve as a valued mentor of numerous trainees.

Dr. Jack Keene: Dr. Keene received his doctorate in microbiology and immunology from the University of Washington in 1974. He was a staff fellow with Robert Lazzarini in the Laboratory of Molecular Genetics, National Institute of Neurological Disease and Stroke (NINDS), NIH, from 1974 until 1978. He was then recruited to the Department of Microbiology and Immunology at Duke. He rose to the rank of professor in ten years, succeeded Dr. Joklik as chairman of the Department of Microbiology in 1992, and was awarded a James B. Duke professorship in 1996. He stepped down as chairman in 2002. His laboratory was very productive, with over one hundred publications in the twentieth century and many more in the subsequent decade.[86-91] He cites four major scientific achievements, as follows:

1) Discovery of the "RNA-Recognition Motif" (RRM) Family of RNA-binding proteins based on biochemical demonstration that it constitutes the functional core of an RNA-binding domain[92-94]

2) Isolation of the first recombinant human autoimmune antigen, elucidation of autoimmune epitopes, and development of a clinical diagnostic test for systemic lupus erythematosus and related diseases[92,95,96]

3) Elucidation of the functions of the ELAV/Hu posttranscriptional regulators (HuA, HuB, HuC, HuD) and their roles in proliferation, neuronal differentiation, and immune response[97-100]

4) Theory of the "Posttranscriptional RNA Operon or Regulon" and modular coordination of RNA gene expression networks in eukaryotes[101,102]

His importance for the understanding and management of infections in the immunocompromised host was in his appointments to his department of faculty with more direct connections to these types of infections.

Wiley Schell, MS: Recruited as the analytical supervisor of the Clinical Mycology and Mycobacteriology Laboratory within the Clinical Microbiology Laboratory, Wiley Schell, MS, became an invaluable asset in the diagnosis, management, and treatment of the increasingly complex patient populations who were being referred to the infectious diseases services. His research proceeded along two lines. First, to improve the diagnosis of fungal infections, he sought to clarify and strengthen morphology-based identifications complemented by histopathologic discoveries. For example, he identified the mechanism (adventitious sporulation) for rapid dissemination of Fusarium infections in neutropenic patients.[103-105] He engaged in the in vitro and in vivo development of antifungal compounds, which led to marketable compounds, posaconazole and pneumocandin,[106,107] and he identified drug-resistant isolates of Candida from HIV-1–positive patients.[108] Schell left the Clinical Microbiology Laboratory in 1990 to establish an independent specialty mycology laboratory within the Adult Division of Infectious Diseases, where he remains today working with adult and pediatric faculty on a productive research agenda.

Dr. Donald Granger: Recruited from the University of Utah, Granger joined the adult Division of Infectious Diseases in 1982, and over the next twelve years, he attended on the medical and ID services at Duke and the VA while he investigated host defenses at the cellular level. Specifically, his work focused on the in vitro fungistatic activity of activated macrophages on *C.neoformans* and collaborated with Drs. Durack and Perfect using a model of cryptococcal meningitis in immunosuppressed rabbits[109-114]. He also served as research mentor for undergraduates, graduates, and ID fellows and published five papers in peer-reviewed journals[115-119]. He also began a long-standing collaboration with Drs. Nick Anstey and Brice Weinberg and others on the role of nitric oxide in malaria,[120] and the details will be discussed in a later publication.

116

Dr. Rytas Vilgalys: Another recruited in the 1980s, Rytas J. Vilgalys, PhD, applied his expertise to the genetics of speciation of fungi and the significance of mating and life cycles. He also worked on determining the genetic structure in wild mushroom species and human pathogenic fungi.

Dr. L. Barth Reller: The establishment of The Cancer Center and other solid organ and stem cell transplant units, the onset of the AIDS epidemic, and the increased referrals of patients with defined immunologic defects to the Pediatric Service expanded the array of individuals defined as immunocompromised hosts. Partially in response to the increased complexity of the immunocompromised patients, Duke hired Dr. L. Barth Reller to be the new director of the Clinical Microbiology Laboratory in 1988. Previously, Dr. Syd Osterhout and subsequently Dr. Dolph Klein directed the laboratory. Dr. Reller previously led the laboratory at the University of Colorado, where he systematically delineated the critical factors for detecting bloodstream infections by culturing the blood. He continued some of that work at Duke. Those factors included inoculum volume, the role of anticoagulants, anticomplement agents, antibiotic inactivation, incubation temperature, and oxygen content.[121-127] In an effort to bring evidence-based clinical microbiology into routine practice, Dr. Reller reorganized the laboratory at Duke.[128-130] He did this while simultaneously addressing the increasing demands and, paradoxically, the decreasing resources that were due, at least in part, to the emergence of managed care. The result was a redefinition of technician responsibilities and changes in the available tests and procedures.

This was a time of great changes in the culture of the Clinical Microbiology Laboratory at Duke and elsewhere. Many of the members of the laboratory staff were experienced but on-the-job trained technicians. New laboratory standardization policies required all technicians to be trained and certified by accredited medical technology programs. At least for a time, morale waned, and several personnel in the laboratory resigned. New policies and practices in the laboratory were not always well received by the clinicians, who were used to different rules, but over time, all came to agree that the laboratory had made improvements in the quality of diagnostic microbiology.

Among the other departures from the clinical laboratory were Wiley Schell, MS, and Thomas Mitchell, PhD.[9] Schell relocated his laboratory to head the Duke Medical Mycology Research Center, a specialized laboratory within the Adult Division of Infectious Disease and International Health. The mission of his specialized laboratory was to facilitate the conduct of research through collaborations with clinical and basic science faculty and industry. An expert in the identification of yeasts and molds, Schell was uniquely qualified to head this laboratory. As evidence of his expertise, Schell was the recipient of the Billy H. Cooper Award for Achievements in Clinical Mycology and Medical Mycological Education in 2006, an award conferred by the Medical Mycological Society of the Americas. Mitchell left his position as director of the Medical Mycology Section of the Clinical Microbiology Laboratory to pursue his research and teaching responsibilities. Upon his return from a sabbatical working with Dr. John W. Taylor at the University of California in Berkeley, Dr. Mitchell switched his research focus to the population genetics and phylogenetics of species of *Cryptococcus* and *Candida*.

Dr. Joseph Heitman: Dr. Heitman, a critical recruit to Duke, received his undergraduate and master's degrees from the University of Chicago and his medical and doctoral degrees from the Medical Scientist Training Program of Cornell University Medical College and the Rockefeller University. From 1989 to 1991, Dr. Heitman was a long-term postdoctoral fellow at the Biocenter in Basel, Switzerland, where the European Molecular Biology Organization sponsored his postdoctoral fellowship. In 1992, Dr. Heitman was recruited to Duke University Medical Center as an assistant professor in the Departments of Genetics, Pharmacology, and Cancer Biology, as well as Microbiology, and Medicine. In addition, he was identified as an assistant investigator in the Howard Hughes Medical Institute. His contributions to the field of fungal genetics have been diverse and extensive. He described his research focus as follows:

> My research focuses on the evolution of sex and fungi and the roles of sexual reproduction in microbial pathogenesis, how cells sense and respond to nutrients and the environment, the targets and mechanisms of action

of immunosuppressive and antimicrobial drugs, and the genetic and molecular basis of microbial pathogenesis and development.[85,131-134]

His initial research focus was the study of nonpathogenic fungi. Later, in collaboration with others at Duke, he embarked on studies of *Cryptococcus neoformans* and *C. gattii*, both of which are pathogenic fungi.[9] Dr. Heitman's scientific productivity in the twentieth century was extraordinary, with numerous publications in high-impact basic science journals. Because of his accomplishments, he was elected to many prestigious organizations and societies. An important contribution of Dr. Heitman was his role as mentor, advisor, and collaborator for other infectious diseases faculty and trainees and basic science faculty. Three of the doctors who benefited from their association with Dr. Heitman were John Perfect, Gary Cox, and Andrew Alspaugh. Dr. Perfect's collaborations with Dr. Heitman were discussed earlier.

Drs. Gary Cox and Andrew Alspaugh: Dr. Cox's studies on the molecular and phenotypic methods to characterize virulence factors in pathogenic fungi were greatly facilitated through collaborations with Dr. Heitman.[135-142] Likewise, Dr. Alspaugh's studies on signal transduction pathways that are associated with fungal pathogenicity were also facilitated through collaborations with Dr. Heitman and others in Heitman's laboratory.[116,136,143-149] Of note, Cox and Alspaugh have been active in the inpatient Immune Compromised Host Service and have attained national and international recognition for their expertise in the basic and clinical science of pathogenic fungi and the diseases caused by those fungi. As a member of DUMRU, the core fungal research group, Dr. Heitman also was able to influence the basic research efforts of others affiliated with that unit.

Drs. Mary Klotman and John Hamilton. In addition to the activities at Duke, research at the VA was exploring the molecular mechanisms that control cytomegalovirus (CMV) latency. This research was done by Dr. Mary Klotman and myself, as well as by trainees Barbara Seaworth, MD, Joan Drucker, MD, Melinda Wharton, MD, Alan Street, MD,

Kenneth Schmader, MD and Kevin Porter, MD. It was driven by the newly recognized emergence of CMV as a cause of major morbidity and mortality in transplant recipients.[150-159] A derivative of this work was the demonstration that CMV was transmitted from a latently infected mouse to an uninfected mouse through transplantation or implantation of sections of kidney.[160-164] Moreover, studies determined that latent virus was preferentially reactivated from the donor tissue rather that the recipient host, a circumstance relevant to the human transplant. This work was supported for over twenty-five years, primarily by an NIH Program Project grant (Paul Klotman, principal investigator) and the VA Merit Review Program (John Hamilton, principal investigator).

Dr. Kenneth Schmader. Dr. Schmader's research activity focused on infections in older adults. Initially, he began working with the VA researchers on studies in aging mice by using murine CMV. He did the following: (a) developed a novel salivary gland biopsy technique that avoided the sacrifice of animals and allowed for longitudinal investigations of murine CMV latency and reactivation, including murine CMV DNA gene amplification;[165] (b) demonstrated that murine CMV reactivation declined with aging;[166] (c) showed that when tissues latently infected with murine CMV were transplanted into severely compromised immune deficient mice, the result was rapid reactivation and dissemination of the virus[167] and (d) collaborated on the development of an isolate of murine CMV that was tagged with an enhanced green fluorescent protein to allow visual localization.[157]

In addition to his work on CMV, Schmader did research on the herpes zoster virus. To understand better the development of herpes zoster reactivation and to improve the prevention and treatment of older persons suffering from herpes zoster illness, especially from postherpetic neuralgia (PHN), Dr. Schmader conducted the first metanalysis showing that antiviral therapies are not effective in preventing PHN. He discovered two new risk factors for herpes zoster reactivation, that of psychological stress and race.[168,169] He also discovered that a positive self-report of shingles has a high positive predictive value and that a negative self-report of shingles has a very high negative predicative value when compared with a diagnosis by a physician.[170] He was instrumental in developing the Zoster Brief Pain Inventory (ZBPI) for the Shingles Prevention Study, the landmark study of the efficacy of the zoster vaccine in older adults. That tool was the first herpes-zoster-specific measure of herpes zoster pain and its impact on functional status and quality of life. Dr. Schmader became one of five lead investigators in the Shingles Prevention Study, the results of which were published in 2005.[171]

Dr. M. Louise Markert: Additional basic science research and extensive clinical programs concerning the immunocompromised host continued through the 1980s and 1990s. Among those making major contributions was M. Louise Markert, MD, PhD. Dr. Markert received her PhD while working in the laboratory of Dr. Peter Creswell on the topic of the human leukocyte antigen, on which her PhD thesis was based. Her house staff training was in pediatrics at Duke, under Dr. Samuel Katz. She subsequently completed her fellowship training in Pediatric Allergy and Immunology with Dr. Rebecca Buckley. Her research during her fellowship was in the laboratory of Drs. Russell Kaufman at Duke and Dr. John Hutton at the Cincinnati Medical Center. In that fellowship, she examined the molecular basis of adenosine deaminase deficiency.[172-175]

In 1992, she was consulted about a child with complete DiGeorge anomaly, a condition noted for a complete absence of the thymus gland, which results in a lack of T cells. To help this child, whose prognosis was grim given his immunodeficiency, Dr. Markert attempted to develop thymus transplantation as a research treatment.[176-181] She based her initial work on reports of thymus transplantation from the 1970s and 1980s. Those early attempts had failed for the most part. Dr. Markert felt that with improved reagents and with the collaboration of Drs. Barton Haynes and Rebecca Buckley, thymus transplantation should be attempted again as therapy. With input from Drs. Haynes and Buckley, Dr. Markert developed improved procedures that allowed for successful engraftment of a donor thymus. Genetically compatible T cells developed in the recipients. By the year 2000, eleven infants with complete DiGeorge anomaly had undergone transplantation, and the number increased to sixty-four by 2012. Of the sixty-four infants, forty-five (70 percent) have survived. After transplantation, most deaths occurred within one year, the time before immune function could be established.

Of the survivors, all but one developed T cells. Currently, Dr. Markert's research remains focused on immune outcomes after thymus transplantation. She also remains active on the Pediatric Allergy and Immunology Service, seeing her patients in clinic.

Dr. Joanne Kurtzberg: In a related but different pediatric population, Joanne Kurzberg, MD, conducted basic research on the preparation of bone marrow to be used in human transplantations.[48] She also characterized the activities of thymocytes[182,183] and studied the early events in human T cell ontogeny.[184] In collaboration with others, she demonstrated that immature human thymocytes can be driven into non-lymphoid linages.[185,186] She led numerous studies of the use of umbilical cord blood as an alternative source of hematopoietic stem cells for transplantation.[187-190] In addition, she contributed to an NIH consensus statement on the ethical issues in the practice of banking umbilical cord blood[191] and reported on the outcomes of placental cord blood transplantations in patients.[192,193] Some of those patients were children with known immune defects, but more than 50 percent of them were children with a malignancy of some sort with an immune deficiency secondary to the malignancy itself or to the chemotherapy that was being used to treat the malignancy. In the early 1990s, CMV was the principal cause of mortality in immunodeficient patients, and later the principal cause was fungi. Later, adenovirus has proven to be a more frequent cause of death. It is fortunate that effective therapies have been developed, at least for the first two of these classes of infectious agents. Also, intensive surveillance for incipient disease and prompt therapy has assisted in reducing attendant morbidity and mortality. Prospective prophylaxis for infectious agents following transplantation is selective, and when indicated, the anti-infective treatment is intensive. Due to the success of Dr. Kurtzberg's program, she was awarded a Translational Cellular Therapy Center by the Robertson Foundation, with her as its director.

Drs. William Peters and Nelson Chao: The first bone marrow transplantations were performed at Duke under the supervision of Dr. William Peters in the early 1990s. Those early transplants were autologous transplants (transplantation of the patient's own cells) for the rescue of the bone marrow following bone marrow ablation therapy for metastatic breast cancer. Dr. Peters left Duke in the mid-1990s, and Dr. Nelson Chao continued his work at Duke. Since autologous or allogeneic bone marrow transplantations were first begun in the late 1980s, the survival of patients undergoing this treatment has dramatically improved, owing in part to better antibiotics, antivirals, and antifungals. In addition, improvements have been made in the drugs used for chemotherapy and in the avoidance of graft-versus-host disease by the use of better immunosuppressive drugs and the use of adjuvants designed to increase the number of cells derived from donors for transplantation.[194-198] Early in 2000, Dr. Chao became the chief of the Division of Cell Therapy in the Department of Medicine. He is currently engaged in studies exploring the possibility of adoptive immunity by the use of T cells targeted to specific pathogens.

Clinical programs in cancer management and treatment began at Duke in the 1940s by Drs. Guy Odom and Barnes Woodhall (Neurosurgery) and expanded with the creation of the Southeastern Cancer Chemotherapy Cooperative Study Group, led by Dr. R. Wayne Rundles in the 1950s.[199] It is not well described what role the infectious diseases program played during those periods when treatments for both the underlying cancers and the complicating infections were undertaken. By the time the Comprehensive Cancer Center was created in 1973, treatments were available for both the cancer and the increasingly recognized serious infections. Since then, Duke has invested heavily in the management of cancer, and several notable cancer specialists have been recruited to the Department of Medicine.

Dr. Joseph Moore: Among the first recruits was Dr. Joseph Moore, who trained at Duke in hematology and oncology and joined the faculty in 1977. His primary responsibility was the care of patients with cancer.[199] He recalls the tremendous advances made in technology (for example, computed tomographic, magnetic resonance imaging, and positron emission tomographic scans); in the availability of more potent cancer drugs and adjuvants; in the better and more specific antibiotics; and in the more effective anti-nausea drugs. Although infections remain ongoing problems in patients with cancer, the major causes of death are progressions of the cancer. Overall, the treatment has shifted from inpatient wards to the outpatient treatment facility, and outcomes generally are better, but the volume of patients has continued to climb, especially in women who have lymphoma or lung cancer. Duke's major contribution to the cancer-related infectious diseases is in the study of the effectiveness of new drugs, done independently or as part of a network of centers.

Dr. Jon Gockerman: A second cancer specialist recruited to Duke was Dr. Jon Gockerman. He particularly treated those with hematologic malignancies. He noted a dramatic improvement in outcomes of cancer chemotherapy since he returned to Duke in 1992. He attributes this improvement to the dramatically improved diagnostic clinical microbiology, to better drugs, to insightful anti-infective clinical trials done in this patient population, and, specifically in the case of HIV-related lymphoma, to improved access to highly active antiretroviral therapy. He has not been enthusiastic about the use of prophylactic antibiotics except in their use for acute leukemia, and he enthusiastically endorses the concept of improved cleanliness, general hygiene, and better hospital infection control as measures to minimize risks to susceptible patients.

To be sure, the Adult Division of Infectious Diseases has been involved in the management of adult patients recognized as immunocompromised from the outset of the kidney transplantation program in the late 1960s and early 1970s. Just how profoundly that newly recognized population would alter the diversity of the clinical infectious diseases was not recognized initially. Experiences elsewhere, however, were increasingly published, and it became evident, certainly by the mid-1970s, that the range of expected infectious disease complications would be dramatically expanded. And certainly this was the case, as reflected by the occurrence of reactivated latent CMV resulting in serious morbidity, such as the occurrence of CMV retinitis and excessive mortality owing in some instances to the occurrence of CMV pneumonia. Throughout the 1970s, 1980s, and early 1990s, faculty within the Division of Infectious Diseases performed consultations on these patients. With the recognition that the increasing numbers of immune-compromised hosts was an unpleasant fact, it became apparent that perhaps a new skill set would be required to provide optimal care. In addition, those who were primarily responsible for the management of patients receiving cancer chemotherapy or the management of recipients of a solid organ, bone marrow, or stem cell transplant generally recognized the need for this new skill set. They also wanted there to be a dedicated subset of the infectious diseases faculty with whom they could relate because they possessed this new skill set. Therefore, in 1994, the division appointed a new faculty member to serve as director of the Transplant Infectious Disease Service.

Dr. Souha Kanj: That new faculty member was Souha Kanj, MD. She had been a Duke Internal Medicine intern and resident and a Duke Adult Infectious Diseases Fellow and had been conducting basic science research for two years in the laboratory of Dr. Laura Davis at the Howard Hughes Medical Institute. She decided, however, to redirect her career from basic science to the clinical support of the new service at Duke. Dr. Kanj began by going to all the weekly transplant meetings (kidney, heart, lung, and liver) and establishing an excellent rapport with all the members of the teams, which, by this time, included surgeons, medical personnel, and the transplant coordinators. Even though she was a junior faculty member, her attention to their needs and the needs of their patients established a strong bond that was essential to the success of the program. In addition to conducting pre-transplant outpatient evaluations, she followed all the inpatients on any of those services and incorporated infectious diseases fellows in those activities when possible. Most of the fellows who worked with Dr. Kanj became interested in transplant-related infectious diseases, including Barbara Alexander, MD, who is now director of the service. During the first several years, Dr. Kanj was on service for eight months of the year, and Drs. John Perfect, Gary Cox, and Andrew Alspaugh filled in for the remaining four months. She was an extremely effective teacher and was nominated for several teaching awards. While at Duke, she published six papers on infectious diseases in transplant patients in refereed journals.[146,158,159,200-202] In the academic year 1998–1999, she returned with her family to her native Lebanon.

Dr. Barbara Alexander: The Division of Infectious Diseases was able to recruit Barbara Alexander, MD, as an associate in Medicine and as director of the Transplant Infectious Diseases and Immunocompromised Host Service in 1999. She was uniquely qualified to assume leadership of this service. In addition to her house staff and fellowship training at Duke, she came with a background in medical microbiology. She had studied mycology in college in the early 1980s, had attended the medical technology school at Duke from 1985 to 1986, and had worked as a mycology technician in the Clinical Microbiology Laboratory at Duke from 1986 to 1989. She then attended medical school at East Carolina University and graduated in 1993. At that point, she joined the Duke house staff as an intern.

As with Dr. Kanj, Dr. Alexander gained the confidence of the attending surgeons through regular interactions at their weekly meetings. She took advantage of an opportunity for a several-month special fellowship in transplant infectious diseases at the Massachusetts General Hospital, which was directed at that time by Drs. Jay Fishman and Robert Rubin. To accommodate the increasing numbers of at-risk patients, Dr. Alexander proposed the expansion of the transplant infectious diseases faculty at Duke. Because her clinical responsibilities were extensive, three other infectious disease faculty with clinical and research interests in fungi assisted her.[116,132,136,143,145-149] In the early 2000s, three additional faculty dedicated to the problems of the immunocompromised host were recruited; they were Drs. Aimee Zaas, Kim Hanson, and Sylvia Costa. Dr. Alexander was thus able to expand the breadth of the Immune Compromised Host (ICH) service to include, along with recipients of transplantations, those patients with cancer who had received aggressive chemotherapy. Her research interests included fungal and viral diagnostics, the genetics of susceptibility to fungal infection, and RNA-based assays for candidemia.[203,204] In the next decade, she successfully applied for K23, K24, and an immunocompromised host training program and published widely in her field, as well as assuming leadership positions in the Infectious Diseases Society of America and other societies.

Photograph 6.5 is a composite of pictures of seven prominent figures in the field of the immunocompromised host at Duke in the twentieth century.

D. Bernard Amos

Rebecca H. Buckley

Norman F. Conant

R. John Perfect

Jack D. Keene

L. Barth Reller

Peter Zwadyk

CHAPTER 7

PROGRAMS IN MICROBIOLOGY, PEDIATRIC, AND ADULT INFECTIOUS DISEASES

> Even with my great personal loyalty to infectious diseases,
> I cannot conceive the need for 309 more infectious disease
> experts unless they spend their time culturing each other.
> —Robert G. Petersdorf, MD,President of the
> Infectious Diseases Society of America,1978[1]

As discussed in greater detail in chapter 1, the recognition and the management of infectious diseases fall into three general time periods: before the nineteenth century, in the nineteenth century, and in the twentieth century. Although practitioners before the nineteenth century dealt with diseases that were later found to be caused by microbes, an understanding of the germ theory of disease was nonexistent. However, the germ theory was recognized in the second half of the nineteenth century, and this concept was then used in the formulation of medical education curricula and medical standards. The twentieth century brought an appreciation of the mechanisms by which infectious agents cause diseases, which led to improved techniques for prevention and treatment and to derivative research that used microbes as valuable tools for the study of various cellular events. In this chapter, the history of Duke's involvement in infectious diseases in the twentieth century will be described in three fairly distinct time periods: before the discovery of antibiotics (between 1930 and 1940); after the discovery of antibiotics (1940s to 1960s); and the final four decades of the century, which I shall call the modern era of infectious diseases. The modern era is characterized by the ability of physicians and scientists to intervene actively in the prevention and treatment of infections in a systematic manner and by the evolution of the infectious diseases specialist and scientist.

THE PRE-ANTIBIOTIC ERA (PRE-1940)

The previous six chapters describe the Duke faculty and the infectious agents with which they are associated and the Duke programs and facilities that existed before 1940. As a brief reminder, the following is a list of those faculty and their principal relationship with infectious diseases: Dr. Wilburt C. Davison was a pediatrician and the founding dean of the School of Medicine and in that role recruited the faculty whose responsibility, in part, was to deal with the prevalent infectious diseases without the benefit of an antimicrobial armamentarium. Dr. Harold I. Amoss, the first chair of the Department of Medicine, came to Duke with experience and training in infectious diseases while at Johns Hopkins and at the Rockefeller Institute, where he had done research on various aspects of infectious diseases. Dr. Frederick M. Hanes succeeded Amoss as chair of the Department of Medicine and in that role supported faculty involved in the management of patients with infectious diseases. Dr. J. Deryl Hart was the first chair of the Department of Surgery and was responsible for surgical interventions necessitated by various infections such as the drainage of infected fluid around the lung resulting from pneumonia (empyema) and surgical repair of a ruptured bowel in the abdominal cavity. Dr. David T. Smith, having had tuberculosis (TB) himself, was recruited to Duke as the chair of the Department of Bacteriology with the additional clinical responsibility to oversee the management of patients with TB and other chronic infections of the lung. Dr. Norman F. Conant, having trained at Harvard in the laboratory of Dr. Hans Zinsser, was recruited as an expert in fungal research.

THE POST-ANTIBIOTIC ERA (1940S–1960S)

Many of the faculty in this era have also been previously described in earlier chapters, but a summary of their particular relationship with infectious diseases follows: Three subsequent department chairs with no special interests in infectious diseases but with responsibility for the departmental management of prevalent infectious conditions were Drs. Eugene Stead (Department of Medicine), Clarence E. Gardner (Department of Surgery), and Jerome S. Harris (Department of Pediatrics). Dr. William B. Tucker was a major figure in the landmark Veterans Administration (VA) Cooperative Study on the Treatment of TB with Streptomycin.[2-4] He was the chief of the Medical Service at the Durham VA Medical Center briefly and was later chief of the Medical Service at the VA Central Office. Dr. Joseph W. Beard, head of the Duke Division of Experimental Surgery, had trained with Drs. Richard E. Shope and Peyton Rous at the Rockefeller Institute and was recognized as one of the earliest retrovirologists.

A somewhat later cohort of contributors to the field included Dr. David C. Sabiston, the third chair of the Department of Surgery, and Dr. James B. Wyngaarden, the fourth chair of the Department of Medicine, both of whom oversaw increasingly sophisticated departments with extensions into the clinical care and research on infectious diseases. Also included were Dr. D. Bernard Amos, a pivotal figure in the development of the science of human transplantation; Dr. Rebecca H. Buckley, a pioneer in the recognition of the specific genetic defects leading to severely compromised immune systems and the management of children so affected; and Dr. A. Derwin Cooper (who himself had had TB), a longtime consultant to Duke and the Durham VA Medical Center on the management of patients with TB.

Four individuals who spanned this era were important in the evolution of the modern era of infectious diseases. They were Drs. J. Lamar Callaway, Herbert O. Sieker, Herbert A. Saltzman, and Suydam Osterhout. Because they were not mentioned previously or only briefly mentioned, brief synopses of their careers and connections with infectious diseases follow.

Dr. Callaway graduated from Duke Medical School in 1932, its first graduating class. He served as an intern and resident in internal medicine at Duke, followed by two years as a fellow and instructor in dermatology at the University of Pennsylvania. He returned to Duke in 1937 and over the next three or four decades became nationally recognized as an expert in dermatology, clinical mycology, and syphilology. As such, he was widely sought as a consultant by the VA, the United States Public Health Service, and the Surgeon General of the United States Air Force. He was a member, and often president, of numerous medical organizations, including, but not limited to, the American Academy of Dermatology, the American Association of Professors of Dermatology, the Southern Dermatology Association, the American Medical Association Section on Dermatology, and the American Board of Dermatology. He published widely in the scientific literature on his areas of expertise and was the coauthor of several books. Attesting to his prominence in clinical and research dermatology, he received numerous honors, including the Gold Medal from the American Academy of Dermatology, and was named a James B. Duke Professor of Dermatology. Dr. Callaway stepped down as chief of the Division of Dermatology (formerly called Syphilology) in 1975, and the J. Lamar Callaway Professor of Dermatology Chair was established. His contributions to the fields of syphilis and fungal diseases of the skin are numerous.[5-13]

Dr. Sieker came to Duke as an intern in 1948 from Washington University in St. Louis, where he completed medical school and initiated some research on histoplasmosis (personal communication). He joined the Duke faculty in the early 1950s as a specialist in cardiopulmonary disease and did research working with Dr. John Hickam. Because of his interests in diseases of the lung, he collaborated with Drs. D. T. Smith and Norman Conant and experienced firsthand the introduction of the use of sulfonamides and the early availability of penicillin. Sieker recalls treating patients with subacute bacterial endocarditis with a mere two thousand units of penicillin per day and the replacement of arsenic injections for syphilis with penicillin under the direction of Dr. Callaway. Dr. Sieker was extremely productive academically, obtaining one of the first training grants (Dr. Rebecca Buckley was the first and arguably the most successful trainee), succeeded Dr. Lige Menafee as the head of the Pulmonary Division, of which he was the only member in 1962, and was appointed assistant dean and chair of the curriculum committee. In the latter role, he, along with Drs. Eugene Stead (chair of the Department of Medicine), Phillip Handler (chair of the Department of Biochemistry), and Barnes Woodhall (dean), revised the medical school curriculum against considerable resistance, resulting in Duke's signature opportunity for third-year medical students to pursue an entire year of research.[14-25]

Dr. Saltzman, also a faculty member in the Pulmonary Division in the Department of Medicine, graduated from the Jefferson Medical School in Philadelphia (personal communication). After serving for several years in the US Air Force, he joined the faculty at Duke. He served as chief of the Pulmonary Section at the Durham VA Medical Center, where he was responsible for pulmonary consultations, but largely he was responsible for the management of patients on the ward designated specifically for patients with active TB. In that capacity, he interacted extensively with Dr. Derwin Cooper, who was the TB consultant for the Durham VA. Of course, house staff and fellows served rotations on that ward. After a number of years, Dr. Saltzman moved his clinical and research programs to Duke South, where he focused on the use of hyperbaric oxygenation for various conditions, including some infections of the bone and brain. In addition, he collaborated with the staff of the Clinical Microbiology Laboratory, who were working on anaerobic bacteria and on animal models of infections with those organisms. The studies were supported by the National Institutes of Health (NIH) and the major military services. Later in his career, he led the national study on acute pulmonary embolism.[26-37]

When still an undergraduate at Princeton, Dr. Osterhout was transferred to Duke Medical School under the auspices of the Naval Training Program in 1945 (personal communication). Upon graduation, Dr. Osterhout served as an intern at the Massachusetts Memorial Hospital in Boston, where he worked under the chairman, Dr. Chester Kiefer, and the head of infectious diseases, Dr. Louis Weinstein, two giants in the field of medicine. After a brief interruption for the Korean War in the early 1950s, Dr. Osterhout returned to Duke for house staff training in the Department of Medicine, including his time as chief resident. At the completion of his house staff training, he joined the faculty at Duke and was supported for study at the Rockefeller Institute to obtain a PhD that was based on his development of a plaque assay for herpes simplex virus.

With his return to Duke in the early 1960s, he assumed responsibility for many infectious diseases consults, although two other faculty members, Drs. Lige Menafee and Samuel Martin, were already seeing most of the inpatients and outpatients with infections. The consults included patients with bacterial pneumonia, meningitis, and skin and gastrointestinal infections. By this time, penicillin and other antibiotics had become available, including drugs for TB. At some point in the early 1960s, Dr. Osterhout was appointed as director of the Clinical Microbiology Laboratory, a position he retained until the mid-1970s. Whether because of the "strike" by Clinical Microbiology Laboratory employees or not, Dr. Osterhout was replaced as head of the laboratory by Dolph Klein, PhD, and was appointed as director of admissions to the School of Medicine. Although he was clinically active up to that point and involved in the Hospital Infection Control (HIC) Program at some level, his new appointment demanded most of his attention.[31,38-46]

By the 1940s, penicillin had become available, albeit in very limited supply outside of the military. Dr. Herbert A. King, the chief resident in medicine from 1944 to 1945 recounts the following episode:

On one occasion, I was taken by surprise regarding the recovery of a very sick patient. This young gentleman had bilateral pneumococcal pneumonia. Obviously, we had no access to penicillin in those days. (All the penicillin was going to the army.) I was unable to type the organism; thus, we couldn't give him the antiserum. Dr. Hanes (then chair of medicine) made rounds one morning with the group and said, "Obviously, this fellow will die within the next forty-eight hours." The patient was a pilot for the civil air patrol. Late that night, his partner came to see me. He inquired, "What's going to happen to my friend?" I replied, "We all think he's going to die shortly." He said, "Is there anything that we might be able to do that would save his life?" "Well," I said, "there is a substance called penicillin that, if we could get any of it, it might help." He looked at me and asked, "Are you going to be on tonight?" I indicated I would be there all night. Two hours later, he came back and brought me a box containing several vials of penicillin. I don't remember how many, but there were quite a few. He indicated to one of the nurses, "I have to have all these vials back by the morning." And I put this penicillin into one of the saline bottles and gave it to the patient. The gentleman indicated that under no circumstances was I to tell anyone that he had given me the medication. Several hours later, the patient's temperature started coming down. In twenty-four hours, he was markedly improved. Dr. Hanes came back a day later and could not believe what he saw. Dr. Hanes remarked, "Well, I don't understand it. I have no idea what happened. This man should have died, but he is obviously getting better." Of course, I had been sworn to secrecy, and I kept my mouth shut. Dr. Hanes asked, "What do you think happened, Dr. King?" I said, "I really don't know. I can't understand this at all." Later, of course, penicillin became a standard therapy for pneumococcal pneumonia and for syphilis under the supervision, usually, of Dr. Callaway.[47]

THE MODERN ERA (1960S TO 2000)

Of particular importance, the Pulmonary Division was the designated home for Dr. Thomas Cate, the first adult infectious diseases specialist recruited to Duke in the late 1960s from the NIH. No division of infectious diseases existed at that time at Duke, although many of the leaders at other major medical centers were specialists in infectious diseases.

Because of the misperception that antibiotics would eliminate the problem with infectious diseases, interest waned both in teaching the principles of microbiology to students and in doing further research to develop additional antimicrobial agents. Indeed, funding through the NIH for such activities also declined. Infectious diseases specialists, however, felt that further support for the field was essential to prepare for unexpected developments. To address that predicted need, support was obtained from the Department of Defense for the creation of the Armed Forces Epidemiologic Board, the mission of which was to examine specific ongoing infectious disease problems such as acute rheumatic fever, influenza, and others and to propose solutions to those problems.[48] Additional support came from a subset of the membership of several prominent scientific societies, namely the American Society for Clinical Investigation (ASCI—also called the Young Turks) and the American Association of Physicians (AAP). Informal meetings of those interested in the ongoing infectious diseases issues were held during the national meetings of the ASCI or the AAP in Atlantic City in the mid-1960s. Members made specific scientific presentations on a diverse array of topics, and subsequent discussions were lively. Led by Dr. Maxwell Finland of Harvard, a planning meeting was proposed to establish a framework for a larger organization, somewhat later to be named the Infectious Diseases Society of America (IDSA). Approximately 250 persons were approached for membership in this organization, and the first meeting of the IDSA was held in 1963 in Airlie, Virginia, with approximately 125 in attendance. Among the charter members were Dr. Samuel Katz, a Harvard faculty member and later the chair of the Department of Pediatrics at Duke; Leighton Cluff, formerly an assistant resident at Duke and later the chairman of the Department of Medicine at the University of Florida; and Dr. Lewis Wanamaker, formerly a medical student and resident in pediatrics at Duke and later professor of pediatrics at the University of Minnesota. There were many other charter members with indirect connections with Duke. Meanwhile, a group of physician scientists more interested in the therapeutic aspects of infectious diseases, including many representatives from industry, formed a separate organization called the Interscience Conference on Antimicrobial Activity and Chemotherapy (ICAAC).

Dr. William G. Anlyan, dean of the Duke School of Medicine from 1964 to 1989, first embraced the concept of infectious diseases specialists when he recruited Dr. Samuel L. Katz from Harvard and the Boston Children's Hospital in 1969 as the third chair of the Department of Pediatrics and Dr. Wolfgang Joklik from the Albert Einstein Medical Center in New York City as the third chair of the Department of Microbiology and Immunology. Both departments were small when Katz and Joklik arrived at Duke, but their collaborations opened many new opportunities for training and research. Those two recruitments set the stage for what would become transformative developments at Duke in the fields of microbiology and infectious diseases over the next three decades. Although activities proceeded largely in parallel, for the purposes of discussion, I will attempt to summarize some of the activities of the faculty in microbiology and infectious diseases in three general areas: first, in the Department of Microbiology and Immunology; second, in the Department of Pediatrics; and third, in the Department of Medicine.

THE DEPARTMENT OF MICROBIOLOGY AND IMMUNOLOGY

As chair of a department with two distinct disciplines, Dr. Joklik concentrated on microbiology, leaving immunology activities to his respected colleague, Dr. D. Bernard Amos. Beginning in 1968 with only six faculty members, the size of the faculty increased to more than thirty over the subsequent twenty-five years. By all accounts, Dr. Joklik was one of the four or five leading virologists in the United States, with more than 150 peer-reviewed publications in prestigious journals focused mostly on his principal interest, the Reovirus family, and he had some fewer numbers of publications on the Rous sarcoma virus and the pox-viruses.[49-58] By the mid-1980s, his department was recognized as one of the top departments of microbiology in the country and ranked in the top three by the National Research Council. Among the many awards and honors conferred upon Dr. Joklik was his election to the National Academy of Sciences. In addition to his extensive record of publications, he served as the editor in chief of the premier textbook of microbiology at that time, named in honor of an eminent microbiologist earlier in the century, Hans Zinsser (*Zinsser's Textbook of Microbiology*). He served in that role for eighteen years, between 1972 and 1990. He served on many national and international scientific committees and founded the American Society for Virology. He also played an important role in determining the fate of remaining stocks of smallpox virus after the completion of the Intensified Program to Eradicate Smallpox, sponsored by the World Health Organization. His role there will be further discussed in the chapter on vaccines.

Among the recruits to the Department of Microbiology was Dr. Jack Keene, who came to Duke in 1978 from the NIH, where he was a staff fellow with Robert Lazzarini in the Laboratory of Molecular Genetics from 1974 to1978. As an assistant and associate professor, he continued his work on molecular genetics, taking advantage of the vastly enhanced technology to study DNA- and RNA-binding proteins and their role in the pathogenesis of autoimmunity. In addition, using viruses and bacteria for his studies, he performed combinatorial analyses to identify novel targets for pharmacologic interventions.

Prior to his becoming chair of the Department of Microbiology, the section on immunology had become a separate department. Dr. Keene lists the following as his major scientific achievements: "First, the discovery in the mid-to-late 1980s of the RNA recognition motif (RRM) family of RNA-binding proteins, which was based on the biochemical demonstration that the RRM family constitutes the functional core of an RNA-binding domain;[59-61] second, the isolation of the first recombinant human autoimmune antigen, elucidation of autoimmune epitopes, and development of a clinical diagnostic test for systemic lupus erythematosus and related diseases;[59,62-64] third [after his appointment as chair of the Department of Microbiology in 1992], the elucidation of the functions of the ELAV/Hu posttranscriptional regulators (HuA, HuB, HuC, and HuD) and their roles in proliferation, neuronal differentiation, and immune response;[65-69] and fourth, proposal of the theory of the 'posttranscriptional RNA operon or regulon' and modular coordination of RNA gene expression networks in eukaryotes."[70,71]

Dr. Keene supported the research by Dr. Thomas Mitchell, a mycologist, and recruited two other mycologists, John McCusker and Joseph Heitman. In addition, he recruited an anthrax specialist with an interest in the immunology of anthrax infection, Philip Hanna, and fostered collaborations with John Perfect and Wiley Schell in the Department of Medicine. Dr. Keene also served as a member of a special committee to consider organizational changes in the graduate school, along with other graduate school faculty Paul Modrich, Thomas Petes, Howard Rockman, Hunt Willard, and Van Bennett.

Another Joklik recruit was Dr. David Pickup, whose work focused principally on the pathogenesis of poxviruses, most specifically on cowpox and vaccinia. His publications were among the first to show that viruses did not simply evade the immune system but actively protected themselves from it by taking control of processes of key importance to it; those publications also helped to establish the understanding that viral pathogenesis was intimately connected to viral interference with immune defenses. His publications and others pointed to ways in which aberrant immune responses might be controlled and were relevant to the design of a poxvirus vaccine because they demonstrated that the immune response to vaccines can be tailored by genetically altering the virus to maximize immunogenicity while minimizing reactogenicity. He also did a number of studies that helped to elucidate the control of gene expression by poxviruses.[55,72-79]

Dr. Joseph Nevins was a PhD student from 1972 to 1976 when he worked with Dr. Joklik on the vaccinia virus. The vaccinia virus was simply a model used for the study of viral replication, in part because vaccinia is one of the few DNA viruses that replicate entirely in the cytoplasm. As a PhD student, Nevins discovered an RNA polymerase in the vaccinia virus.[80-82] After receiving his PhD degree in 1976, he moved to the Rockefeller University as an assistant and then as an associate professor of molecular cell biology, where he worked in the laboratory with Dr. J. Darnell until 1987.[83] At that time, he was recruited to Duke as a professor of microbiology and an investigator in the Howard Hughes Medical Institute. In 1990, he was named professor of genetics and head of the Section of Genetics until 1994, when he was appointed as chair of the Department of Genetics, where he remained until 2002. At that point, he was named as chair of a newly created Department of Molecular Genetics and Microbiology. The author of nearly three hundred peer-reviewed publications, his appointment to numerous professional advisory committees, and his numerous awards and honors all attest to his highly productive and successful professional career. He cites as his major scientific achievement the early identification of the E2F transcription factor in nuclear digests of the adenovirus.[84,85]

Specifically, he and others found that an adenovirus protein, E1A, inactivates the retinoblastoma (RB) tumor suppressor. The consequence of that inactivation is that the tumor suppressor activity of RB, which normally controls the E2F transcription factor, leads to disruptions in the normal cell cycle and potentially causes cells to convert to the uncontrolled growth of specific cell types or cancer.[86] The subsequently recognized implications of those findings led to his change in focus from virology and genetics to the importance of genetics in cancer.[87] Scientific curiosity certainly had led him in that direction, but he also was tempted by an evolving interest in some application of his scientific achievements to the discovery of possible interventions in the disease process of cancer.

Mariano Garcia-Blanco obtained his MD and PhD degrees at Yale, did a postdoctoral fellowship with Phillip Sharp at Massachusetts Institute of Technology and was recruited to Duke in 1990 with appointments in the departments of microbiology, molecular cancer biology, and medicine. Following up on his work on the biochemical activity of the HIV-1 protein Tat and its RNA binding site, the TAR element, he collaborated on the development of an in vitro transcription system that recapitulated Tat-dependent transactivation.[88] At Duke, he collaborated with Dr. Brian Cullen in describing the molecular basis of latency in pathogenic human viruses.[89] In his further research, he determined the structure of the terminal loop of the RNA element by using biochemical structure probing[90] and two-dimensional nuclear magnetic resonance spectroscopy[91,92]. Using the in vitro transcription system, he collaborated on the identification of cellular proteins that interacted with Tat and TAR, and, using affinity chromatography and mass spectrometry, he and colleagues discovered a protein originally termed CA 150 (now TCERG1) that is involved in transcript elongation and processing.[93-97]

With a trainee in the adult infectious diseases training program, he explored the use of RNA decoys as potential anti-HIV-1 tools and used them to explore the mechanism of Tat transactivation.[98,99] During that time he also served as co-scientific director of the VA Research Center on AIDS and HIV Infection (with Dr. Thomas Palker), the goal of which was to bring the Adult Division of Infectious Diseases within the Department of Medicine closer together with the Department of Microbiology (now known as the Department of Molecular Genetics and Microbiology). Relationships between these departments have matured with time, bringing dividends to both departments. One example is Garcia-Blanco's mentoring of an adult ID trainee, who later became a faculty member (Kimberly Hanson) and developed the molecular skills needed to oversee the Clinical Microbiology Laboratory's section on molecular diagnostics.

Dr. Joseph Heitman was recruited to Duke in 1992 after obtaining his MD and PhD degrees from Cornell and Rockefeller Universities. He was the recipient of many awards and honors, including his election to the Institute of Medicine and an appointment as a James B. Duke Professor of Medicine. He describes his research as follows:

[My] research program focuses on model and pathogenic fungi addressing sexual reproduction of microbial pathogens, cellular sensing of nutrients and environmental cues, targets and mechanisms of action
of immunosuppressive/antimicrobial drugs, and the
molecular basis of microbial pathogenesis and development. [Our] studies reveal general biological and genetic
principles. These include pioneering studies in Baker's
yeast to discover FKBP12 and Tor1/2 as the conserved
targets of the immunosuppressive/anti-proliferative
drug rapamycin,[100] defining the roles of calcineurin in
governing fungal virulence,[101] morphogenesis, and antifungal drug action,[102] deciphering how cells sense and
respond to nutrients via cell surface receptors and Tor
signaling,[103,104] and elucidating mechanisms via which
fungi sense environmental signals such as light and CO_2
and infect the host.[105]

THE DIVISION OF INFECTIOUS DISEASES IN THE DEPARTMENT OF PEDIATRICS

Dr. Katz, a Harvard-trained pediatrician, had worked in the laboratory
of Dr. John Enders as a house officer and as a faculty member at the
Boston Children's Hospital. Dr. Enders was a basic virologist and in
1954 was a corecipient of the Nobel Prize in Medicine, along with Drs.
Thomas Weller and Frederick Robbins, for work on the in vitro isolation of the poliovirus. That work led ultimately to the development of
effective polio vaccines, perceived at the time as a national emergency.

Dr. Katz, however, worked on the measles virus and the measles virus
vaccine and other infectious diseases.[106-109] His legacy at Duke resides in
his pivotal contributions to the development of the measles vaccine while
at Harvard, his role in the definition of worldwide vaccine policies with
the World Health Organization, the United Nations, and the Advisory
Committee on Immunization Practices (ACIP), the latter sponsored by the
Centers for Disease Control (CDC).[110-114] In addition, he oversaw the dramatic expansion of the pediatric faculty. See figure 7.1 and tables 7.1 and 7.2
for synopses of faculty and fellow developments over the next three decades.

Figure 1. Faculty Leadership Positions in Pediatric Infectious Diseases

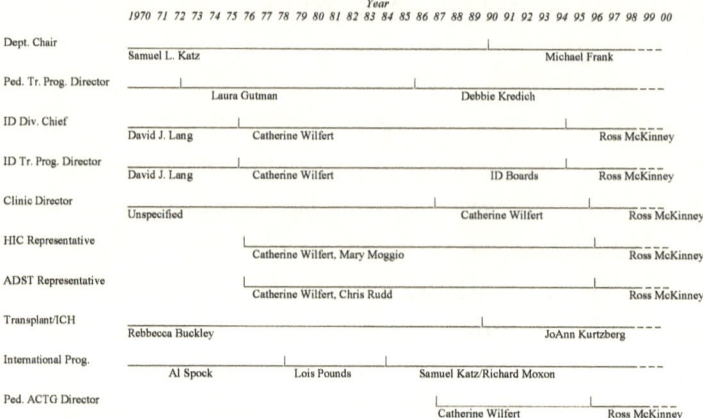

Abbreviations are defined as follows: Ped., Pediatric; Tr. Prog., Training Program; ID, Infectious Disease; Div., Division; HIC, Hospital Infection Control; ADST, Antibiotic Decision Support Team; ICH, Immunocompromised Host; ACTG, AIDS Clinical Trials Group.

Table 7.1. Pediatric Infectious Diseases Faculty

	1970's	1980's	1990's
Dept. Chair	S. Katz	S. Katz	M. Frank
ID Division Chief	D. Lang	C. Wilfert	R. McKinney
Div. Faculty	4	7	12
Clinical Services	Duke ID	Duke ID	Duke ID
Clinical Emphasis:			
General	All until HIV		
ICH			
HIV/viruses	2	6	7
INH		1	
TB	1		
Clin. Micro.			
Vaccine	1		1
Research Emphasis:			
Bacter., including TB	1		
Viruses, including HIV	2	4	7
Fungi			
Parasites			
Rickettsiae	1		
Hosp. Infect. Control			
Immunology	1		
Awards			
Golden Apple			
Basic	1	0	1
Clinical	4	1	0
Hs	0	1	1
S. Katz			2
M. Frank			
Williams			
Master Clin			
Palumbo			
New Programs		Inpatient & Outpatient Pediatric AIDS Services	
Funding Sources		Ped. AIDS Clin. Trials Group (NIH) Rickettcial Epidemiology (NIH)	Vaccine Trials Unit
Publications	CV's		

Abbreviations are defined as follows: ID, Infectious Disease; ICH, Immunocompromised Host; HIV, human immunodeficiency virus; INH, international health; TB, tuberculosis; Hs, housestaff

142

Table 7.2. Pediatric Infectious Diseases Fellows

	1970's	1980's	1990's
Fellows:			
No.	9	9	11
MD's	9	7	11
MD/PhD's		2	---
Post fellowship:			
Academic	2	5	7
Public Health	0	0	1
Industry	2	1	1
Private Practice	4	1	2
Military/NIH/CDC	1	1	1
Fellow funding:			
Clinical year	Department	Department	Department
Research	Division	Division	Division
K award within 3 yrs	0	0	2
Current position:			
Academic	3	5	8
Public Health	1	0	1
Industry	1	1	1
Private Practice	4	1	2
Military/NIH/CDC	0	1	0

New recruits to the faculty in the late 1960s to 1970s: Three specialists in infectious diseases were also recruited from Harvard soon after Katz's arrival at Duke. One was Dr. David Lang, the first chief of the newly created Pediatric Division of Infectious Diseases within the Department of Pediatrics and a protégé of Dr. Enders. Dr. Lang's research focused on cytomegalovirus, a newly recognized opportunistic viral infection in transplant patients and in pregnant women.[115-124] Another new recruit was Dr. Catherine Wilfert, who had also trained in the Enders laboratory. Upon her arrival at Duke, she chose to focus initially on enteroviral research[125,126] and on diagnostic virology while also rounding on the newly formed inpatient Infectious Diseases Service. Somewhat later, Dr. Wilfert developed a serious clinical research interest in diseases caused by rickettsia,[127] and later still, her energies were consumed by pediatric AIDS (acquired immunodeficiency syndrome),[128-133] which I discuss in greater detail in chapter 4). The other recruit from Harvard was the neurologist Dr. John Griffith, who had an interest in herpes simplex virus infections of the brain and Subacute Sclerosing Pan-Encephalitis.[125,134,135]

Four years after his arrival at Duke, Dr. Katz recruited Dr. Laura Gutman from the University of Washington. Her background was in TB. She became one of the two local authorities on pediatric TB and HIV; she later focused on child abuse.[136-140] She also was instrumental in establishing a novel treatment protocol for herpes simplex virus infections in the brains of newborns and the very young.[141-143] That protocol extended acyclovir therapy from three weeks to two years with beneficial results.

(Appendix A lists the entire faculty in the Department of Pediatrics in the twentieth century, including their academic appointments, and Appendix B lists all the fellows, their current locations, and their positions if known.)

Although not yet a program requiring accreditation, pediatric infectious diseases fellows were recruited to ID. They included Drs. Robert Snowe, Robert Greenberg, Mickael Kannan, Ousama Tomeh, Gerald Aronheim, Ziad Idriss, James Waller, Jeffrey Davis, Betty Raffin, Herbert Lassiter, Cornelia Dekker, and Sandra Lehrman.

New recruits to the faculty in the 1980s. With the explosive evolution of pediatric AIDS in the mid-to-late 1980s, additional recruits to the faculty and fellowship program were essential to deal with the crisis. Among those recruited to the fellowship program who ultimately joined the faculty were the following individuals, with brief biographical descriptions of their principal activities: Drs. Ross McKinney, Sandra Lehrman, Robert Drucker, and Dennis Clements.

Dr. McKinney obtained his MD degree from the University of Rochester, was a pediatric resident and infectious diseases fellow at Duke, and joined the faculty in pediatrics in 1985 (personal communication). His first grant, an IDSA award, was to study the immune response to enteroviral infections in children with x-linked agammaglobulinemia.[126] He later became extensively involved in the inpatient and outpatient management of children with the human immune deficiency virus (HIV) infection, and much of his research involved a pivotal early clinical trial of zidovudine in children with HIV infection.[128,132,144-148] Following the publication of the Pediatric AIDS Clinical Trial Group (PACTG) 076[149] and the implementation of its conclusions in the development of the formal program titled "Prevention of Mother-to-Child Transmission (PMCT)," new cases of pediatric AIDS declined dramatically, and the previously infected children were placed on effective suppressive therapy. In 1998, Dr. McKinney became the third chief of the Division of Pediatric Infectious Diseases and presently is director of the Trent Center for Bioethics Humanities and the History of Medicine.

Dr. Sandra Lehrman obtained her MD from Brown University, was an intern and resident in pediatrics at the Massachusetts General Hospital, and came to Duke as a fellow in pediatric infectious diseases in 1980. Prior to attending medical school, Dr. Lehrman worked in the virology laboratory overseen by Dr. Robert Chanock at the NIH from 1969 to 1972. Because of the strength in virology at Duke, she chose to come to Duke for a fellowship working with both Drs. Lang and Wilfert. In 1984, she was recruited by Dr. David Barry to work on antiviral resistance at the Burroughs Wellcome Company. When HIV became recognized as a retrovirus, and with Wellcomes's experience with the first effective antiviral drug acyclovir, Dr. Lehrman and others at Wellcome became increasingly active in the development of zidovudine, the first effective antiretroviral agent[128,144,150-155]. From that point on, she became increasingly involved with the biotechnology industry and drug development, and she is currently the global director for Scientific Affairs at Merck.

Dr. Robert Drucker obtained his MD at Duke, working in his third year in Dr. Wilfert's laboratory and subsequently doing house staff training at the Massachusetts General Hospital. He began a pediatric infectious diseases fellowship in 1983. He joined the faculty in pediatrics in 1986 and worked for a period on poxviruses in Dr. David Pickup's laboratory before serving as chief resident in pediatrics. He continued to participate in various clinical research projects.[156-164] Recognizing that his strengths lay in the clinical and educational arena, he redirected his career in those directions, becoming associate residency director and became one of four faculty who were advisory deans.[165] He continues to serve in a consulting capacity on the Pediatric Infectious Diseases Service.

Dr. Dennis Clements obtained his MD from the University of Rochester, was a pediatric intern and resident at Duke, and went into private practice in pediatrics in Durham, North Carolina, after a two-year stint with the US Air Force (personal communication). He remained in practice until 1986. At that time, he returned to Duke as a pediatric infectious diseases fellow. Following this, he did another fellowship, this time in the Department of Microbiology and Infectious Diseases at the Royal Children's Hospital in Melbourne, Australia. During this period, he conducted epidemiologic studies on *Haemophilus influenza* group B epiglottitis and meningitis and used those data as the basis for his MPH and PhD.[166,167] When he returned to Duke in 1990, he started the Duke Pediatric Vaccine Unit, where he conducted more than thirty clinical trials over a period of approximately five years, mostly phase I trials capitalizing on access to day care centers.[168-170] Subsequently, he initiated a new global health program in Honduras involving medical and nursing students, who prepared for the three-week expedition through formal coursework.

Other fellows recruited in the 1980s included Drs. John Frank, Stephen Eppes, Jeffrey Snedeker, Emmanuel Walter, Emilia Rivadeneira, Gareth Tudor-Williams, and Ghassan Dbaibo.

New recruits to the faculty in the 1990s: Dr. Michael Frank was recruited to Duke from his position as the clinical director at NIAID in 1990 to succeed Dr. Katz as the third chair of the Department of Pediatrics. Dr. Frank came with impressive credentials, having trained in Boston and Baltimore in medicine and pediatrics and eventually assumed positions of increasing responsibility in the Laboratory of Clinical Investigation at the NIH from 1966 to 1990. While there, he focused most of his research efforts on the role of complement in the immune response to infectious diseases and published widely prior to coming to Duke. Elected to membership in many prestigious organizations and societies, including the ASCI and AAP, he was also the recipient of numerous awards, including the European Lifetime Achievement Award, precipitated by his discovery of the cause and treatment of hereditary angioedema, a cause of four to five hundred deaths annually.

Upon arriving at Duke, he encountered numerous financial and personnel challenges, the most pressing of which was a limitation in departmental space to accommodate what he perceived as the need to expand the breadth and depth of clinical expertise. When he arrived, the department had sixty-three faculty members and most divisions within the department had only one or two faculty, with the exception of Infectious Diseases, which by that time had nine or ten faculty. With the assistance of able personnel serving as business manager (Richard Leitwig), fund-raiser, and administrative assistant (Diane Crayton), Dr. Frank embarked on the recruitment of faculty whose primary emphasis was to enhance the clinical profile of the department to the exclusion somewhat of faculty whose focus was on research. He is pleased with his success in the expansion of the faculty to 120 by the time he stepped down as well as leading the construction of a new children's building to accommodate the outpatient activities of the department. His research output declined somewhat as chairman, but in the decade to come after stepping down, he again began to investigate novel roles that the complement system may contribute to the development of effective vaccines, including HIV.

Dr. Kenneth Alexander received his MD and PhD from the University of Washington in Seattle. While at Duke, he proved to be an outstanding clinician and superior teacher, while simultaneously conducting research at a basic level on the human papilloma virus.[171-176] He remained at Duke for approximately thirteen years and was then recruited to the University of Chicago as professor of pediatrics and chief of the Division of Infectious Diseases.

Dr. Emmanuel "Chip" Walter joined the faculty in 1990 after completing medical school and house staff training at the University of Maryland and a fellowship in pediatric infectious diseases at Duke. As a fellow, Dr. Walter worked in Dr. Kent Weinhold's laboratory on various aspects of cell-mediated immune phenomena in children. Because he found that laboratory work was not sufficiently rewarding, he shifted his focus to clinical research. In preparation for this, he pursued a degree in public health and completed the requirements for the MPH, supported in part by the Pediatric AIDS Foundation. Although the first effective HIV treatments were becoming available in the early 1990s, Dr. Walter moved his primary appointment to the Division of General Pediatrics in 1992. In that role, he has been less involved with ongoing treatment trials for HIV. Instead, he has been primarily involved in the Pediatric Vaccine Unit in partnership with Dr. Clements on phase I trials of new vaccines that potentially would be useful in both children and adults.[169,177-184]

Other fellows recruited in the 1990s included Catharine Moffitt, Miguela Caniza, Cynthia Jackson, Pamela Palasanthiran, Paul Adholla,

Kenji Cunnion, Alicia Johnson, Danny Benjamin, EmiliaRivadeneira, and Kathleen Vozzelli.

Several long-serving and invaluable support staff in the Department of Pediatrics included Edna Royal and Diane Crayton. Two others served as administrators in the Division of Infectious Diseases, namely Lydia Richards and Ruth Robinson. All the current and former faculty and fellows remember them fondly and gratefully.

Clinical services: There was a single inpatient Pediatric Infectious Diseases Service. The attendings for the Service rotated among the faculty who supervised the fellows, residents, and medical students on the team. There was no dedicated outpatient clinic until the arrival of pediatric HIV infection. At that time, a subset of the faculty assumed responsibility for oversight of the care provided by the fellows and residents. Prior to the availability of effective antiretroviral therapy, the clinical management of these patients was extremely difficult and a major strain on all the health-care providers.

Teaching and training program: The increasing numbers of faculty members with clinical and educational interests contributed to the teaching of trainees at all levels. Although able to support a single new fellow every year, the Division of Infectious Diseases did not have NIH support for the training program until later.

Research themes: There were a number of successful research programs within the Division of Infectious Diseases. These included clinical research on HIV, epidemiologic research on rickettsia, research on enteroviruses, basic research on human papilloma viruses, *Staphylococcus aureus,* and fundamental immunology, as well as operational research on child abuse.

Finances: The Department of Pediatrics, patient care revenue, the viral diagnostic laboratory, and a portion of the proceeds from the Children's Classic golf tournament were the mainstay of financial support in the 1970s and 1980s. Additional revenue became available to support division activities related to HIV clinical trials from the NIH and the Non-HIV Vaccine Trials Unit in the 1990s.

New programs: The major new programs revolved around the inpatient and outpatient Pediatric AIDS Services

THE DIVISION OF INFECTIOUS DISEASES IN THE DEPARTMENT OF MEDICINE

Dr. James B. Wyngaarden, chair of the Department of Medicine from 1967 to 1983, was both a rheumatologist and a basic science investigator in the area of metabolic diseases. As chair, his major contribution to infectious diseases at Duke was to establish a new Division of Infectious Diseases and to name David Durack, MD, DPhil, as its first division chief in 1977. Later, as director of the NIH, his involvement with infectious diseases exploded with the recognition of AIDS in 1981 and the numerous controversies of that era, his appointment of Dr. Anthony Fauci as director of the National Institute of Allergy and Infectious Diseases (NIAID), and the initiation of the Human Genome Project, all of which had major implications for the discipline.

Pre-division status: Prior to the creation of the Division of Infectious Diseases in 1977, the faculty who dealt with the clinical infectious diseases or did research on those diseases or their causative agents were affiliated with the Pulmonary Division in the Department of Medicine or with the Department of Microbiology. Dr. James B. Wyngaarden recruited Dr. Thomas Cate in the late 1960s to serve as the first newly recruited infectious diseases specialist at Duke. Dr. Cate received his medical degree and house staff training at Vanderbilt University and subsequently joined the laboratory, overseen by Vernon Knight, at the NIH. The focus of that laboratory was viral research, particularly influenza. While there, Dr. Cate focused his laboratory research on various aspects of influenza.[185-190] Upon arrival at Duke in 1968, Dr. Cate joined Dr. Suydam Osterhout (who was also head of the Clinical Microbiology Laboratory) as the other clinician caring for the majority of the patients with infectious diseases at Duke Hospital. In addition, Dr. Cate continued his virologic research on influenza and served as a resource for another new faculty member in pediatrics, Dr. Catherine Wilfert, who worked initially in Dr. Cate's laboratory. Dr. Cate remained at Duke until 1976, when he was recruited to the Influenza Research Center located at the Baylor College of Medicine in Houston.

(Appendix C lists the entire faculty affiliated with infectious diseases, and Appendix D lists all the fellows in the twentieth century.)

In 1968, I was serving as an Epidemiology Intelligence Service officer of the CDC, assigned to the State Board of Health in Raleigh, where I was responsible for investigations of infectious disease outbreaks in the state of North Carolina. I had completed a year of training in infectious diseases at the Cleveland Metropolitan General Hospital under the mentorship of Drs. Charles Rammelkamp and Emanual Wolinsky. After a brief conversation with Dr. Wyngaarden, chair of the Department of Medicine at Duke, I was requested to serve as a consultant in infectious diseases at the Durham VA Hospital while still serving as a federal employee. I did this with the approval of my superior in Raleigh, Dr. Martin Hines, the state epidemiologist, until I joined the faculty in 1971 as an assistant professor in the Pulmonary Division and as head of the Infectious Diseases Section at the VA. For the next six years, I oversaw the VA clinical service, was responsible for teaching the medical students and residents rotating on the Infectious Diseases Service, and conducted research initially on antibiotic susceptibility and resistance and on the treatment of cryptococcal meningitis, for which I used a murine model.[191] In the academic year 1977–1978, I joined an immunogenetics laboratory at the Academisch Ziekenhuis in Leiden, the Netherlands, overseen by Professor Jon J. Van Rood, a co-discoverer of the HLA-B locus for a yearlong sabbatical. The laboratory specialized in transplantation immunology, and I gained considerable experience in that area.[192-194]

The Clinical Microbiology Laboratories at Duke and the VA were critical to the support of the increasingly complex patients being followed by the clinical services. Dolph Klein, Gail Hill, and, somewhat later, Barth Reller led the laboratory at Duke and were briefly discussed in chapter 6. Peter Zwadyk, PhD, directed the laboratory at the VA. He obtained his PhD in microbiology from the University of Iowa in 1971 and came to Durham shortly thereafter. A diplomat of the American Board of Medical Microbiology, he also rose to the academic rank of associate professor of Pathology as a Duke faculty member. He was a prolific contributor to the premier textbook on microbiology, *Zinsser's Microbiology*, but his passion is best reflected in a quotation by a colleague as follows: "He [Zwadyk] understood, better than almost anyone, that students are our legacy and that teaching allowed us to touch the future."[195]

At Duke, other infectious diseases specialists had previously joined the faculty and assumed some of the clinical responsibilities in my absence and as the clinical demands increased. They were Drs. Harry Gallis and Charles Ellenbogen. When I returned to the VA and Duke in 1978, some welcome changes had taken place.

Post-division status: Among the most important changes that occurred was the creation of the new Division of Infectious Diseases in the Department of Medicine and the appointment of the first division chief, Dr. David T. Durack. An Australian by birth, Dr. Durack received his medical degree and house staff training at Oxford University, where he was a Rhodes Scholar. While at Oxford, he worked with Dr. Paul Beeson, the regius professor of medicine, and Dr. Robert Petersdorf, then-chair of the Department of Medicine at the University of Washington in Seattle. This experience working on in vitro and in vivo animal models of infective endocarditis presaged the trajectory of Durack's research career.[196-201] At the conclusion of his experience at Oxford, rather than returning to Australia as he had planned, Dr. Durack was recruited to the University of Washington to serve initially as the chief resident in the Department of Medicine in 1974 and subsequently as a member of the faculty. Seeking a chief for the newly created Division of Infectious Diseases at Duke, Dr. Wyngaarden was referred to Dr. Petersdorf as the possible source of leading candidates. Dr. Durack, although then only thirty-seven years old, was highly recommended, and Dr. Wyngaarden initiated the process of what became a successful recruitment. Dr. Durack arrived at Duke in the 1977–1978 academic year and established his office and laboratory in the Stead building in Duke South to begin a productive career as division chief, which lasted seventeen years (personal communication).[202-205]

The division underwent dramatic changes over the subsequent decades of the twentieth century. One of the most notable changes was the dramatic increase in the numbers of faculty and trainees, as illustrated in figure 2. The leadership and organizational structure of the division also became more complicated over time, as illustrated in figure 3.

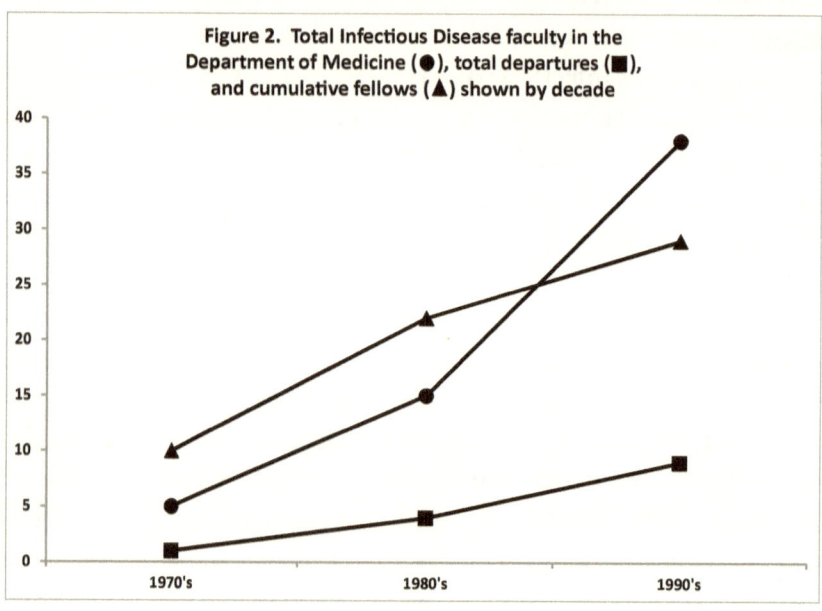

Figure 2. Total Infectious Disease faculty in the Department of Medicine (●), total departures (■), and cumulative fellows (▲) shown by decade

Figure 3. Faculty Leadership Positions in Adult Infection Disease Division

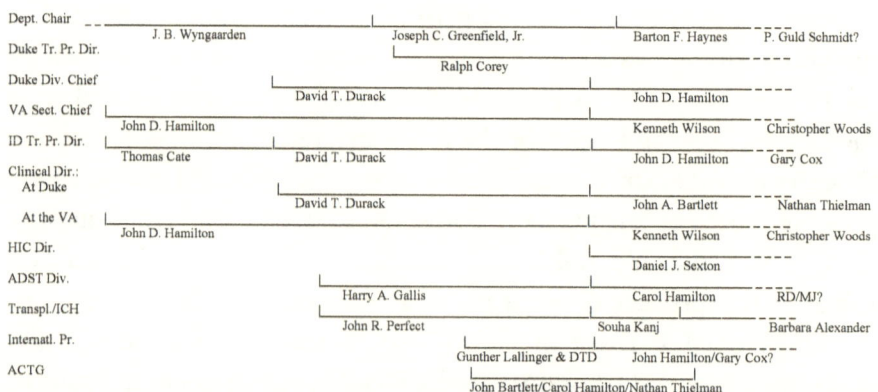

Abbreviations are defined as follows: Tr. Pr. is Training Program; Dir., Director; Div., Division; Sect., Section; HIC, Hospital Infection Control; ADST, Antibiotic Decision Support Team; Transpl., Transplant; ICH, Immunocompromised Host; Internatl., International; ACTG, AIDS Clinical Trials Group; DTD, David T Durack

152

In the period between 1970 and the beginning of the twenty-first century, the department chairs made enormous contributions to the field of infectious diseases at Duke and elsewhere. Dr. Wyngaarden's contributions were discussed earlier.[206]

After completing his house staff training and rising through the academic ranks at Duke, Dr. Joseph C. Greenfield, Jr., became chairman of the Department of Medicine at about the same time as GRID was discovered and helped to formulate local policies to deal with the manifold health-care issues that arose at Duke. A cardiologist by training, his research career was most productive.[47,207-218] He also provided financial support for the international health program in Tanzania, which allowed students, residents, fellows, and faculty to rotate for various periods in Dar es Salaam at the Muhimbili Medical Center. The program focused initially on clinical care and the training of Tanzanian trainees and colleagues. In addition, Dr. Greenfield supported the expansion of the number of ID Division faculty, appointed me as Dr. Durack's successor in 1994, and supported the initiation of the Duke Infection Control Outreach Network. This network proved to be a remarkably successful program that provided both an academic platform for research and an exceptionally effective and productive collaboration with community hospitals. And finally, his appointment of Ralph Corey, an ID Division faculty member, as the program director for the Internal Medicine Training Program provided a virtual highway for promising medicine residents to enter the field of infectious diseases.

Dr. Greenfield's successor, Dr. Barton F. Haynes, also a former house officer and Duke faculty member, was a classically trained immunologist and expert in infectious diseases at the NIH. As such, he brought considerable expertise to the study of the pathogenesis and immune response to HIV infection through his earlier work at the NIH, particularly on B and T-cell biology. As noted in chapter 4, Dr. Haynes's laboratory isolated human T-cell leukemia virus (HTLV 1) from only the second recognized case of that disease. In addition, in collaboration with colleagues at the NIH, including Dr. Robert Gallo, Dr. Haynes provided blood specimens from patients with hemophilia who were later determined to be infected with HIV. He acquired the specimens through his collaborative work with Dr. Gilbert White at the University of North Carolina at Chapel Hill. Those specimens were some of the earliest from which the HIV virus was isolated. In subsequent years, Dr. Haynes continued to make numerous fundamental contributions to the understanding of HIV.

Others making important contributions to the field of infectious diseases in the last two decades of the twentieth century included the following: Dr. Ralph Corey, who had been an ID fellow at Duke, was program director for the Internal Medicine Training Program at Duke, in which trainees were exposed to the management of infectious diseases and other illnesses that afflict adults. Dr. Corey served in this capacity from 1984 to 2002, influencing countless students, residents, and fellows, and likely will be the one individual most well remembered by former trainees. A consummate clinician, he was also academically productive as evidenced by the following citations, which do not include his many papers on bacteremia and endocarditis that are cited in chapter 10.[219-226]

Dr. Donald Granger was a member of the division from 1982–1994, worked on the clinical services at Duke and the Durham VA, and was very active in the laboratory, working on fundamental aspects of host cellular defenses, including macrophage antifungal activity in-vitro and with an experimental model of cryptococcal meningitis in immuno-suppressed rabbits.[227-231] He also served as research mentor to fellows, medical students, and undergraduate students as well as published a number of papers along the lines of fundamental cellular immunology, the citations for which follow.[232-236] In 1996, he coauthored the first of a series of papers on the role of nitric oxide in falciparum malaria work that has continued in the twenty-first century.[237]

As section chief of infectious diseases at the VA from 1995 to 2005, Dr. Kenneth Wilson played significant roles in both the clinical and research spheres of infectious diseases.[238-246] In parallel, faculty members Dr. Harry A. Gallis[247-253] and Dr. John R. Perfect[254-258] made seminal contributions, particularly in the understanding and management of patients whose immune systems were compromised. Drs. Barth Reller and Lizzie Harrell made important contributions to the Clinical Microbiology Laboratory and to the training of clinical microbiology trainees, some of whom were adult and pediatric infectious diseases fellows. Dr. Peter Zwadyk served in a similar capacity as head of the

Clinical Microbiology Laboratory at the VA. Dr. Souha Kanj made the immune-compromised host her primary clinical and laboratory interest and established the Transplant Infectious Diseases Service.[222,259-262] Dr. Carol Dukes-Hamilton adopted TB as her primary focus, and somewhat later, in the 1990s, assumed responsibility as the North Carolina state TB controller and, later still, became an international leader in the field.[263-272]

After being in private practice for a number of years, Daniel J. Sexton, a former Duke fellow in infectious diseases, returned to Duke as a faculty member to assume responsibility for an expanded role of the Hospital Infection Control Program. This led ultimately, in the latter part of the 1990s, to the creation of a far-flung network focused on infection control, called the Duke Infection Control Outreach Network (DICON). The network reached nearly fifty distant hospitals within and outside the state of North Carolina.[273-281] Somewhat in parallel, an international program was launched, first in Dar es Salaam, led by Dr. Durack and Dr. Gunter Lallinger.[223,282-285] Then, beginning in 1994, after a move from Dar es Salaam to Moshi, Tanzania, I led the program.

Finally, the clinical and clinical research efforts on HIV/AIDS within the division were led by Drs. John Bartlett,[286-294] Charles Hicks,[288,295-297] and Nathan Thielman,[284,285,298-300] with support from the NIH, foundations, industry, and philanthropy. Others who made significant contributions to the Division of Infectious Diseases, some of whom are mentioned in greater detail in other chapters, include (alphabetically) Drs. Barbara Alexander, June Almenoff, Andrew Alspaugh, Miriam Cameron, Peter Cegielski, Gary Cox, Vance Fowler, Richard Frothingham, Alison Heald, Charles Hicks, Charles van der Horst, Jerome Kim, Katherine Kirkland, Mary Klotman, Diego Miralles, Mark Perkins, Kevin Porter, Nathan Thielman, and Hetty Waskin.

Table 7.3 shows the clinical and research emphasis, clinical teaching and research awards, new programs, and funding sources. Of note, the number of consultation services expanded from two in the 1970s to four in the 1990s. Although general infectious diseases dominated clinical interests from 1960 to 1980, new emphasis was given to a new array of possibilities, including the immunocompromised host and transplantation patients, HIV and AIDS patients, international infectious diseases, TB, and clinical microbiology. Research emphasis included research programs on bacteria (including TB), viruses (including HIV), fungi, rickettsiae, and hospital infection control. Awards made by national organizations for clinical and research achievements were given to five faculty members in the Division of Infectious Diseases. Awards for clinical, teaching, and research by Duke University Medical Center were given to 15 Division faculty.

Table 7.3. Adult Infectious Diseases

	1970's	1980's	1990's
Dept. Chair	Wyngaarden	Greenfield	Haynes
Events	ID Division Durack, chief		ACGME Accredited: JDH, Chief
Division faculty	2 - 6	7 - 18	19 - 36
Clinical services	Duke (1) & VA (1)	Duke (2) & VA (1)	Duke (3) & VA (1)
Clinical emphasis: General ID Transplant ID HIV/AIDS International Tuberculosis Clinical Microbiology	5	11 2 3 2	19 5 7 3 1 1
Research emphasis: Bacteria, including TB Viruses, including HIV Fungi Rickettsiae Infection Control Immunology	2 2 1	5 8 3	9 15 5 1 3

Awards:
Golden Apple:
  Basic
  Clinical
  House Staff
 	
Duke Scholar Teacher
 	
Gorgas Medal
 	
Young Investigator
 	
Stead
	
	2/1
	-/4
	1/-
	
	1/-
		
		--
		--
		-/1
		
		4/1
			
			--
			2/1
			--
			
			1/-
			
			1/-
			
			3/-
			
			4/-

New programs	Division of ID	Endocarditis Service HIV clinic ACTU RCAHI Duke/Tanzania Collaboration ACGME accreditation	Transplant ID Antibiotic Stewardship Sexually Transmitted Dis Infection Control Outreach Network Clinical Research Network Mycology Center of Excellence

Other than creation of the new Division of Infectious Diseases in the 1970s, no additional new programs were developed in that decade. In the

1980s, however, new programs emerged, with faculty champions in infective endocarditis, inpatient, and outpatient management of patients with HIV infection, the AIDS Clinical Trials Unit (ACTU) of the NIH, the Research Center on AIDS and HIV Infection (RCAHI) at the Durham VA Medical Center, the Duke International Program in Tanzania, and the official accreditation of the Infectious Diseases Training Program by the Accreditation Council for Graduate Medical Education (ACGME). In the 1990s, new programs included the Immunocompromised Host and Transplantation Service (ICH-TX), the Antibiotic Decision Support Team (ADST, later called the Antibiotic Evaluation Team (AET), the Sexually Transmitted Diseases (STD) Program, the Hospital Infection Control (HIC) Program, the Duke Infection Control Outreach Network (DICON), a clinical research network for antibiotic studies, and a Duke University Mycology Research Unit (DUMRU). Funding to support research expanded from clinical revenues, industry, VA and NIH grants to funds from the Health Resources and Services Administration (HRSA) and the National Institutes for Environmental Health and Safety (NIEHS), the National Foundation for Infectious Diseases (NFID), and other foundations (see table 7.4).

Table 7.4. Funding sources for the Adult Infectious Diseases

1970's	1980's	1990's
Department PDC	Department PDC	Department PDC
VA Merit	National Institutes of Health: R01s PPG	National Institutes of Health: R01s, U01, P01 ACTU Roadmap K23s, 24s T32s Fogarty International Center MIRT
VA Career Development	VA Merit	HRSA
Industry	Industry	NIEHS
	Foundation	NFID
		VA: Merit CSP

Abbreviations are defined as follows: PDC, Private Diagnostic Clinic; VA, Department of Veterans Affairs (Merit Review; Cooperative Studies Program); National Institutes of Health (R01s; PPG; U01; P01; ACTU; K23; K24; T32), FIC, Fogarty International Center (MIRT).

The fellowship program within the Adult Infectious Diseases Service increasingly focused on trainees with career trajectories directed toward academics or public health while increasing the number of trainees as well. Table 7.5 shows the numbers. In the 1970s, eleven trainees were accepted into the training program. In the 1980s, during the time the training program became ACGME accredited, twenty-five trainees were accepted into the program; in the 1990s, thirty-four trainees were accepted into the program. This increase in trainees can be attributed to the increased burden of patients with infectious diseases in an increasingly diverse array of circumstances and to an enhancement of the training options offered to applicants.

Table 7.5. Fellows in the Department of Medicine

	1970's	1980's	1990's	Since 2000
Fellows:				
Number	10	22	29	42
Board Pass Rate				100%
MD's	10	22	26	
MD/PhD's	--	--	3	
Postfellowship appt (immediate):				
Academic	7	8	18	39
Public Health		1	3	
Industry	--	1	3	
Private Practice	2	10	4	
NIH	1	--	1	
Military		2	--	
Fellow funding:				
Clinical, *years*	Duke/VA	Duke X2/VA	Duke X3/VA	
Research, *years*				
HIC	--	--	11	
Training grants	--	--	10	
Clin. Micro.	--	--	7	
Dept./Division	9*	10*	3	
Mentor/Industry	2*	15*	3	
K Award within 3 years	Unk	Unk	Unk	11
Current position:				
Academic				37
Public Health				
Industry				
Military				
Private Practice				

* Estimates
**Abbreviations are defined as follows: NIH, National Institutes of Health; HIC, Hospital Infection Control; Clin. Micro., Clinical Microbiology.

The format of the fellowship from the early 1990s was an initial clinical year in which the first-year fellows rotated on three distinct clinical services—two at Duke and one at the VA. Somewhat later, there were four distinct services: two Duke services, one Duke Immunocompromised Host and Transplant Service, and one VA service. First-year fellows also attended a half day of clinic per week, during which they followed up on patients whom they had seen as inpatients. During the subsequent two years, fellow responsibilities generally continued for a half day per week in the outpatient clinic. The majority of their time, however, was dedicated to the training for, and conduct of, specific research in their areas of interest, which was arrived at through conversations with faculty members. During all three years, fellows were required to attend three weekly conferences: Infectious Diseases Grand Rounds, Journal Club, and Case Conference. The goal of these prescribed activities was to prepare trainees for successful careers as clinical investigators in academic departments.

Table 7.4 summarizes the early choices made by fellows completing the training program. In the 1970s, most trainees entered academic fields; in the 1980s, approximately equal numbers entered academic or public health-related activities as did those entering private practice; and in the 1990s, by far the majority of trainees who completed the fellowship entered academic or public health fields. Funding for the clinical year of fellowship was derived from the Duke Department of Medicine, Duke Medical Center, or from the Department of Veterans Affairs. During the research years, funding came from the Department of Medicine, the Division of Infectious Diseases, the trainee's mentor, or, in some cases, from industry. Similar sources of funding were accessed in the 1980s, whereas in the 1990s, the majority of funding came from the Hospital Infection Control Program, training grants—including grants for clinical research—T 32, K 12, and the Clinical Microbiology Training Program. In table 7.4, the source of funding largely reflects where the trainee was spending his or her effort. Examination of the careers of the forty-two trainees who completed the program since 2000 reveals the following salient facts: 100 percent passed the infectious diseases boards; thirty-nine (or 93 percent) went directly into an academic spot; thirty-seven (or 88 percent) currently remain in academics; eleven (or 26 percent) obtained a K award within three years of completing the fellowship, thanks largely to the training program director, Dr. Gary Cox, showing that basic research activities continued to thrive in the twentieth century.[301-307] Although no objective measures of infectious diseases fellowship programs exist to compare Duke with other schools around the United States, Duke surely ranks within the top twenty and possibly the top ten.

The administrative staff supporting all of these activities in the division was incredibly important and greatly appreciated. Duke staff included Olive Sherman, Janet Routten, Mary Ann Howard, Jackie Robbins, Mary Moggio and Mary Oden. Staff located at the Durham VA in the Section on Infectious Diseases included Margaret Arnold, Kay McKeel, Pat Spivey, Pam Stewart, Mary Moorefield, and Joyce Frederick.

A number of prior fellows wrote brief summaries or gave information about their fellowship experiences in the following statements:

FIRST FELLOW

Richard V. McCloskey wrote:

> In 1965, when I finished as a Senior Assistant Resident and took on an immunology/ID [infectious diseases] training program with Dr. Sieker, Dr. Amos, and Dr. Zmijewski, there wasn't actually a formal ID section at Duke, but I did certainly assume responsibility for the ID Service originally at the VA Hospital, which actually turned out to be part of the training program. It probably comes as no surprise to you that Dr. Stead felt that people who are interested in acquiring skills would design their own training program which is, in fact, what happened (personal communication).[308-318]

RETURNED FROM OR EMBARKED ON A PRIVATE PRACTICE

Daniel J. Sexton (medicine) wrote:

> My fellowship at Duke was a transformative experience in my career. In many ways, my experience was quite different [from] the experience of subsequent fellows who trained at Duke in the 1990s and beyond. For example, I was on call fifty of fifty-two weekends. I saw all the

consults at Duke and the VA every day. I did a teaching conference on IDs [infectious diseases] to medical residents every two weeks, and the teaching obligation was educational in the true sense of the word. And I did things that would strike modern fellows as quite unique, such as Gram stains and urinalysis. Some of the most valuable experiences during my fellowship [were] not recognized as valuable until much later. Because the average length of stay was substantially longer in the mid-1970s than now, I had the wonderful opportunity to observe infections in hospitalized patients over time. This opportunity to learn by doing during my fellowship remained central to my beliefs about the imperative for lifelong learning and served me well in private practice and now back at Duke in leading the Duke Infection Control Outreach Network (personal communication).[273-281]

ON AN ACADEMIC TRAJECTORY

Wendy Keitel (medicine) wrote:

As I recall, I was the only clinical fellow in 1980–81 at Duke. [I] completed my fellowship training at Baylor, and subsequently joined the faculty there after having two kids. At Duke, I participated in the original acyclovir studies on genital herpes and herpes zoster in immunocompromised hosts, and I vividly remember the first biopsy proven case of herpes simplex virus encephalitis at Duke and a patient with a brain abscess caused by Cladosporium. And, of course, our small journal club during which we reviewed the first descriptions of AIDS thought then to be caused by amyl nitrate. My experience as a fellow was great, and I never wanted to leave. It did get me started on the path to clinical trials, which has served me well and has led to my appointment as

161

professor of molecular virology and microbiology and medicine at Baylor.[319-324]

Stephen Eppes (pediatrics) wrote:

> As a pediatric ID [infectious diseases] fellow at Duke from 1984 to1987, I spent my first year doing clinical consultations. Most of the next two years I spent in Dr. Joklik's laboratory, researching the mechanism of antiviral activity of ribavirin. When Duke became involved with the initial clinical trials of AZT in children, pediatric HIV infection became my passion and took most of my time. Under Cathy Wilfert's supervision, I believe I was the first physician to treat a child with an antiretroviral medication. I am currently a professor of pediatrics at the Jefferson Medical College (personal communication).[141,144,325,326]

CAREERS IN PUBLIC HEALTH

Dr. Elizabeth Talbot (medicine): Dr. Talbot knew she wanted to be an infectious disease specialist since she was thirteen years old. She embarked on her professional trek with a liberal arts foundation in biology and philosophy at Mount Holyoke College, then "loved *almost* every moment" of medical school at the University of Medicine and Dentistry of New Jersey, in her home state. She confirmed her childhood ambition during formal education at the London School of Hygiene and Tropical Medicine. Drawn to a full view of health and healing, Dr. Talbot completed a combined psychiatry-medicine internship at the University of Iowa and began at Duke as a junior resident in 1993. Like many others with an interest in global health, she was drawn to Duke because of the unique international research opportunities, such as the TB research she accomplished in Vitoria, Brazil, as a senior resident. There, she fell in love with the many facets of TB control and returned to Duke as an infectious diseases fellow in global mycobacterial research, with Richard Frothingham as her mentor. She continued on this professional trajectory through the CDC's

Epidemic Intelligence Service and as a medical officer in Botswana with the International Activities of the Division of TB Elimination.

Dr. Talbot remains engaged in international TB control through clinical projects, research, and consultation. She is currently an associate professor at Dartmouth College and is federally funded as the deputy state epidemiologist for New Hampshire. She is a medical scientist for the Foundation for Innovative New Diagnostics in Geneva, where she participates in efforts to identify and implement novel approaches to the diagnosis of poverty-related diseases such as TB. She consults for PATH, USAID, and TB CARE II for global TB control activities, especially in Tanzania, Haiti, and the Democratic Republic of Congo. Dr. Talbot has participated in the writing of national and international TB guidelines and has authored more than sixty peer-reviewed publications, with additional studies of TB diagnosis ongoing. She has collaborated with other Duke faculty members throughout her career, including Drs. Mark Perkins, Peter Cegielski, Carol Dukes-Hamilton, and Kathryn Kirkland. She is grateful for her training at Duke that allowed her to realize her aspirations in infectious diseases and international TB control (personal communication).[327-334]

Dr. Erin Staples (pediatrics): With an MD and PhD in microbiology and immunology from the State University of New York, Dr. Erin Staples was an intern and resident in pediatrics and a fellow in pediatric infectious diseases at Duke. Her prior academic achievements included election to Phi Beta Kappa and Alpha Omega Alpha. After leaving Duke in 2006, she joined the CDC as an Epidemic Intelligence Service officer and was assigned as a medical epidemiologist in the Arboviral Disease Branch of the Division of Vector-borne Infectious Diseases. In that capacity, she served as a consultant on arboviral diseases, conducted outbreak investigations, designed and implemented research studies addressing gaps or key questions related to arboviral epidemiology, oversaw domestic surveillance of arboviral diseases, and supervised preventive medicine residents. She received the director's recognition award in 2010 for her work on a yellow fever outbreak in Uganda and for work on a West Nile virus outbreak in Arizona.

She is an author or coauthor of 42 peer-reviewed publications, on five of which she is first author and on ten of which she is the senior author. In addition, she is a coauthor on numerous additional non-refereed publications concerning vector-borne diseases, as well as scientific presentations.

INTERNATIONAL FELLOWS

David Murdoch (medicine): A native New Zealander, Dr. Murdoch did much of his medical and microbiological training at the University of Otago in Christchurch, New Zealand, punctuated by several years' experience working in Nepal as a resident physician with the Himalayan Rescue Association or in the District of Solukhumbu in Nepal before coming to Duke as a fellow in the Division of Infectious Diseases in the Department of Medicine in 1998. With his experience, he brought a unique perspective both to the practice of medicine and infectious diseases and to life. He published eleven peer-reviewed articles resulting from his work as a fellow or from collaborations with Duke faculty and fellows. His academic achievements since leaving Duke are numerous, in terms of publications in peer-reviewed journals, chapters, reviews, editorials, and a thesis on the diagnosis of Legionella. Upon returning to the University of Otago in 2000, he rapidly rose through the academic ranks to be named professor and head of the Department of Pathology at the University of Otago. He notes the following: "My Duke fellowship was a wonderful experience. Particularly notable are the excellent mentorship, an environment that provided opportunities to bring out the best in people, and the lasting friendships that have arisen from this time."[335-345]

Ghassan Dbaibo (Pediatrics): Another international fellow was Ghassan S. Dbaibo, who came to Duke as an intern in pediatrics in June of 1986 from the American University of Beirut School of Medicine. He completed his house staff training at Duke, including his position as chief resident, and he completed his fellowship in pediatric infectious diseases and the appointments in pediatrics until he returned to Lebanon in 1996. At American University of Beirut, Dr. Dbaibo has risen to professor of pediatrics and head of the Division of Pediatric Infectious Diseases. He has published over seventy-five peer-reviewed papers and is first author on thirteen with his collaborators in the Department of Medicine, Drs. Youssef Hunun and Lina Obeid. Of the fifteen papers published on work done at Duke, Dr. Dbaibo was first author on eight and focused on the molecular mechanisms of the effects of cytokines and apotosis. He contributed to many book chapters, abstracts, and presentations. Upon his return to Lebanon, he led over twenty-five antibacterial, antiviral, and vaccine clinical trials.

INTERNATIONALLY FOCUSED FELLOWS

Peter Cegielski (medicine): Dr. Cegielski's interest in international health began early in his career. As a medical student at the University of California at San Diego, he pursued a project entitled "Clinical Malaria, Falciparum Parasitemia, and Malnutrition in Urban and Rural Kenyan Children" as the topic for his thesis, requiring him to work for five months in Kenya. The topic of his senior assistant residency talk at Duke was "AIDS in Africa," and as a fellow in infectious diseases at Duke, Dr. Cegielski lived and worked in Dar es Salaam, Tanzania, as part of the Duke-Tanzania collaboration. He joined the faculty at Duke in the Division of Infectious Diseases in 1990. Over the next four years, while a Duke faculty member, he pursued a MPH degree at the University of North Carolina; the topic of his thesis was "TB and Malnutrition in a Nationally Representative Cohort of Adults in the United States." In 1994, the University of Texas at Tyler, Texas, recruited him to oversee the TB clinics in thirty-five separate counties in northeastern Texas. In 1996, he was recruited to the Department of Epidemiology as an assistant professor at the Johns Hopkins University's Chang Mai Program to conduct large cohort studies of populations at risk of HIV infection. Two years later, he joined the Division of Tuberculosis at the CDC as a medical epidemiologist, where he was responsible for TB activities in Russia, at the United States-Mexico border, and in foreign-born persons in the United States. In 2001, he was named the team leader for drug-resistant TB, where he has remained to the present. He has been active in public health circles, received numerous awards, and been published widely, including twenty-five peer-reviewed publications for work done at Duke or with Duke faculty collaborators. Finally, he has been successful in competing for extramural funds from USAID, WHO, Rockefeller, Family Health International, and the NIH.[282,346-353]

Dennis Clements (pediatrics): A pediatrician by training, Dr. Clements spent the decade between 1976 and 1986 in the US Air Force (two years) and in the private practice of pediatrics (eight years). In 1986, he began a fellowship in pediatric infectious diseases at Duke, and following its completion, he spent another two years, from 1988 to 1990, as a fellow in the Department of Microbiology and Infectious Diseases at the Royal Children's Hospital in Melbourne, Australia. His work in Australia was on *Haemophilus influenza* group B (HIB) epiglottitis and meningitis, which was funded by Merck. Dr. Lynn Gilbert oversaw his work and helped to facilitate the creation of a *Haemophilus influenza* group B vaccine network.[354,355] Upon his return to Duke in 1990 as a faculty member in the Department of Pediatrics, he assumed inpatient and outpatient clinical responsibilities and extended the work started earlier by Dr. Catherine Wilfert in the creation of the Duke Vaccine Unit. Over the next five years, he conducted over thirty phase 1 trials of novel vaccines, including HIB vaccine and one successful later phase study of varicella vaccine in day care centers[167-170]. Occasional fellows and residents worked with him on those studies for short periods, and he attracted a continuous string of medical students pursuing their third year in the clinical research track. Although he continued his interest in vaccines, he passed along responsibility for the Vaccine Unit to Dr. Emmanuel Walter and began to focus his career on the development of educational modules, exploring medicine and other cultures, and on the establishment of a program in Honduras to provide medical care staffed by volunteers, most of whom were undergraduate or medical students or house officers. This popular program has continued to the present day.

CAREERS AT THE NIH, THE CDC, OR THE MILITARY

Sanjay Desai MD, PhD (Medicine): Dr. Desai is one of only a few fellows in the Adult Infectious Diseases Program who came to the program already having obtained both an MD and a PhD degree, both of which were obtained at Washington University in St. Louis. He entered the Duke House Staff Training Program in Medicine on the Clinical Investigator Pathway Program in 1992, completed two clinical years in medicine, followed by three years of basic research on fundamental aspects of malaria and then a year in the Clinical Infectious Diseases Fellowship. Already being an experienced investigator, Dr. Desai functioned independently but with continued collaborations with investigators at Washington University and later collaborations with Dr. Robert Rosenberg, a scientist at the University of North Carolina in Chapel Hill. He published his work in prestigious journals as a fellow, was the recipient of an NIH-funded KO8 grant in 1994, and received the Young Investigator Award from the American Society for Microbiology in 1998.[356-358] He was appointed a research fellow in the Laboratory of Parasitic Diseases at the NIAID in 1998 and continued his productive research program focused on nutrient transport channels in infected erythrocytes. He discovered two novel ion channels and received external funding that was based on the possibility that these ion channels could serve as targets to neutralize infection by the malaria parasite.

Jerome Kim, MD (Medicine): Dr. Kim trained in medicine at Duke and joined the infectious diseases fellowship in 1987. In addition to demonstrating outstanding clinical skills, he also proved to be an excellent investigator in basic science laboratory work on HTLV-I and -II as a senior fellow and a junior faculty member, as described in chapter 4.[359-366] He left Duke in 1991 to accept a position as assistant professor of medicine at the Uniformed Services University, followed later by positions at the Institute of Human Virology at the University of Maryland and as chief of the Virology Section in the Department of HIV Vaccine Research at the Walter Reed Army Institute of Research.

Shannon Hader MD (Pediatrics): Dr. Hader trained in pediatrics and infectious diseases at Duke and went to the CDC as an Epidemiology Intelligence Service officer. Subsequently, she directed the CDC's Zimbabwe Global AIDS Program, worked with children and adults with HIV infections in Brazil and Jamaica, and later assumed a President's Emergency Plan for AIDS Relief position as a senior scientific advisor for the (PEPFAR). She was director of the District of Columbia's HIV/AIDS administration from 2007 to 2012.

FELLOWS WITH BOTH MD AND PHD DEGREES

Paul Bohjanen, MD, PhD (Medicine): Dr. Bohjanen received his MD and PhD from the University of Michigan in 1993. He completed his house staff training in medicine and a fellowship in infectious diseases in the Clinical Investigator Pathway Program at Duke in 2000. His major research focus was on T-cell activation, and he found a good balance between his basic research and clinical work during this period. He completed his final year of clinical training in the last year of his fellowship, which was challenging.[98,99,367-369] During the research years, he worked with Dr. Mariano Garcia-Blanco and Dr. Jack Keene, with support from an NIH K award. In the final year of his fellowship, he successfully applied for a NIH RO1 and accepted a position on the faculty at the University of Minnesota, where he continued his productive research career and participation in the clinical activities of the Division of Infectious Diseases.[370-374] Dr. Bohjanen cites three general strengths of the Division of Infectious Diseases Training Program: clinical care and training, the opportunity for useful collaborations with other basic scientists, and independence.

Danny Benjamin, MD (Pediatrics): Dr. Benjamin elected to come to Duke as a pediatric infectious disease fellow in 1998 because of the strength of the pediatric faculty working on HIV-AIDS. It became apparent relatively soon after his arrival that a career in pediatric AIDS was not viable given the impact of PACTG 076 and that a career focused on laboratory-based research was not to his liking. Instead, he embarked on clinical research, an interest initially stimulated by a patient with a Candida kidney infection, which led him to pursue additional training, leading to both the MPH and soon thereafter the PhD.

INDUSTRY-BASED CAREERS

Cindy Jackson, MD (Pediatrics) wrote:

> My experience at Duke in the Pediatric Infectious Diseases Fellowship program was transformational. I

had the opportunity to learn from and directly interact with giants in the field of Peds ID [pediatric infectious diseases], including Samuel L. Katz and Catherine M. Wilfert as well as others in multiple departments that were also seen as key leaders in their fields.[375-377] There was also a deliberate effort to foster interactions between the adult and pediatric ID fellowship programs, which was very beneficial to both. I fondly remember Micro Rounds led by Barth Reller, combined Journal Club led by David Durack, and the weekly adult case conferences and pediatric patient rounds. I was given a unique opportunity to continue my fellowship research project in the laboratory of Marty St. Clair at Burroughs Wellcome, one of the patent holders of AZT. In Marty's lab, I learned scientific method by one of the most rigorously meticulous people in HIV research at the time. My work at BW was supported by an NIH HIV training grant led byJohn Hamilton. This experience gave me an insider look at pharmaceutical drug development. Marty was never hesitant about including me on all discussions relating to a specific compound and planning clinical trials. Much of the work done in the lab at the time was centered on the antiretroviral combination of AZT and 3TC and resistance mutations seen in patients, which was groundbreaking information at the time. Coupling the lab work at BW with the PACTG clinic work at Duke, led by Ross McKinney, the PACTG PI, I became interested in the world of clinical trials, where I remain today as head of the pediatric group at Quintiles.[378-380] None of this would have been possible without my time at Duke and the support of too many individuals to mention.

Corrie Dekker (Pediatrics): Dr. Dekker trained in pediatrics and infectious diseases at Duke and embarked on a career in industry with an appointment as a senior clinical research scientist at the Wellcome

Research Laboratories in the Research Triangle Park, North Carolina. She then worked on HSV pathogenesis in the Laboratoire d'Oncologie Virale at the Fondation Pour la Recherche Medicale, INRS, Villejuif, France. First as deputy director of clinical research at Lederle Biologicals in Pearl River, New York, and later as medical director of the Biocene Company, she was responsible for the global clinical development of vaccines. In 1992, she was appointed vice president of Clinical Research and Medical Affairs at the Chiron Corporation in Emeryville, CA. After a brief stint as a consultant with Bayer, she was appointed medical director of the Stanford LPCH Vaccine Program and a research professor in the Stanford School of Medicine, where she remains.

FORMER DUKE MEDICAL STUDENTS AND HOUSE OFFICERS

Robert Purcell, MD: Dr. Robert Purcell received his MD from Duke University in 1962 and, as a medical student, worked in the laboratory with Dr. Joseph Beard, subsequently did a year as an intern in the Department of Pediatrics, followed by appointments at the CDC and the NIH, and finally was appointed head of the Hepatitis Viruses Section of the NIAID. There, he oversaw two other individuals with Duke connections—Drs. Ticehurst and Cohen. A prolific scientist with over five hundred publications, those papers that he coauthored with Drs. Ticehurst and Cohen are cited below.

Jeff Cohen (Medicine): Dr. Jeffrey Cohen attended medical school at Johns Hopkins, was an intern and resident in Medicine at Duke from 1981 to 1984, and joined the program at the NIH in Dr. Robert Purcell's laboratory working on hepatitis. Specifically, he worked on the hepatitis A virus from 1984 through 1987 and published a number of papers,[381-386] after which he was a clinical infectious diseases fellow at Brigham and Women's in Boston from 1987 until 1989. Then he joined Elliot Kieff's laboratory at Brigham, working on the role of EB virus in tumorigenesis.[387-389] In 1991, Dr. Cohen moved to the NIH in Steven Strauss's laboratory, where his research was mostly directed to the fundamental virology of the Varicella Zoster virus.[390-392] He has remained productive scientifically, and he became the chief of Medical Virology within the Laboratory of Infectious Diseases at the NIH.

John Ticehurst (Medicine): Dr. John Ticehurst graduated from Duke Medical School was a resident at the University of North Carolina and then a fellow for two years in the Infectious Diseases program at Duke, leaving in 1979 also to join Dr. Purcell's laboratory at the NIH. There, he worked and published mostly on hepatitis A but also on hepatitis B and E.[393-400] In 1987, he was recruited to the Walter Reed Army Institute Research, where he became more involved with hepatitis E.[401-403] He completed a fellowship in laboratory medicine at Hopkins in 1995 and was then recruited to the FDA and later recruited back to Hopkins, where he was responsible for the development of point-of-care diagnostic devices for the detection of assorted viruses.

Fred Ruben (Medicine): Dr. Ruben received his undergraduate and medical degree from Duke and trained at Cleveland Metropolitan General Hospital (Case Western affiliate) under Dr. Charles Rammelkamp, infectious diseases at CMGH under Dr. Emanuel Wolinsky, and the University of Illinois in Chicago under Dr. George Jackson. He then joined the CDC as an Epidemiology Intelligence Service officer, where he assisted in the eradication of smallpox. In 1972, he joined the ID Division at the University of Pittsburgh, where he remained until retiring as professor of medicine in 1997 to become the director of Scientific and Medical Affairs for Sanofi Pasteur. He was productive academically, with numerous peer-reviewed publications focused on vaccines, especially for influenza.[404-410] Dr. Ruben was active in several professional societies such as IDSA and ATS and served on advisory committees for the CDC (ACIP), the FDA (Anti-Infective Drugs Advisory Committee), and the NIH (intramural and extramural advisory committees). He retired from Sanofi in 2008.

Photograph 7.6 is a composite of pictures of six prominent department chairs at Duke in the twentieth century.

David C. Sabiston

Wolfgang "Bill" Joklik

Samuel L. Katz

Michael Frank

Joseph C. Greenfield

James B. Wyngaarden

CHAPTER 8

INTERNATIONAL ACTIVITIES

> What are the long-term implications of what we do to-
> day? They are endless and therefore do important things.
> Consider a time one thousand years ago, when some of
> the great cathedrals of the world were being built by
> artisans who would never see the magnificent fruits of
> their labors. We in medicine are a version of these same
> artisans, building what medicine will become.[1]

In 1982, Dr. David Durack, then head of the Adult Division of Infectious Diseases at Duke, met Tanzanian Professors Mahalu, Masselle, and Shao at a meeting in Cairo, Egypt, and initiated discussions about a possible collaboration between Duke and the Muhimbili Medical Center, located in Dar es Salaam, Tanzania. The following year, Dr. Durack recruited Dr. Gunther Lallinger to launch the collaborative program in Africa after Lallinger completed a Duke fellowship in infectious diseases. He completed the fellowship in 1987. In that same year, Dr. Lallinger; his wife, Rupa Redding-Lallinger (a pediatrician); and their infant son moved to Dar es Salaam, funded by the Department of Medicine. Initially, Duke's mission, led by Dr. Lallinger, was to provide inpatient clinical care and to teach Tanzanian students and house staff. In 1988, Lallinger was joined by Dr. Peter Cegielski, the first infectious diseases fellow recruited to the international track, a new program created in 1988, called the Duke Infectious Diseases International Program. Prior to going to Dar, Dr. Cegielski had gained some experience in clinical microbiology at Duke and in pulmonary medicine through a rotation at the Asheville VA Hospital, where he had learned to perform bronchoscopies, a skill he later put to use in studies of opportunistic infections in patients with HIV infections in Africa. With the expectation that he would be doing research as well as clinical care, he obtained warehoused equipment, including centrifuges and microscopes, and miscellaneous supplies to take with him. Also, Dr. Carl Ravin, then chair of radiology at Duke, donated a fiber-optic bronchoscope. Without specific research funding, Dr. Cegielski sought funding from pharmaceutical firms and initiated various clinical studies focused on prevalent conditions related to HIV infections. Specimens were processed locally in Dar as well as returned to Duke for more sophisticated analyses, which were supported in part by a three-thousand-dollar award from the Duke Medical Center Women's Auxillary to perform electron microscopy on selected specimens and to test for *Mycobacterium tuberculosis* (TB). Over time, increasing emphasis was placed on clinical research and funds were obtained from the Carnegie Foundation to build a stand-alone laboratory to support the following research projects: (a) the causes of anemia in adults and children, conducted by Dr. Rupa Redding Lallinger; (b) the effect of nitric oxide in cerebral malaria, conducted by Dr. Nicholas Anstey and funded by the American Society for Tropical Medicine and Hygiene; and (c) an early study of mother-to-child transmission of HIV, conducted by Drs. Michael Greenberg and Christine Hahn and supported by the National Institutes of Health (NIH). The laboratory was actually dedicated in 1991 by the minister of health at the time, Professor Philemon Sarungi, along with other dignitaries.

At its peak, three full-time Duke faculty and fellows, as well as medicine and pediatric residents from Duke, lived and worked in Dar and were supported largely by the Department of Medicine at Duke. The Lallinger family, with four children by then, remained in Dar es Salaam at the Muhimbili Medical Center for seven years before returning to North Carolina in 1994. Other Duke faculty joined the Lallingers for periods of up to two years, including Duke's Dr. Malcolm MacDonald and the University of California at San Francisco's Drs. Charles Daly and Peter Small. Numerous fellows and residents, including Drs. Peter Cegielski, William Miller, Nicholas Anstey, Carol

Dukes, and Jeremy Sugarman also contributed to the growing research enterprise. Together with Muhimbili faculty and staff, including Drs. McLarty, Swai, Palangyo, Mwaikambo, Mgusi, Mbaga, Kinabo, and Mwakusa, Duke staff published important validation studies of clinical case definitions for HIV-AIDS in Africa.[2] In addition, they proposed that delayed-type hypersensitivity along with total lymphocyte counts[3] and beta-2 microglobulin levels[4] might be useful for following AIDS patients longitudinally. Others from Duke in Tanzania published on the use of fusidic acid in patients with AIDS[5], the epidemiology of AIDS in Africa,[6] the pulmonary complications of HIV disease—including TB[7] and pericardial effusions[8-10]—diarrheal diseases,[11-13] and skin manifestations of HIV-infected persons in Tanzania.[14] Aside from HIV topics, collaborators from Duke published studies on malaria,[15-18] hepatitis,[19] rickettsial disease,[20] and research ethics.[21]

In 1995, Professor John Shao, one of Duke's principal collaborators, left Dar es Salaam for Moshi, Tanzania, to lead the creation of the second medical school in Tanzania, at the Kilimanjaro Christian Medical Center (KCMC). Although the Duke effort in Dar es Salaam was productive, I, chief of the Adult Infectious Diseases Division at the time, moved our program to Moshi for a variety of reasons, the most important of which was the pledged support offered by Professor Shao, and the KCMC-Duke collaboration began. From that point forward, the "Research with Service" mission of our collaboration evolved to what it became as of 2010. Professor Shao delivered on his promise to help the Duke faculty implement a robust clinical research program in Moshi, assisting in overcoming the myriad of regulatory, administrative, personnel, and practical obstacles encountered in the development of a credible research enterprise. Medical education was a central component of the KCMC-Duke relationship: fourteen Tanzanians received training at Duke before 2000. In addition, over fifty Duke residents, students, and fellows trained in Tanzania between 1995 and 2000, supported by the Department of Medicine, the Adult Division of Infectious Diseases, the American Society of Tropical Medicine and Hygiene, and various philanthropic organizations. Under the auspices of the NIH-funded Minority International Research Training (MIRT) Program overseen by Dr. Redding-Lallinger and me, sixty-five minority students from Duke, UNC, and NCSU trained in seven different countries: Colombia, Costa Rica, Hungary, Slovak Republic, Ghana, Tanzania, and Brazil. Moreover, it became evident that the health of the populations in developing nations was central to the stability of those nations. That, combined with the desire to attack the overwhelming diseases associated with poverty, led the United States government and other organizations to support operational research that would address the health problems prevalent in countries like Tanzania. Faculty in the adult and pediatric divisions of infectious diseases was extraordinarily successful in capturing those newly available funds.

In the mid-1980s, Dr. Ralph Corey led a program in the Department of Medicine at Duke that was to evolve from his vantage point as program director of the Internal Medicine Training Program. With the support of Dr. Joseph C. Greenfield, Jr., the chair of medicine, three-month rotations were instituted for interested residents in a number of foreign sites, including Tanzania, Kenya, China, Thailand, Vietnam, Brazil, and the United Kingdom. The purpose of those rotations initially was to provide residents the opportunity to experience the practice of medicine in another culture. In general, the residents were incorporated into the inpatient and outpatient health-care systems to serve as both providers and teachers. Dr. Corey chose the sites because he knew native residents of those countries who had once trained at Duke and because of his personal visits, he was convinced that the residents would have a valuable experience. As judged by evaluations provided by the medical residents while abroad and when they returned home and by the reports from the supervising on-site mentor, this program was an enormous success. Funding to support the residents and to replace them at Duke during the time when they were away was derived from the Department of Medicine and from the residents themselves. At its peak before the beginning of the twenty-first century, ten to fifteen internal medicine residents took advantage of this opportunity in any one year. The residents were selected based on their interest in international health and their excellence as house officers at Duke. The residents worked closely both with their local counterparts and with infectious diseases fellows and faculty who were in some instances also living and working abroad. During the period from 1985 to 2000, over 170 residents benefited from the experience. Although some unpredicted personal and professional challenges arose, by all accounts those challenges were not insurmountable and, in the larger sense, contributed substantially to the experience for the trainees. From a separate vantage point, the opportunity to serve three-month rotations in a foreign site was a remarkably potent recruitment tool for the house staff and for the infectious diseases fellowship program, and ultimately even for the faculty.

Dr. Corey's interest in international health began with his own experience working at a small rural hospital in Kenya in 1985, not too long after he joined the Duke faculty in internal medicine. Because of his position as the director of the Internal Medicine Training Program, he was able to influence the agenda for the Training Program, which was principally exposing residents to clinical medicine at a foreign site. With time, however, the mission for the residents evolved to include collaborative research activities both to enhance their academic aspirations and to support the ongoing research activities. It had become apparent that the cultures within which the residents worked was not well versed in the incorporation of research into the clinical management of patients. Nevertheless, in time, an impressive research portfolio of support was captured, which allowed an extension of training opportunities both for the Duke trainees and for the other foreign trainees, with whom research collaborations were extremely common and productive. The support obtained for the conduct of research also allowed expansion of the critical local support staff, ranging from clinical trials coordinators to research nurses, data management, and personnel in the clinical laboratory. Beginning with no staff in 1995 when our program moved from Dar es Salaam to Moshi, Tanzania, the numbers of the faculty, fellows, residents, and medical students living and working in Tanzania increased gradually. By the beginning of the twenty-first century, it had increased to five to seven full-time individuals. In the first decade of the twenty-first century, those aligned with the Duke-KCMC collaboration had at least quadrupled. The topics of research were dominated initially by studies of HIV and HIV-related issues, including clinical trials of antiretroviral drugs in adults and children, TB, and social and behavioral science studies. Later the research agenda expanded dramatically, owing in part to the recruitment of the initial cohort of Duke personnel, who subsequently focused their careers on international health.

The Duke faculty in pediatrics also became more involved in international activities. For example, Dr. Catherine Wilfert's contributions to infectious diseases, including HIV, while an active faculty member at Duke were extraordinary. But after her retirement from Duke in 1996, her contributions to the field were magnified manyfold by her appointment as scientific director of the Elizabeth Glaser Pediatric AIDS Foundation, as discussed in chapter 4.

Dr. Richard Frothingham entered medical school at Duke, trained in medicine and pediatrics at the University of Rochester, and attended the Columbia International University seminary for one year with the full intention of becoming a medical missionary to serve the needs of people outside the United States. He met and married a pediatrics intensive care unit nurse who had a similar missionary spirit, and they applied for a nondenominational mission in Haiti. The family included Richard; his wife, Margaret; a twenty-month-old child, and a four-month-old baby. They lived in a remote village on the southern peninsula of Haiti from 1987 to 1990, and Dr. Frothingham served as he had desired, as a medical missionary. They were required to provide all their own compensation, lived in one of the cinderblock dwellings close to the hospital, shared responsibilities with two other American physicians and three Haitian physicians, provided care to inpatients and outpatients in what he describes as an "open" hospital facility, by which he meant not closed in by doors and windows. Although medical supplies and other resources were limited, Dr. Frothingham felt that quality care was provided to the population in that remote village located four to five hours from the capital of Port-au-Prince. Prostate surgery was compensated by the government and served as the major source of support for the remainder of the hospital. Routine blood work could be done locally, but HIV tests were referred to the Red Cross, and a US pathologist read Pap smears. HIV-AIDS was rampant, as were TB, malaria, typhoid fever, tetanus, and measles. Margaret Frothingham stayed busy taking care of their now three children and helping with community health programs related to safe child delivery. The high point of the three-year experience was the spiritual environment that integrated their faith with medicine. Although formal religious activities were not required of the staff, the providers commonly prayed with their patients. Later their children viewed the experience variously. After three years, they, along with their parents, were ready to return to the United States. Gradually it had become apparent that working as a medical missionary did not meet Dr. Frothingham's professional needs, although the experience in Haiti was still quite rewarding in many respects. Transitioning back to the United States had its own challenges, but professionally he chose to pursue a career in academic medicine focused on adult infectious diseases in the Duke Division of Infectious Diseases, where he first trained as a fellow. He later joined the faculty and pursued clinical medicine at the Durham VA as well as conducted both clinical and basic laboratory research.

Dr. Mark Perkins, a Duke faculty member, was a productive investigator who lived and worked in South America for five years between 1993 and 1998.[22-29] He gained considerable insight into the factors that are critical for the success of a collaboration between a foreign and a US academic institution. Dr. Perkins was the codirector with his Brazilian collaborator, Dr. Reynaldo Dietz, of an infectious diseases unit located at the Universidade Federal de Espirito Santo (UFES) in the island city of Vitoria, Brazil. An informal collaboration had been initiated a number of years earlier between Drs. Dietz and Ralph Corey of Duke. Their goal was to foster an exchange of a small number of medical residents. In 1992, however, the Duke Department of Medicine endorsed a plan to expand this interaction to involve Duke faculty on-site and to set up a research laboratory in support of fundable scientific projects. The collaborating institution in Vitoria (UFES) was a 280-bed public hospital that provided health care for indigent Brazilians. The infectious diseases unit began with only the two codirectors but grew over the five successive years to include six doctoral level faculty, five physicians in training, nine technologists, twelve nurses, as well as secretaries, a driver, and assorted students. Within five years, the unit had been awarded five grants for projects in mycobacteriology and parasitology and had achieved a level of financial sustainability and independence. In a report presented to a group of Duke alumni in April 1998, Dr. Perkins identified three factors critical for the success of such a collaboration: scientific credibility, stability, and integration.

As for *scientific credibility*, there must be scientifically credible research goals, adequate disease prevalence to support the research, interest on the part of the collaborators to pursue the projected goals, and sufficient technical expertise to accomplish the stated aims.

As for *stability*, the lack of political and financial stability is a problem in much of the world, perhaps especially the developing world, and clearly was an issue in the early years of the Brazil project, which saw runaway inflation and repeated strikes of public health workers, bus drivers, university professors, and others for as long as up to six months. Serial changes in senior leadership led to uncertainty about the country's priorities and caused problems for prospective planning for the infectious diseases unit. This disarray posed difficulties not only for the infectious diseases unit but also for the US-based sponsoring entity, especially as it related to success in competing for extramural funds, providing support for the US personnel, and ultimately in agreeing on modified benchmarks for faculty advancement because the traditional academic benchmarks were not applicable.

As for *integration*, the mission of the infectious disease unit must fit into the long-range institutional goals, and clear lines of communication must be established. A common problem that needs to be recognized and dealt with proactively for foreign trainees working in US-based institutions is the training of individuals for positions that do not exist in the host country to which the trainee will return. For US institutions, collaborating with foreign sites requires an investment in the site as well as in the project. Individuals must also make their own investment by learning the language, teaching in the university, and helping to find bridging funds to sustain the site between grants. To assure Duke's success with international collaborations, Dr. Perkins identified the following elements: clarify institutional goals, identify only a select number of overseas sites, identify faculty-level personnel dedicated to the sites, and have a clear and understood career trajectory for those faculty. After leaving Duke in 1998, Dr. Perkins pursued a successful career in international public health with the World Health Organization and, later, as scientific director of the Foundation for Innovative New Diagnostics (FIND). The mission of FIND is to identify novel and improved diagnostics for TB and other infectious diseases.

Another successful program on malaria initiated first in Dar es Salaam, Tanzania, by Duke Drs. Nicholas Anstey, Brice Weinberg and

Donald Granger in 1994, focused on nitric oxide in Tanzanian children with malaria.[17,30-38] Subsequently, further studies were performed with additional collaborators in Indonesia, resulting in seven to ten high-impact publications. The details of these and later studies will be discussed in chapter 10.

Duke's involvement in Honduras dates back to the early 1990s. Missions to Honduras were organized initially by the Duke Chapel and consisted of students and faculty who traveled to Honduras to work with the Christian Commission for Development to provide support for refugees. The dean of the chapel, Dr. Will Willimon, recruited Dr. Marvin Hage, an obstetrician and gynecologist, in an effort to engage medical personnel on these missions. Dr. Hage developed a "house course" for Duke undergraduate students to provide background information on the culture, demography, and health status of the native Hondurans. The intention was to prepare selected students to participate in a mission to a rural site in Honduras. The first mission in 1993–1994 was led by Dr. Hage, and he was accompanied by Dr. Michael Sheets (a Duke neurobiology scientist), undergraduate and medical students, house staff, and other faculty members.

Dr. Linda Lee was recruited to accompany successive missions beginning in 1997. Her role was to provide a formal evaluation of the program to enhance its academic credibility. Mission participants financed their own expenses, which amounted to approximately one thousand dollars per person. Not long thereafter, Dr. Hage left Duke and Dr. Lee recruited Dr. Dennis Clements, a faculty member in the Department of Pediatrics in 2000, to take on the responsibility of organizing and expanding the core course, which had become increasingly popular. Dr. Clements had pursued a PhD degree from 1988 to 1990 in the Department of Microbiology and Infectious Diseases at the Royal Children's Hospital in Melbourne, Australia, under the supervision of Dr. Lynn Gilbert. His thesis had focused on the epidemiology of *Hemophilus influenza* group B epiglottitis and meningitis.[39-43]

Dr. David Walmer became involved in international health activities because of an opportunity in 1993 to participate in a mission to Haiti, sponsored by his church, the Triangle Presbyterian Church. He previously had trained in obstetrics and gynecology and had conducted molecular biologic research at Duke. He elected to accompany the second group of parishioners to explore the needs of the populace in the capital city of Haiti, Port-au-Prince, and potentially to identify areas where his church might provide expertise and support. Being the only physician in the group, Dr. Walmer was encouraged to interact with health-care personnel there and to establish a relationship with a local obstetrics and gynecology physician (Dr. Jean-Claude Fertillien) whose practice was connected with the Hospital St. Croix, a regional referral hospital located some twenty-two miles from the capital. He quickly established excellent working relationships and, after a relatively brief period, came to understand that the top health-care priority for obstetricians and gynecologists in Haiti would be a dramatically enhanced program in the diagnosis, prevention, and treatment of cervical cancer.

Over the next seven years, Dr. Walmer periodically traveled to Haiti, bringing with him the supplies and equipment that were essential to the development of an enhanced program directed at cervical cancer. Those efforts were extremely well received, and progress toward the ultimate goal was measurable, at least in small increments. Members of his church pursued other activities, of course, but this collaborative project took on a life of its own, resulting in the development of a non-profit organization, Family Health Ministries. Family Health Ministries addressed an increasing spectrum of issues related to the cervical cancer program, including the recruitment of local staff, identification of consultants, involvement of additional personnel, including students, from the United States, training of technical and medical personnel to provide enhanced diagnostic and treatment modalities, and, of course, fund-raising. A dramatic increase in the capability of this ongoing project occurred in the first decade of the twenty-first century, and this project, which began very much as a grassroots effort, succeeded far beyond the expectations of those who began it.

Seven Duke faculty members have served in various capacities in the Epidemiology Intelligence Service (EIS) of the Centers for Disease Control and Prevention prior to joining the faculty, and five had international experiences. This service is described on the CDC's web site as follows:

[The Epidemiology Intelligence Service] is a unique two-year postgraduate training program of service and on-the-job learning for health professionals interested in the practice of applied epidemiology. Since 1951, over [three thousand] EIS officers have responded to requests for epidemiologic assistance within the United States and throughout the world. EIS officers are on the public health front lines, conducting epidemiologic investigations, research, and public health surveillance, both nationally and internationally.

Dr. Barth Reller was aligned with the Enteric Section of the Bacterial Diseases Branch at the CDC in Atlanta from 1968 to 1970. During that interval, he served in Bangladesh for two months, focusing on oral rehydration therapy for persons infected with cholera; in El Salvador for one month, focusing on the epidemic of *Shiga-Bacillus* dysentery; and, for the remainder of his two years as an EIS officer, investigating various domestic outbreaks of diarrheal disease. I also served as an EIS officer from 1968 to 1970, based at the State Board of Health in Raleigh, North Carolina. My international experience included a four-month period in which I joined a team in India consisting of a representative of the Indian Ministry of Health, the director of public health in Chandigar, India, and a malaria expert from the World Health Organization to investigate the progress being made on a nationwide malaria-eradication program. Our role was to travel by car from village to village and to evaluate critically the accuracy of blood smears on which were based estimates of malaria prevalence. Domestically, I investigated outbreaks of Shiga bacillus dysentery in an institution for mentally handicapped persons and outbreaks of the Hong Kong flu and botulism.

Dr. Daniel Sexton was appointed to the State Board of Health in Jackson, Mississippi, where he investigated various outbreaks of measles, rubella, and echovirus encephalitis. He also worked in the Rocky Mountain Laboratory in Hamilton, Montana, investigating the epidemiology of Rocky Mountain spotted fever under the direction of Dr. Willie Burgdorfer, and worked on a health survey in northern Alaska.

Others with specific international experience included Dr. John

Crump, who was also assigned to the Foodborne and Diarrheal Disease Branch of the CDC, investigating outbreaks of typhoid fever in Egypt, Uzbekistan, and Geneva and conducting a large project to assess the efficacy of a novel product used for water purification in the Kisumu region of Kenya. He later conducted seminal studies on the transmission of *E. coli* 0157 directly from zoo animals to humans when they pet them, for which he received the prestigious James Steele award. Dr. Christopher Woods was assigned to the Special Pathogens Branch of the CDC, where he worked from 2000 to 2002. He worked on an outbreak of human anthrax in Kazakhstan, on Rift Valley fever in northeastern Kenya, on meningococcal meningitis in Ghana, on unexplained encephalitis in the Maldives, and on studies involving various bacterial and viral vaccines. Two other Duke faculty members (Drs. John Falletta and Conrad Fulkerson) were also EIS officers but had no international experience.

The International Collaboration on Endocarditis (ICE) began at the 1999 International Society for Cardiovascular and Infectious Diseases (ISCVID) in Amsterdam, when Dr. Ralph Corey of Duke asked Dr. José Miro for a copy of his extensive database on endocarditis. Along with Dr. Chris Cabell, a cardiologist, Dr. Corey also asked for copies of databases from Drs. Bruno Hoen, Susan Eykyn, and Lars Olaison, all of whom agreed with the merger of their databases with the plan to conduct retrospective analyses. The merger of those databases posed enormous challenges for a variety of reasons, including that the records were neither in a standardized format nor in a common language. In all, 2,268 cases of definite endocarditis were included in what was called the "merged database." Although those analyses revealed a number of important new pieces of information about endocarditis, it was perhaps most helpful in generating hypotheses for future testing.[44-55] A prospectively created database on endocarditis was implemented in June of 2000.

Photograph 8.7 is a composite of pictures of five prominent internationalists at Duke in the twentieth century.

Catherine M. Wilfert

David T. Durack

G. Ralph Corey

Richard Frothingham

J. Peter Cegielski

CHAPTER 9

VACCINES

> If a house is on fire, no one wastes time putting water on nearby houses just in case the fire spreads. They rush to pour water where it will do the most good— on the burning house. The same strategy turned out to be effective in eradicating smallpox.
>
> —Anonymous[1]

For centuries, many have tried to prevent or to modify the course of disease, perhaps most famously by the practice of vaccination or variolation. Edward Jenner is credited with the first scientific effort to modify the course of smallpox by cowpox inoculation in 1798. Eighty-seven years later, Pasteur used an attenuated strain of the rabies virus for the first human rabies vaccination. That work depended heavily on his prior work in the attenuation of the chicken cholera bacterium. Prior to the beginning of the twentieth century, Metchnikoff, Ehrlich, and others made seminal contributions to the understanding of the immune basis for vaccination by developing three other bacterial vaccines—for typhoid, cholera, and plague. Before 1945, additional vaccines were developed, including the first toxoids for diphtheria and tetanus; the BCG vaccine against tuberculosis (TB); the yellow fever virus vaccine; the first live influenza A virus vaccine; vaccines against Rocky Mountain spotted fever, typhus, and Q fever; and the pertussis vaccine.[2]

It is somewhat surprising that an early Duke faculty member made fundamental discoveries in the creation of the first vaccine for the prevention of animal and human Eastern and Western equine encephalitis. In the early 1930s, Dr. Joseph Beard, a fully trained surgeon, was studying rabbit papillomatosis, caused by a tumor virus, with Peyton Rous at the Rockefeller Institute in Princeton, New Jersey, when Duke recruited him to the Department of Surgery in 1937. Through what appear to be a series of serendipitous connections, Dr. Beard embarked on experiments leading to the discovery of a commercially feasible method of purifying sufficient Eastern and Western equine encephalitis virus to produce an effective virus vaccine, preventing an epidemic in horses, particularly thoroughbred horses.[3,4] The disease in horses, commonly called the "blind staggers," would have led inevitably to their demise and, if unchecked, would have decimated much of the thoroughbred horse population. Although effective also in humans, the vaccine had the major side effect of a rather severe immediate pain at the site of injection, which was likely due to residual formaldehyde in the preparation.[5-9] Nonetheless, the success of the vaccine represented a major commercial opportunity for Lederle Laboratories, a subsidiary of the American Cyanamid Corporation, and led to a prolonged period of financial support for the Beard laboratory at Duke, where he continued to study avian tumor viruses. In later years, well after his discovery of the second animal tumor virus, the avian myeloblastosis virus (AMV),[10] he became the dominant commercial source of RNA-dependent DNA polymerase (Reverse Transcriptase) from the AMV, for which he ultimately established a company in south Florida.

Another viral disease for which a vaccine was urgently needed, this one unique to humans, was poliomyelitis. It was clinically recognized throughout the United States in the early part of the twentieth century and was reported in North Carolina as well. Dr. Carl V. Reynolds, the chairman of the North Carolina State Board of Health, stated that the average number of cases of acute poliomyelitis in North Carolina was seventy-one. In 1935, however, 675 cases were reported, of which 67 were fatal.[11] The pathogenesis of the disease was poorly understood, although the disease was clearly caused by a virus.[12] The disease also had political ramifications as reflected by the following statement made by one of the attendees at a meeting of the Medical Society of the State of North Carolina, Dr. Lessesne Smith:

I was in hopes that the North Carolina Medical Association would have a meeting without discussing the subject, because the more we discuss it, the more popular it becomes; the more the phobia becomes increased in people's minds. It wasn't anybody else's fault. I lay the whole trouble to the egotism of the State Board of Health; that is, that they wanted to make out that they were doing a whole lot of stuff, and I say it without any rancor. They wanted to advertise themselves, and they thought that was a good way to advertise themselves, by putting it in the paper.

In response, Dr. Reynolds, the Chairman of the State Board of Health, stated the following:

I think North Carolina has the reputation today throughout the United States that when they write to the State Board of Health of North Carolina they're going to get the truth and they can depend on it, and they will not go into any town or any part of the state with fear because they do not know the facts; and so as long as I am health officer, they're going to get the truth.

Dr. Root, also a member of the North Carolina Medical Society, added:

It was their (the State Board of Health) duty to let the people know just where they stood. It did create hysteria, and excitement, but it saved a whole lot of sickness and probably a good many lives.

Discussions of the clinical disease, laboratory manifestations, prevention, and treatment reflected the limited understanding of the pathogenesis of acute poliomyelitis, and further clarification awaited discoveries related to the development of effective vaccines as described by the first dean of the Duke School of Medicine, Dr. Wilburt Davison.[12]

189

The development of the technology to grow human viruses outside of a living host revolutionized vaccinology for the remainder of the twentieth century. John Enders and his colleagues led the way with the cultivation of a strain of the poliovirus in human embryonic tissue cultures.[13-17] For that achievement, Enders, Thomas Weller, and Frederick Robbins were awarded the Nobel Prize. That advance led to the development of an effective polio vaccine.

One of Enders's protégés, Dr. Samuel Katz, who later became the chairman of the Department of Pediatrics at Duke (1968 to 1990), played a pivotal role in the development and evaluation of an attenuated measles virus vaccine. Drs. Enders and Peebles described the cultivation of this attenuated virus in tissue cultures in 1954. Subsequent experiments at Harvard by Katz and colleagues demonstrated that after sequential passages in human kidney cells, human amnion cells, the amniotic sac of the fertile hen's egg, and finally in monolayer cultures of trypsinized chick embryo cells,[18] the virus showed a significant reduction in virulence when injected into monkeys. This less virulent virus, however, conferred resistance to superinfection when the animals were challenged with the wild-type virus. (The experiments leading to this less virulent virus were instrumental in the development of complement fixing and neutralizing antibodies.) Katz then led studies in children and found that compared with the effects of the naturally acquired disease, the attenuated virus caused substantially less fever and, in some, less of a rash in the few who experienced a rash.[19,20] Moreover, when used as a vaccine, it elicited a prompt antibody response in 95 percent of the recipients and prevented superinfection in most of those who were subsequently exposed to the natural disease.

In the early 1960s, naturally acquired measles was a major cause of morbidity and mortality worldwide, with over a million deaths. In an effort to address this epidemic, the Red Cross organized a consortium consisting of representatives from the Centers for Disease Control (CDC), the World Health Organization (WHO), the United Nations Foundation, and the United Nations Children's Fund (UNICEF) to deliver the newly discovered attenuated measles virus vaccine worldwide. Seven separate companies manufactured this vaccine, but ultimately Merck was awarded the patent, and Dr. Katz led the measles vaccination program in Nigeria.[21] Although measles was not eradicated, there was a reduction in mortality to several hundred thousand per year worldwide. Dr. Katz was appointed to the CDC's Advisory Committee on Immunization Practices in 1982 and served as its chairman from 1985 to 1993. As a member of that committee, he was well aware of the anti-vaccine movement, the modern birth of which occurred on April 19, 1982, when the *DPT: Vaccine Roulette*, written and produced by Lea Thompson, was aired on an NBC affiliate station.[22] The program seemed to galvanize the call to action of the parents of children who felt their children had suffered irreparable harm following the administration of a vaccine, particularly the combined diphtheria, pertussis, and tetanus (DPT) vaccine. One of the most vocal activists was Barbara Loe Fisher, who together with other parents with similar experiences formed what initially was called the Dissatisfied Parents Together (a play on the initial letters of the target vaccine). The name of the organization subsequently changed to the National Vaccine Information Center and became the most powerful anti-vaccine organization in the United States. Most scientists who have examined a coincident occurrence of a vaccination and a serious medical condition consider it just that, a coincidence. Moreover, these same scientists see the undesirable repercussions of the anti-vaccine movement, including the ever-larger numbers of unvaccinated children and a return of the naturally acquired disease with its attendant morbidity and mortality. Dr. Katz served in various capacities related to pediatrics and vaccines and as a consultant to numerous organizations over the course of his career.[23-26]

Recruited to Duke at the same time as Dr. Katz was Dr. Wolfgang Joklik, who came as chair of the Department of Microbiology with expertise in the basic science of virology, including the poxviruses. This was only a year after the WHO embarked upon the Intensified Smallpox Eradication Program, which had the goal of eliminating smallpox worldwide. The basic research done by Joklik and his colleagues Drs. David Pickup and Joseph Nevins revolved around related poxviruses, not smallpox per se, and did not specifically contribute to the broader public health goal. Nevertheless, their counsel was frequently sought by those dealing with the worldwide epidemic, which in the 1960s was the cause of death in millions around the globe. An effective vaccine, live vaccinia virus, already existed, but the problem was understanding how it worked and knowing what its limitations were. Joklik, Pickup, and Nevins contributed to the strategic decisions about how best to use the vaccine.[27-45] In addition, Dr. Joklik served on several international committees that convened to discuss the status of endemic and epidemic viral diseases, including smallpox. The committees included the WHO's "Informal Consultation on Monkeypox and Related Viruses" and the International Association of Microbiology Societies' Pox Virus Study Group. Led by public health officials at the WHO, the CDC, and all the countries in which smallpox was epidemic, the "Intensified Program on Smallpox Eradication" was eminently successful. The last case of naturally acquired smallpox was detected in Somalia on October 26, 1977.

In the next year, two laboratory cases were identified in England, raising a related issue as to the need or desirability of retaining stocks of the smallpox virus (inasmuch as the virus had been eliminated in humans and no alternative reservoir was known). A concern was that residual stocks of virus could pose threats for accidental or intentional infections. Dr. Joklik, accompanied by other distinguished virologists, defended the case in the journal *Science*, claiming that remaining stocks of smallpox virus should not be destroyed but instead retained only in highly secure laboratories for future research.[46] This was the action ultimately undertaken by senior public health officials, and stocks are said to remain only in Atlanta (in a P4 containment facility at the CDC) and at the Moscow Research Institute for Viral Preparation in the USSR (later renamed the State Research Center of Virology and Biotechnology in Koltsovo, Novosibirsk Oblast, Russia). Dr. Joklik and his colleague Dr. Y. Ichihasi worked on the development of radio precipitation and monoclonal antibody tests to allow more rapid testing for specific poxviruses should the need arise.[47] The alternative view, that all remaining stocks of smallpox virus should be destroyed, accompanied the piece by Joklik in *Science*.[48] It is, of course, impossible to tell whether surreptitiously collected samples of live smallpox virus exist elsewhere in the world, but no subsequent detection of clinical smallpox has been recognized since 1978. (Note that unauthorized vials purportedly containing the smallpox virus were discovered frozen in a laboratory in the United States as recently as 2014)

Chapter 5 describes the epidemic of hepatitis B that occurred in

Duke employees in the 1970s owing to their contact with patients under-going dialysis or the dialysis equipment or with patients who underwent organ transplantation.[49,50] I participated in four separate clinical studies related to the prevention of acute hepatitis B, as follows: (1) the administration of hyperimmune hepatitis B immune globulin (HBIG) following an accidental needlestick exposure;[51] (2) the administration of plasma-derived hepatitis B vaccine (Heptavax B) to hemodialysis patients[52] and their attending staff;[53] (Szmuness1982); and (3) the administration of hepatitis B vaccine (HBV) to patients progressing toward the need for hemodialysis.[54] Of these three trials, HBIG was more successful than immune serum globulin (ISG) was in the prevention of acute hepatitis B; Heptavax B significantly reduced new cases of hepatitis B in the medical staff in dialysis units but did not significantly reduce new cases of hepatitis B in the hemodialysis patients themselves; and the administration of HBV to patients with progressive kidney disease before dialysis provided a better immunologic response than historical controls on dialysis had. Finally, (4) a cost-effectiveness analysis of HBV vaccine administered to pre-dialysis patients demonstrated that pre-dialysis immunization can prevent additional acute hepatitis B infections, but the cost is high.[55]

Chapter 5 describes the epidemic of hepatitis B that occurred in Duke employees in the 1970s owing to their contact with patients un-dergoing dialysis or the dialysis equipment or with patients who under-went organ transplantation.[49,50] I participated in four separate clinical studies related to the prevention of acute hepatitis B, as follows: (1) the administration of hyperimmune hepatitis B immune globulin (HBIG) following an accidental needlestick exposure;[51] (2) the administration of plasma-derived hepatitis B vaccine (Heptavax B) to hemodialysis patients[52] and their attending staff;[53] (Szmuness1982); and (3) the administration of hepatitis B vaccine (HBV) to patients progressing toward the need for hemodialysis.[54] Of these three trials, HBIG was more success-ful than immune serum globulin (ISG) was in the prevention of acute hepatitis B; Heptavax B significantly reduced new cases of hepatitis B in the medical staff in dialysis units but did not significantly reduce new cases of hepatitis B in the hemodialysis patients themselves; and the administration of HBV vaccine to patients with progressive kidney disease before dialysis provided a better immunologic response than historical controls on dialysis had. Finally, (4) a cost-effectiveness analysis of HBV vaccine administered to pre-dialysis patients demonstrated that pre-dialysis immunization can prevent additional acute hepatitis B infections, but the cost is high.[55]

Chapter 4 describes many of the accomplishments of the Duke faculty who worked on HIV-1 and HIV vaccine. The following direct quote from a report prepared by Dr. Barton Haynes provides a background description of the Duke Human Vaccine Institute (HVI)[56]:

> The roots of the Duke Human Vaccine Institute (DHVI) began in 1985, soon after the discovery of HIV-1. In order to tackle this newly emerging infectious disease, Dani Bolognesi and I formed a working group at Duke University Medical Center with Kent Weinhold, Tom Matthews, and Tom Palker. Not everyone at Duke was supportive of AIDS research efforts, but one of the greatest supporters of this effort at Duke was David Sabiston, the chair of Duke Surgery from 1964 to 1994. Sabiston raised $5 million to build the Surgical Oncology Research Facility in 1986, where all AIDS research at Duke was performed, and supported both Bolognesi and me throughout those early and difficult years. In 1990, Bolognesi and I established the Duke Human Vaccine Institute (DHVI) to support interdisciplinary efforts across Duke to develop vaccines and therapeutics for HIV and other emerging infections that threatened the health of our nation and our world.... Neither the early work of the Duke Center for AIDS Research and DHVI nor any of the other AIDS basic research at Duke could have been performed without the support of David Sabiston.

Dr. Haynes goes on to say:

> Some of the most elegant work on both HTLV-1 and HIV-1—regarding how these human retroviruses

194

overcame the host cells and "hijacked" the machinery of the host cell to the viruses' benefit—was the work of Drs. Warner Greene and Bryan Cullen in the Division of Rheumatology and Clinical Immunology of the Department of Medicine at Duke.[57] Their work played a major role in helping the field understand how human retrovirus genes worked to subvert the host immune system.

Since the beginning of the AIDS epidemic, the ongoing major focus of the Duke HVI investigators has been the development of a safe and effective HIV vaccine. Dani Bolognesi, recruited to Duke by Dr. Joseph Beard in 1966, led the initial efforts in HIV vaccine development. He was working on a cancer vaccine when the AIDS epidemic was recognized, and he convinced the National Cancer Institute (NCI) to change his grant in midstream to a grant for HIV vaccine development. When that grant went up for re-competition, Tom Palker and Haynes joined Bolognesi, Kent Weinhold, Tom Matthews, and Alphonse Langlois and successfully obtained the new grant in 1986. Later David Montefiori joined the team and became a key national leader in the area of neutralizing antibody work.

Early on, Bolognesi had teamed with Peter Fischinger and Robert Gallo at the NCI to focus first on the outer coat protein of HIV, gp120. They showed the ability of gp120 to be a vaccine candidate because it could induce antibodies and prevent infection by the laboratory-grown strains of HIV that were being used at that time.[58] Bolognesi teamed with Scott Putney of Repligen Incorporated and Haynes's group to define an early target for a vaccine, the V3 loop of the envelope gp120.[59,60] With Tom Palker, Haynes converted this region into a vaccine candidate, which was first synthesized by Dr. Tony Moody and Richard Scearce in his laboratory. Dr. John Bartlett in the Division of Infectious Diseases at Duke carried out the first clinical trial of this HIV vaccine, which was developed in a basic science laboratory and taken directly to the bedside. This trial confirmed the safety and immunogenicity of that vaccine.[61] Soon thereafter, Bolognesi, Matthews, and others in the field found that the strains of HIV grown in the laboratory were not like the strains infecting people in real-life settings, prompting the realization that the field was using the wrong strains of HIV-1 for testing vaccines—a major setback.

Over the years, the HIV-1 vaccine development effort has been long and complicated owing to the extraordinary nature of HIV's diversity and its ability to evade host immune responses. Nonetheless, HIV vaccine developers have been encouraged by studies demonstrating the ability of cytotoxic T cells (killer T cells) to control HIV replication, as well as the ability of high levels of neutralizing antibodies to protect against HIV infection (by providing sterilizing immunity). These studies suggest that an immunogen that induces broadly reactive killer T-cell responses and B cell neutralizing antibodies that work against the type of viruses circulating in the community (so-called primary isolates) can be helpful in stemming the AIDS epidemic. To this end, the Duke HVI development team has a broad-based program to design a number of experimental immunogen constructs for testing in animals and in humans.

In the late 1990s, although the Duke investigators demonstrated that the V3 loop could induce antibodies that were effective against a few primary isolates, they were discouraged by the realization that the V3 loop vaccines were not going to be effective against the majority of primary HIV-1 strains. Nonetheless, this work led to the discovery that the V3 loop was a key component of binding the HIV-1 envelope to the HIV-1 coreceptors CCR5 and CXCR4 and that its structure determines the type of coreceptor to which HIV-1 binds.

Drs. Dennis Clements and later Emmanuel "Chip" Walter initiated the Duke Pediatric Vaccine Unit in the late 1980s, following some early vaccine trials by Drs. Katz and Wilfert. Studies done before 2000 included follow-up of varicella vaccine recipients and the effects on naturally acquired disease, the influence of immunization of neonates with influenza A vaccine on otitis media, and studies of combination vaccines with acellular pertussis, diphtheria, tetanus toxoid and Hemophilus influenza group b. After 2000, they conducted many more trials.[62-73] Most recently, they joined a national network of Vaccine and Treatment Evaluation Units, and they continue collaborating with Duke vaccine trialists in select and unique populations.

Dr. Kenneth Schmader's research activity focused on herpes zoster, a Herpes group virus, and vaccines in older adults. The herpes zoster virus (HZV) causes chicken pox and can reactivate later and cause disease. The aims of his work were to understand better the development of herpes and post-herpetic neuralgia (PHN) in older adults; to improve the prevention and treatment of older persons suffering from herpes zoster and PHN, including the use of the zoster vaccine; and to develop and to deploy better vaccines for older adults. Dr. Schmader conducted experiments with the mouse cytomegalovirus (also a Herpes group virus) animal model to understand the nature of aging and herpes virus latency. With this model, he accomplished the following: He developed a novel salivary gland biopsy technique that avoided the sacrifice of animals and allowed for longitudinal investigations of MCMV latency and reactivation including MCMV DNA gene amplification;[74] showed that MCMV reactivation declined with aging;[75] showed that transplantation of latently infected MCMV tissues into the severely compromised immune deficient (SCID) mice resulted in rapid reactivation and dissemination of the CMV virus;[76] and assisted with studies to insert a green fluorescent protein as a marker for localizing murine CMV infection.[77] On the human level, he conducted the first meta-analysis showing that antiviral therapies are not effective for the prevention of PHN;[78] discovered two new risk factors for herpes zoster, that of psychological stress and race[79]; and discovered that a positive self-report of shingles has a high positive predictive value and that a negative self-report of shingles has a very high negative predictive value compared to physician diagnosis.[80] He was instrumental in developing the Zoster Brief Pain Inventory (ZBPI), the first herpes-zoster-specific measure of herpes zoster pain and its impact on functional status and quality of life.[81] Dr. Schmader is a lead investigator in the Shingles Prevention Study, the landmark study of the efficacy of the zoster vaccine in older adults; its planning began in the mid-to late-1990s.[82] The singular accomplishments of that study were revealed in that publication in 2005 and will be discussed in a later publication.

Photograph 9.8 is a composite of pictures of five prominent vaccinologists at Duke in the twentieth century.

Kenneth E. Schmader

Kent J. Weinhold

David C. Montefiori

Emmanuel Walter

Dennis Clements

CHAPTER 10

SERIOUS BACTERIAL, VIRAL, FUNGAL, RICHETTSIAL, AND PARASITIC INFECTIONS

It is certainly a one-sided opinion—even though generally adopted at the moment—that all infectious agents which are still unknown must be bacteria. Why should not other micro-organisms just as well be able to exist as parasites in the body of animals?

—Robert Koch 1881[1]

SERIOUS BACTERIAL INFECTIONS

BACTEREMIA/SEPSIS AND ACUTE AND SUBACUTE BACTERIAL ENDOCARDITIS

Two of the most serious bacterial infections seen at Duke University Medical Center were bacteremia and acute and subacute bacterial endocarditis. Although these infections surely were seen earlier than 1940, that is the year when Dr. Eugene A. Stead, an early Duke faculty member, and his coauthor, Dr. Richard Ebert, published some of the first work describing the effects of acute bacterial infections on the peripheral circulation. In that work, they concentrate on the physical signs relevant to the peripheral circulation, namely body temperature, radial and digital pulses, capillary pulsations in the fingernail bed, femoral and brachial bruits, the temperature and color of the skin, pulses and arterial blood pressure, and estimates of venous pressure.[2] They describe how these signs may be affected during an acute infection, as follows: elevated body temperature, full bounding pulses, warm and flushed skin, femoral bruits, dilated veins in the extremities, and either low or normal blood pressure. In discussing five illustrative cases, they note that not all the physical signs are necessarily present. In a subsequent paper, they describe their conclusions about the circulatory effects, most notably circulatory failure. Of the eight patients studied, most were described as having lobar pneumonia, most commonly caused by the pneumococcus. The pneumonia was accompanied by circulatory failure and characterized by a decrease in peripheral blood flow and a fall in arterial pressure.[3] They found no evidence of hemoconcentration or diminished blood volume, and neither transfusions nor elevation of the foot of the bed improved the blood pressure. The blood pressure improved only when the infection was brought under control. Notably, six out of eight were bacteremic with the pneumococcus (n = 4), hemolytic streptococcus (n = 1), or *Staphylococcus* (n = 1). None had evidence by physical examination of endocarditis, although the results of autopsies, done in most, were not described. The next paper by Duke faculty members on bacterial endocarditis appeared in 1965, now in the post-antibiotic era. It was the first successful demonstration of the value of heart valve excision and replacement in a patient who had conclusively failed to be cured by antibiotics.[4]

Although Dr. Stead made further comments on endocarditis as a journal editor,[5] it is not evident that he conducted further research on the topic. Surely, as the chairman of the Department of Medicine at Duke from 1947 to 1967, he was aware of a dramatic change in the outcome in patients with bacteremia, acute or subacute endocarditis, owing to the discovery and use of powerful antibiotics.

Drs. Durack, Corey, Sexton, and Fowler shared an interest in bacteremia and endocarditis. Thus they are commonly coauthors on publications. As a Rhodes Scholar at Oxford from 1969 to 1973, David Durack, MD, DPhil, conducted a series of experiments by using a model of bacterial endocarditis in rabbits. He placed a polythene catheter in the right side of the rabbit heart and collected serial blood samples for culture. Normal animals demonstrated rapid clearance of injected bacteria. Animals with a preexisting, nonbacterial vegetation, however, had substantial numbers of bacteria adhering to the vegetation, where the bacteria rapidly reproduced in a manner similar to that observed in vitro.[6] Using a labeled metabolite measured by autoradiography, Dr. Durack demonstrated differences in activity between the superficial and the more deeply located bacterial colonies.[7] Further studies demonstrated the sequence of events in the developing infected vegetation, including the creation of layers of fibrin and platelets along with bacterial colonies, and the predictable fatal outcome in animals with left-sided endocarditis in contrast with right-sided endocarditis.[8,9] In the same animal model, vancomycin was found to be the only single antibiotic that was uniformly successful in preventing the development of bacterial endocarditis.[10-12] Tetracycline, a bacteriostatic agent, failed to prevent the development of experimental streptococcal endocarditis.[13] In studies of the treatment of experimental streptococcal endocarditis, Dr. Durack demonstrated a synergism when penicillin and streptomycin were combined.[14] Along with others, Dr. Durack developed new and significantly improved criteria for the diagnosis of infective endocarditis by using specific echocardiographic findings in studies of 405 consecutive cases collected from 1985 to 1992.[15] Using data from the same 405 episodes of endocarditis, Duke colleagues demonstrated a negative predictive value of at least 92 percent by the new Duke clinical criteria for endocarditis.[16] In another study of sixty-one patients with endocarditis owing to penicillin-susceptible streptococci, a two-week course of combination therapy with ceftriaxone and gentamicin was found to be as effective and safe as a four-week course of ceftriaxone alone.[17] As an acknowledged expert on bacterial endocarditis, Dr. Durack contributed to a consensus statement on the treatment of most microbiological causes of endocarditis.[18]

Drs. G. Ralph Corey and Daniel Sexton were coauthors with Dr. Durack on the papers describing the new Duke criteria for diagnosing endocarditis and its negative predictive value.[15,16] They contributed to papers on the role of echocardiography in evaluating patients with *Staphylococcus aureus* bacteremia[19] and patients with bacterial endocarditis who were receiving hemodialysis treatment.[20,21] Further studies demonstrated the following: an improved diagnostic sensitivity of the Duke criteria compared with other criteria[22]; an association between acute renal failure and a fatal outcome in patients with bacterial endocarditis;[23] similar clinical presentations of native valve endocarditis in elderly patients and younger adult patients;[24] an improved clinical outcome for patients who had S. aureus bacteremia and who did not follow the advice of the infectious disease specialist;[25] an improved prognosis in patients whose vegetation was visualized by only transesophageal echocardiography compared with only transthoracic echocardiography;[26] and a finding that recurrent episodes of S. aureus bacteremia are due primarily to a relapse of endocarditis.[27]

The S. aureus bacteremia Group (SABG), initiated in 1994 by Dr. Vance Fowler when he was a first-year Duke internal medicine resident, continuously enrolled Duke patients with bacteremia throughout the 1990s. By the end of the first decade of the twenty-first century, SABG had accrued one of the world's largest prospective biorepositories of bloodstream isolates, patient serum, and DNA, as well as outcomes data on patients with S. aureus bacteremia and gram-negative bacteremia. Descriptions of the clinical course of patients with bacteremia reveal the extraordinary excess morbidity and mortality that attend this condition.[26] Research using this resource focused on the role of echocardiography and transesophageal echocardiography in the evaluation of patients with S. aureus bacteremia.[19,28] Further, it defined patients on hemodialysis and the elderly as specific risk groups,[21,29] described the usefulness of pulsed field gel electrophoresis in patients with recurrent S. aureus bacteremia,[27] and described the outcome of S. aureus bacteremia according to compliance with the recommendations of the infectious diseases specialist.[25] Those discoveries revolutionized the management of patients with S. aureus bacteremia and allowed Dr. Fowler to apply successfully for numerous NIH grants, including ROIs, K23, K24, and others. The SABG team consisted of a study nurse, study coordinator, research assistant, postdoctoral fellows, international scholars, and undergraduates, all sharing responsibility for screening patients daily, providing written informed consent for the collection of detailed clinical data, collecting the bacterial isolate, and collecting the human DNA and serum.

Trained in both infectious diseases and pulmonary/critical care medicine, Dr. Karen Welty-Wolf embarked on basic and translational research investigating organ dysfunction as a consequence of the maladaptive host response to bacterial sepsis. She translated the results of critical animal studies to the clinical arena, investigating the role of mitochondrial injury and biogenesis in the pathogenesis and recovery from sepsis-induced organ injury in studies published in the 1990s.[30-37] Further discoveries in the early part of the twenty-first century will be discussed in more detail in a later publication.

TUBERCULOSIS IN INFANTS AND YOUNG CHILDREN

In 1994, Duke pediatric faculty member Dr. Laura Gutman published a paper in which she writes the following: "Tuberculosis occurring in a child is a public health emergency, because it represents recent acquisition of disease from a contagious adult, who is also a health hazard to other persons."[38] Before coming to Duke in the early 1970s, Dr. Gutman had worked in public health on tuberculosis (TB) in Seattle, Washington, and therefore had firsthand knowledge of the implications of new infections with TB in children. Although the implications were relatively well known in pediatric circles, many health care personnel, particularly those working in county health departments, were less familiar with the urgency of prompt treatment, both for the child's prognosis and for actual and potential contacts. A paper published much later outlined the scientific literature on the natural history of intrathoracic TB in children in the pre-chemotherapy era, from approximately 1920 to 1950.[39] This work provided concrete reasons to support the need for prompt action when a child less than ten years of age became infected and especially so when the child was less than two years old.

Nine relevant studies conducted during that time, each enrolling in excess of 950 subjects with a follow-up of approximately ten years, served as the basis for the analysis of the natural history of TB in children.[40-48] Comparison of the results of those studies allowed for a careful delineation of the expected timetable for progression from infection to disease in the immune-competent host. The distinctions made based on the age of the host were fairly reproducible. Table 10.6, abstracted from the Marais paper, describes the average age-specific risk for disease development following a primary infection.

Table 10.6

Age at primary infection	Immune-competent children (dominant disease entity shown in brackets)	Risk of disease following primary infection (%)
Less than 1 yr	No disease	50
	Pulmonary disease [Ghon focus, lymph node, or bronchial]	30-40
	TBM* or military disease	10-20
1-2 yrs	No disease	70-80
	Pulmonary disease [Ghon focus, lymph node, or bronchial]	10-15
	TBM or military disease	2-5
2-5 yrs	No disease	95
	Pulmonary [lymph node or bronchial]	5
	TBM or military disease	0.5
5-10 yrs	No disease	98
	Pulmonary disease [lymph node, bronchial, effusion or adult-type	2
	TBM	Less than 0.5
More than 10 yrs	No disease	80-90
	Pulmonary disease [effusion or adult-type]	10-20
	TBM	Less than 0.5

*TBM = Tuberculous meningitis

These data demonstrate dramatic changes in the progression of disease in those infected in the first year of life compared with those infected later. Those infected later had less pulmonary disease and less TB meningitis and military disease. This underscores the need for prompt attention to children whose primary infection occurs before the third year of life, especially those less than one year of age—a practice used by Dr. Gutman years before.

When children are coinfected with TB and HIV and are at the stage when significant compromise of the immune system has occurred, they will demonstrate progression of their TB in the same manner as children less than two years of age do. Certainly the superimposition of the HIV epidemic on endemic TB has contributed significantly to the unexpected rise in the number of new cases of TB and the fatalities domestically and abroad.[49-51] The reasons include increased susceptibility to TB when also infected with HIV; the concentration of these diseases in marginalized populations, thus facilitating transmission; and the relatively increased prevalence of both diseases in foreign-born individuals.

Gutman served as the lead investigator in a national sample of infants and children with TB exposure, infection, or disease who were within a larger sample of children also exposed to, or infected with, HIV and being followed in one of the seventy Pediatric AIDS Clinical Trials Group (PACTG) clinics. Gutman and her colleagues collected data derived from 14,038 of those patients.[38] From 1981 to 1992, they identified 75 total active cases of TB disease in coinfected children. They determined that another forty children had asymptomatic infection with TB and that another seventy-one had had significant exposure to TB. All study children who were identified to have been diseased, infected, or exposed to TB received anti-TB therapy according to standard recommendations at that time. A survey completed by each of the PACTG sites allowed the annualized case rates to be calculated. The rates ranged from 58 per 100,000 cases from the sites that originated in the early to mid-1980 to 478 per 100,000 cases at the sites that were initiated from 1990 to 1992. Because of the increase in annualized case rates across PACTG sites, the authors posited that the rates of infection were increasing over time. In addition, this study demonstrated substantial numbers of isolates of *Mycobacterium tuberculosis* resistant to both isoniazid and rifampin in both the pediatric patients (20 percent) and their adult contacts (15 percent).

SERIOUS VIRAL INFECTIONS

ENTEROVIRUSES IN PATIENTS WITH HYPOGAMMAGLOBULINEMIA

A deficiency of gamma globulins is not usually associated with a failure to control or eradicate viral infections, but reports beginning in 1973 described eight patients who had chronic (a duration of two months to three years) enteroviral infections of the central nervous system (CNS) and whose gamma globulin levels were low or absent. Duke pediatrician Dr. Catherine Wilfert assembled the data for five of these patients who were diagnosed at Duke, Harvard, or Virginia, all of whom shared the following characteristics: their lymphocytes lacked surface immuno-globulin, their lymph nodes lacked cortical follicles, and their T cell function was normal.[52] In other words, they could be described as being agammaglobulinemic. Their clinical courses varied. Few had signs of acute CNS infection, but of the five Duke-reported cases, three had a dermatomyositislike syndrome with peripheral lymphocytes that re-acted with anti-human leukemia-specific primate and rabbit serums in a cytotoxicity assay. In a follow-up review of forty-two patients in 1987, most were shown to have X-linked agammaglobulinemia, and their clin-ical course varied as well, but the dermatomyositislike syndrome was again represented.[53] Some succumbed to this infection, but somewhat surprisingly, some had a total absence of symptoms. McKinney et al. postulated that a clinical scenario of hypogammaglobulinemia accom-panied by pleocytosis and an elevated cerebrospinal fluid protein should raise the differential diagnostic possibility of a concomitant Enteric Cytopathic Human Orphan (ECHO) viral infection of the CNS and that treatment with IV immunoglobulins at regular intervals appears to alter the clinical course favorably. He also tested a novel antiviral drug in suckling mice infected with another enterovirus, Coxsackie A 9 virus (McKinney, personal communication).

HERPES SIMPLEX VIRUS INFECTIONS IN NEONATES

In 1980, Whitley and colleagues described the natural history of herpes simplex virus (HSV) infection in pregnant women and newborns.[54] Typically, the women had prior histories of genital herpes or of an expo-sure to a sexual partner with presumed HSV lesions but were themselves asymptomatic at the time of delivery. Premature labor was recognized in approximately 50 percent of the mothers. In those infants with localized CNS HSV disease, skin rash was the most frequent site of peripheral clinical disease. The spinal fluid was infrequently positive by culture for HSV, and in some cases, a brain biopsy was necessary for diagno-sis. The long-term outcomes were very poor for the reported infants, with some dying and the majority of survivals with extensive debil-itating and chronic neurologic disabilities. Although treatment with ARA-A or ARA-C improved outcomes somewhat, the results remained

206

discouraging. In 2001, the natural history of neonatal HSE in the acyclovir (ACV) era was reported and again found frequently to include extensive disabilities with only some improvement in outcomes from the earlier results.[55]

In 1986, Duke pediatrician Dr. Laura Gutman and colleagues reported that the clinical courses and late results of cerebrospinal fluid analysis of surviving infants provided sufficient evidence to hypothesize that "HSV encephalitis in infants may result in the late persistence or recurrence of disease of the central nervous system."[56] In addition, the medical literature had already included several case reports of recurrent symptomatic HSE disease in infants who had survived neonatal HSE disease and had received standard acute anti-viral treatment when the disease was first diagnosed. Most of those recurrences had occurred before the infant reached two years of age. During this period, acyclovir became available in both intravenous and oral preparations. Standard recommendations for therapy of neonatal HSE began to include IV acyclovir as well as ARA-A, with a treatment duration of two to three weeks total. However, because of concern about the frequent dermal recurrences that were experienced by many infants, as well as the evidence of CNS persistence and/or recurrence in some, studies were begun at Duke on a treatment regimen that included following the IV therapy with courses of oral ACV therapy.

In 1994, Rudd and colleagues published the first study of ACV serum concentrations taken one to two hours following various oral doses in infants in order to determine the optimal antiviral dose.[57] The approximate distribution of ACV from serum to cerebrospinal fluid (CSF) was known, and a serum concentration of three micrograms per millimeter or greater was chosen to achieve the desired CSF level. The information from the Rudd study formed the basis for oral dosing decisions where the objective was to achieve CSF levels sufficient to suppress viral replication. The serum levels following oral doses that were being widely recommended would be expected to fail in that regard. Consequently, Gutman's colleagues initiated a treatment regimen using high-dose oral therapy during which serum concentrations were periodically monitored and appropriately increased to achieve specific serum concentrations . A duration of two years was selected because of the pattern of symptomatic recurrences in the literature, all in younger infants. These clinical cases were monitored for safety, efficacy, and feasibility of prolonged oral high dose therapy to improve the outcomes. Between 1990 and 2003, sixteen children had been managed through age two years with this regimen and were reported in 2005.[58]

Assessments of the children included periodic determination that the target serum concentration of ACV was being achieved and dosing was accordingly increased as needed. At one and at two years of age, the child's mental and motor neurodevelopmental status was assessed using Bayley Scale for Infant Development scores. For the sixteen reported children, there were no CNS recurrences, and two transient instances of dermal blisters that did not yield virus cleared within several days without alteration in therapy, and whose etiology was uncertain. All children were nutritionally independent, were developing language abilities, and were independently ambulatory, with none having recurrent seizure disorder or ocular paresis. The Bayley Scale score for mental development was eighty or above for fifteen of the sixteen, and the motor score was eighty or above for thirteen of sixteen, both results exceeding those expected from historical controls.

In 2011, Kimberlin and Whitley reported a study through the National Institute of Allergy and Infectious Diseases (NIAID) Collaborative Antiviral Study group.[59] Seventy-four neonates had received parenteral acyclovir for fourteen to twenty-one days, following which half received up to six months of suppressive oral doses using much lower doses than the Duke program, and half received placebos. Serum concentrations were not assessed. Those who received suppressive therapy had a better outcome, but measurement of mental and motor outcomes were not reported so as to allow comparison with the Duke outcomes.

The precise duration of suppressive oral therapy that will benefit an infant following neonatal HSE remains uncertain, but Dr Gutman deserves credit for exposing the outcomes and limitations of short-term parenteral therapy alone, for recognizing that HSE acts as do other herpes virus infections, with a predilection for recurrences, and for providing a therapeutic program to control the medical threat that those recurrences bring to infected infants.

CYTOMEGALOVIRUS INFECTION IN TRANSPLANT RECIPIENTS

My colleagues and I published clinical case reports[60-62] and clinical research studies on the seroepidemiology of cytomegalovirus (CMV),[63] CMV pneumonia,[61] and CMV disease in transplant recipients.[64] I also published a monograph on CMV and immunity[65] that summarized the state of knowledge of clinical disease and the importance of normal or abnormal immune systems.

When a variety of clinical circumstances led to immunosuppression, it was found that previously latent viruses, including CMV, had reactivated to result in clinical disease. An understanding of the basis for the process of viral latency and its reactivation was not known. I therefore embarked on a series of experiments to investigate the cellular and molecular events associated with viral latency and reactivation. I, along with others, published three papers addressing various host immune and viral phenomena demonstrating the complexity and interdependence of the immune defense mechanisms and the viral replicative processes.[66-68] A year's sabbatical in the immunogenetics laboratory of Professor Jon J. van Rood at the Academisch Ziekenhuis in Leiden, Netherlands, greatly enhanced my knowledge of the complexity of a rapidly evolving field and provided me with new insights regarding likely productive research on viral-host interactions. My work there demonstrated the clonality of human cytotoxic T cells by utilizing a monolayer absorption technique that I developed. In addition, in collaboration with others, I demonstrated that anti-self-HLA was also likely to be expressed clonally.[69]

Subsequent publications, with colleagues as co-authors, demonstrated the following: the transmission of murine cytomegalovirus (MCMV) from a latently infected mouse to an uninfected mouse in a kidney transplant model;[70] the reactivation of latent MCMV from the kidney of latently infected mice;[71] the source of latent virus reactivated in the transplant model (donor versus recipient);[72,73] the identification of the viral genome in kidney [74]; the evidence for viral transcription in latently infected mice [75]; a suppressive effect of MCMV on macrophage and granulocyte-macrophage progenitor cells;[76] and evidence of later stages of viral replication in latently infected mice.[77]

Also, I initiated studies to further characterize the state of activity by using an extremely useful recombinant MCMV expressing "Green Fluorescent Protein" (GFP) constructed in my laboratory.[78] The development of further powerful technologies has allowed additional investigations focusing on the reservoir of latent virus and the impact of MCMV infection on innate and adaptive immunity.[79,80] These experiments will be discussed in a later publication.

SERIOUS FUNGAL INFECTIONS

CRYPTOCOCCUS NEOFORMANS FUNGEMIA AND MENINGITIS

Cryptococcus has been the most common cause of infectious meningitis at Duke University Medical Center for many years. This was especially true when effective treatments for HIV were not available. A group of eight to ten faculty members in the Departments of Medicine, Pediatrics, and Molecular Genetics and Microbiology have devoted their careers to the study of the *Cryptococcus* organism and its effect on humans. Some have engaged in clinical research and others in basic research. Most of these faculty members have been discussed in earlier chapters, but for the purpose of this section, I will focus on the contributions of Dr. John Perfect. His work is most conveniently summarized by dividing his contributions into one of five general categories.

First, throughout his career, which began in 1978 when he came to Duke as a fellow in infectious disease, he has played a prominent role in teaching medical students, house staff, and fellows, and in the care of patients on the general medicine service on the infectious disease consult service and on the transplant service. Second, he coauthored two comprehensive books on *Cryptococcus*, one published in 1998[81] and the other in 2011.[82] Third, he has worked as an experimental pathologist in the study of animal models for fungal pathogenesis and treatment. His *in vivo* models bolstered by *in vitro* studies allowed him to understand old, new, and combination antifungal agents. These insights were then taken to the clinic for translational studies.[83-86] Fourth, Dr. Perfect retrained in molecular biology to allow him to study the molecular pathogenesis of *Cryptococcus neoformans*. His laboratory defined the boundaries of cryptococcal virulence by using models in rabbits, mice, worms, grubbs, and zebrafish, and the tools of molecular yeast manipulation. In addition to his studies on the basic mechanisms of fungal pathogenesis, he conducted clinical research on the epidemiology of cryptococcal disease, newer diagnostics, novel treatment strategies, the pharmacokinetics of antifungal agents, and the genetics of antifungal resistance, and he developed databases and complementary biorepositories and created study drug protocols.[87-97] Fifth, as a member of advisory groups for the NIH and pharmaceutical industries, he has contributed to the development of guidelines for the use of antifungal agents.[98,99]

SERIOUS RICHETTSIAL INFECTIONS

ROCKY MOUNTAIN SPOTTED FEVER

Only the state of Oklahoma reports more cases per year of Rocky Mountain spotted fever (RMSF) than does the state of North Carolina (Dr. Daniel Sexton, personal communication). In North Carolina, RMSF is most prevalent west of Durham, and emergency rooms and urgent care centers expect to see numerous critically ill patients with the infection in the spring, summer, and fall, some patients with and some without histories of a tick bite. With the history of a tick bite, a febrile illness with a characteristic, though not pathognomonic, rash will always merit empiric therapy, even without confirmatory diagnostic serology tests, immunofluorescence microscopy, or polymerase chain reaction tests. Depending on the stage of the disease and on the presence of hemodynamic instability, the prognosis is very good with appropriate therapy. Duke Infectious Diseases faculty and, later, Assistant Professor Dr. Kathy Kirkland and colleagues, however, reported six cases that documented gangrene as one very serious complication of the acute disease.[100] Dr. Kirkland and colleagues also reported on the dire consequences of a delay in the initiation of effective therapy.[101] Other complications include renal failure, hypotension and vascular collapse, electrolyte imbalance, seizures, stupor, and coma.[102] Complicating the prompt initiation of effective therapy is the existence of variations on the classical clinical presentation, namely Rocky Mountain "spotless and almost spotless" fever.[103]

During a one-year gap between having a busy private practice in Oklahoma City and joining the faculty at Duke, Dr. Daniel Sexton worked in Australia as a clinical consultant while also pursuing his interest in the spotted fever group of infections in humans[104] and dogs.[105] In the latter, he demonstrated an unexpectedly high seroprevalence. Following along on this line of investigation, Dr. Sexton, along with one of the legends in rickettsial research (Dr. Willy Burgdorfer), conducted a survey in Mississippi for spotted fever antibodies in dogs and for *Rickettsia rickettsii* in dog ticks.[106] Sexton also described a new endemic region for spotted fever in the Espirito Santo region of Brazil[107] and was the first to describe, along with Dr. Burgdorfer, a new spotted fever group of *Rickettsia* in the United States, transmitted by *Rhipicephalus sanguineus*.[108] In studies aimed at improving the sensitivity of laboratory tests for the diagnosis of RMSF, one study compared four serologic methods by using sera samples referred to the North Carolina Division of Health Services. The complement fixation test was found to be less sensitive than were the concordant microimmunoflorescence, microagglutination, and hemagglutination tests.[109] Attempts to use polymerase chain reaction of blood and urine to detect *Rickettsia rickettsii* in ten patients with suspected RMSF demonstrated that this method appeared to be not very sensitive.[110]

Dr. Catherine Wilfert, a pediatric faculty member, collaborated with personnel at the State Board of Health in Raleigh, North Carolina, to survey 1,976 residents of Cabarrus, Rowan, and Granville counties to define the prevalence of antibody to *R. rickettsii* (sera samples were tested by indirect hemagglutination). These counties were chosen because of the excess number of case reports to the State Board of Health. Of the samples tested, 28.7 percent had antibody titers greater than or equal to 1:8, 4.0 percent had titers greater than or equal to 1:64, and 70 percent had no detectable antibody. Indirect immunofluorescence tests were done on a subset of these sera samples that demonstrated significant differences from those tested by indirect hemagglutination. The results suggested that "the antibodies measured may not be specific for

R. rickettsii or that the antibody levels wane with time or both" and that "it is probable that unrecognized infection occurs, but the true incidence or prevalence cannot be determined by available serological tests."[111,112]

SERIOUS PARASITIC INFECTIONS

MALARIA

Another robust collaboration between fellows and faculty in the division of infectious diseases was initiated by the research on malaria conducted by Dr. Nicholas Anstey, a fellow from 1993 to 1996, during which time he lived and worked in Dar es Salaam, Tanzania. Now the Head of the Global Health Division at the Menzies School of Health Research in Darwin, Australia, Dr. Anstey began his research with pathophysiological observations of impaired levels of nitric oxide (NO) production in children and adults with severe malaria in collaboration with, among others, Drs. Donald Granger and Brice Weinberg, both faculty members at Duke in the 1990s. Eight publications on topics related to malaria ensued before 2000[113-120]

Photograph 10.9 is a composite of pictures of six prominent leaders in the field of serious infections at Duke in the twentieth century.

Vance G. Fowler

Laura T. Gutman

J. Brice Weinberg

Nicholas M. Anstey

Donald L. Granger

Ross E. McKinney

AFTERWORD

This book attempts to capture the history of the people and programs that proved so important in the emergence of the subspecialty of Infectious Diseases at Duke University in the 20th century. Beginning with a modest commitment focused on the local scourges prevalent at the time, Duke University-- a fledgling institution in the 1930s in the years after Mr. Duke's bequest--evolved to become one of the premier institutions of higher learning with an academic portfolio that undoubtedly exceeded Mr. Duke's expectations.

Initially, the disease profile was dominated by recognizable contagious diseases, such as tuberculosis and the severe and often fatal bacterial and parasitic diseases. The Duke staff devoted themselves to a better understanding of the diseases and to improving methods that would prevent, manage and treat them. Over the period of the 20th century, the faculty, students and staff contributed mightily to this mission, supported by an ever more complex infrastructure and the requirement for increasing sophistication of the underlying science and sources of financial support.

To the extent that this book describes the elements of progress over the 20th century selectively, it is my view that Duke University and Medical Center has contributed substantially to the still evolving infectious threats to humankind. The momentum has accelerated dramatically in the first part of the 21st century, and an anticipated subsequent publication will address these new developments.

CITATIONS BY CHAPTER

CITATIONS FOR CHAPTER 1

1. Strauss MB, ed *Familiar Medical Quotations*. Boston, MA: Little, Brown and Company; 1968. Strauss MB, ed.
2. Gehlbach SH. *American Plagues: Lessons from Our Battles with Disease*. New York, NY: McGraw-Hill; 2005.
3. Sherman IW. *Twelve Diseases That Changed Our World*. Washington, DC: American Society for Microbiology; 2007.
4. McNeill WH. *Plagues and Peoples*. New York: Anchor Books; 1998.
5. Barry JM. *The Great Influenza: The Epic Story of the Deadliest Plague in History*. London, England: Penguin Books; 2004.
6. Florey HW. General Pathology, Based on Lectures Delivered at the Sir William Dunn School of Pathology, University of Oxford. In: Florey HW, ed. *General Pathology*. 3rd ed. London, England: Lloyd-Luke Medical Books Ltd; 1964.
7. Magner LN. *A History of Infectious Diseases and the Microbial World*. Westport, CT: Praeger Publishers; 2009.
8. Long D, ed *Medicine in North Carolina: Essays in the History of Medical Science and Medical Service*. Raleigh, NC: North Carolina Medical Society; 1972. I.
9. Long D, ed *Medicine in North Carolina: Essays in the History of Medical Science and Medical Service*. Raleigh, NC: N.C. Medical Society; 1972. II.
10. Stern AM, Markel H. International Efforts to Control Infectious Diseases: 1851 to the Present. *JAMA*. 2004; 292:1474-1479.
11. Notifiable Diseases: Prevalence During 1913 in Cities of 10,000 to 100,000. *Public Health Rep*. 1914; 29:2883-2902.
12. Bowman JL. The Medical Profession and Notifiable Diseases: Why Reports Should Be Made. *Public Health Rep*. 1920; 35:2503-2509.
13. Lee HSJ, ed *Dates in Infectious Diseases*. New York, NY: The Parthenon Publishing Group; 2002.

14. Rifkind D, Freeman GL. *The Nobel Prize Winning Discoveries in Infectious Diseases*. Amsterdam: Elsevier Academic Press; 2005.

15. Centers for Disease Control. National Notifiable Diseases Surveillance System. 2010:2. http://www.cdc.gov/ncphi/disss/nndss/nndsshis.htm.

16. Rankin WS. *First Annual Report of the Bureau of Vital Statistics, 1914 of the North Carolina State Board of Health*. Raleigh, NC: The Bureau; 1915.

17. Anderson JB. *Durham County: A History of Durham County, North Carolina*. Durham, NC: Duke University Press; 1990.

18. Flexner A. Medical Education in the United States and Canada: A Report to the Carnegie Foundation for the Advancement of Teaching. *Bulletin*. 1910; 4.

19. Arena JM, McGovern JP, eds. *Davison of Duke: His Reminiscences*. Fulton, MO: The Ovid Bell Press, Inc; 1980.

20. Gifford J. *The Evolution of a Medical Center: A History of Medicine at Duke University to 1941*. Durham, NC: Duke University Press; 1972.

21. Robertson GD. Pardon Sought for Impeached 1871 N.C. Gov. Holden. *Raleigh News and Observer* 2012.

22. Baker RB, Washington HA, Olakanmi O, et al. African American Physicians and Organized Medicine, 1846-1968: Origins of a Racial Divide. *JAMA*. 2008; 300:306-313.

23. US Supreme Court. Plessy V Ferguson. 1896.

24. Reynolds PP. *Watts Hospital of Durham, North Carolina, 1895-1976: Keeping the Doors Open*. Durham, NC: The Fund for Advancement of Science and Mathematics in North Carolina; 1991.

25. Reynolds P. *Durham's Lincoln Hospital*. Charleston, SC: Arcadia Press; 2001.

26. Stephens J. Asheville: The Tuberculosis Era. *NC Medical Jour.* 1985; 46:455-465.

CITATIONS FOR CHAPTER 2

1. Gifford J. *The Evolution of a Medical Center: A History of Medicine at Duke University to 1941*. Durham, NC: Duke University Press; 1972.

2. Arena JM, McGovern JP, eds. *Davison of Duke: His Reminiscences*. Fulton, MO: The Ovid Bell Press, Inc; 1980.

3. Durden R. *Lasting Legacy to the Carolinas: The Duke Endowment, 1924-1994*. Durham, NC: Duke University Press; 1998.

4. Flexner A. Medical Education in the United States and Canada: A Report to the Carnegie Foundation for the Advancement of Teaching. *Bulletin*. 1910; 4.

5. Reynolds PP. *Watts Hospital of Durham, North Carolina, 1895-1976: Keeping the Doors Open*. Durham, NC: The Fund for Advancement of Science and Mathematics in North Carolina; 1991.

6. Long D, ed *Medicine in North Carolina: Essays in the History of Medical Science and Medical Service.* Raleigh, NC: North Carolina Medical Society; 1972. I.
7. Long D, ed *Medicine in North Carolina: Essays in the History of Medical Science and Medical Service.* Raleigh, NC: N.C. Medical Society; 1972. II.
8. Anderson JB. *Durham County: A History of Durham County, North Carolina.* Durham, NC: Duke University Press; 1990.
9. Paul JR. A Memorial. 1956.
10. Rankin WS, Hannaford HE, van Arsdall HP. The Small General Hospital: Prepared for the Trustees of the Duke Endowment. *Bulletin.* 1932; 3.
11. Campbell WE, High M, Kennedy EJ. *Foundations for Excellence : 75 Years of Duke Medicine.* Durham, N.C.: Duke University Medical Center Library; 2006.

CITATIONS FOR CHAPTER 3

1. Haas F, Haas SS. The Origins of Mycobacterium Tuberculosis and the Notion of Its Contagiousness. In: Rom WN, Garay SM, eds. *Tuberculosis.* Boston, MA: Little, Brown and Company, Inc; 1996:3-19.
2. Murray JF. A Century of Tuberculosis. *Am J Respir Crit Care Med.* 2004; 169:1181-1186.
3. Davis AL. History of the Sanatorium Movement. In: Rom WN, Garay SM, eds. *Tuberculosis* New York: Little, Brown and Company, Inc; 1996.
4. Williams LE. 50th Anniversary Historical Data of the North Carolina Tuberculosis Association, 1906-1956. 1956.
5. Von Ruck K. A Practical Method of Prophylactic Immunization against Tuberculosis. *J Amer Med Assoc.* 1912; 58:1504-1507.
6. Goldstein SE. After the Sanatorium What? *J Outdoor Life.* 1914; 11:266-268.
7. Mitchell RS. Mortality and Relapse of Uncomplicated Advanced Pulmonary Tuberculosis before Chemotherapy: 1504 Consecutive Admissions Follwed for Fifteen to Twenty Five Years. Part I. *Am Rev Tuberc.* 1955; 72:487-501.
8. Stephens MG. Follow-up of 1,041 Tuberculosis Patients. *Am Rev Tuberc.* 1941; 44:451-462.
9. Watson S. Rankin Interviewed by James F. Gifford, January 11, 1965. Durham, NC: Duke Medical Center Archives, Biographical Files Collection; 1965.
10. Rankin WS, Hannaford HE, van Arsdall HP. The Small General Hospital: Prepared for the Trustees of the Duke Endowment. *Bulletin.* 1932; 3.
11. Watson S. Rankin Obituary. *Charlotte News* 1970.
12. Amberson JB. A Clinical Trial of Sanocrysin in Pulmonary Tuberculosis. *Amer RevTuberc.* 1931; 24:401-435.

13. Schatz A, Bugie E, Waksman SA. Streptomycin, a Substance Exhibiting Antibiotic Activity against Gram-Positive and Gram-Negative Bacteria. *Exp Biol Med.* 1944; 55:66-69.

14. Feldman WH, Hinshaw HC, Mann FC. Streptomycin in Experimental Tuberculosis. *Am Rev Tuberc.* 1945; 52:269-298.

15. Hinshaw HC, Feldman WH. Streptomycin in the Treatment of Clinical Tuberculosis: A Preliminary Report. *Proc Staff Meet Mayo Clin.* 1945; 20:313-318.

16. Tucker WB. The Evolution of the Cooperative Studies in the Chemotherapy of Tuberculosis of the Veterans Administration and Armed Forces of the U.S.A. An Account of the Evolving Education of the Physician in Clinical Pharmacology. *Bibl Tuberc.* 1960; 15:1-68.

17. Hays MT. *VA Research 1925-1980.* Bloomington, IN: Authorhouse; 2010.

18. Harris HW. Chemotherapy of Tuberculosis: The Beginning. In: Rom WN, Garay SM, eds. *Tuberculosis.* Boston, MA: Little, Brown and Company; 1996:745–749.

19. British Medical Research Council. Streptomycin Treatment of Pulmonary Tuberculosis: A Medical Research Council Investigation. *Br Med J.* 1948; 2:769-782.

20. Ferebee SH. *Report on the Usphs Streptomycin Study Program.* Chicago, IL: Veterans Administration Branch Office;1948/49.

21. Bernheim F. The Effect of Salicylate on the Oxygen Uptake of the Tubercle Bacillus. *Science.* 1940; 92:204.

22. Lehmann LS. Para-Aminosalicylic Acid in the Treatment of Tuberculosis. *Lancet.* 1946; 1:15.

23. Abernathy RS. David Tillerson Smith. *N C Med J.* 1990; 51:473-481.

24. Collected Papers of David T. Smith, Trent Collection. Durham, NC: Duke University; 1930-1970.

25. Smith DT, Abernathy RS, Smith GB, Jr., Bondurant S. The Apical Localization of Reinfection Pulmonary Tuberculosis. I. The Stream Flow Theory. *Am Rev Tuberc.* 1954; 70:547-556.

26. Smith DT, Abernathy RS, Smith GB, Jr., Bondurant S. Basic Considerations in Tuberculosis: The Apical Localization of Reinfection Pulmonary Tuberculosis. *Transactions of the annual meeting National Tuberculosis Association.* 1954; 96:189-195.

27. Smith DT, Abernathy RS. Selective Localization of Pulmonary Emboli: An Explanation for the Apical Localization of Reinfection Tuberculosis. *Trans Am Clin Climatol Assoc.* 1949; 61:191-220.

28. Abernathy RS, Smith GB, Jr., Smith DT. The Apical Localization of Reinfection Pulmonary Tuberculosis. II. Selective Localization of Experimental Emboli. *Am Rev Tuberc.* 1954; 70:557-569.

29. Harris J. Memorial for David T. Smith. 1981.

30. Tucker WB. Comparison of the Effect of Four Variables in the Antimicrobial Therapy of Pulmonary Tuberculosis. I. Report of the Cooperative Study of the Veterans Administration, Army, and Navy, April, 1949 to January 1951. *Am Rev Tuberc.* 1955; 72:718-732.

31. Tucker WB. A Review of the Current Status of the Chemotherapy of Tuberculosis. *Ann Intern Med.* 1953; 39:1045-1061.

32. Tucker WB. Evaluation of Streptomycin Regimens in the Treatment of Tuberculosis; an Account of the Study of the Veterans Administration, Army, and Navy, July 1946 to April 1949. *Am Rev Tuberc.* 1949; 60:715-754.

33. Tucker WB. Streptomycin in the Treatment of Various Tuberculous Conditions. *J Lancet.* 1948; 68:282-292.

34. Falk A. U. S. Veterans Administration-Armed Forces Cooperative Study on the Chemotherapy of Tuberculosis. Xii. Results of Treatment in Miliary Tuberculosis: A Follow-up Study of 570 Adult Patients. *Am Rev Respir Dis.* 1965; 91:6-12.

35. Falk A, Stead WW. U.S. Veterans Administration-Armed Forces Cooperative Studies of Tuberculosis. V. Antimicrobial Therapy in the Treatment of Primary Tuberculous Pleurisy with Effusion: Its Effect Upon the Incidence of Subsequent Tuberculous Relapse. *Am Rev Tuberc.* 1956; 74:897-902.

36. Falk A. U. S. Veterans Administration-Armed Forces Cooperative Study on the Chemotherapy of Tuberculosis. 13 Tuberculous Meningitis in Adults, with Special Reference to Survival, Neurologic Residuals, and Work Status. *Am Rev Respir Dis.* 1965; 91:823-831.

37. Falk A. A Follow-up Study of the Initial Group of Cases of Skeletal Tuberculosis Treated with Streptomycin, 1946-1948; the United States Veterans Administration and Armed Forces Cooperative Studies of Tuberculosis. *J Bone Joint Surg Am.* 1958; 40-A:1161-1168.

38. Hammersten. Memorial for Warren B. Tucker. 1979.

39. Hamilton JD. Medical Service Records at the VA. 2010.

40. Karpik R. Tuberculosis. In: Mandell H, Spiro H, eds. *When Doctors Get Sick.* New York, NY: Plenum Medical Book Company; 1987:401-404.

41. Gutman LT, Moye J, Zimmer B, Tian C. Tuberculosis in Human Immunodeficiency Virus-Exposed or -Infected United States Children. *Pediatr Infect Dis J.* 1994; 13:963-968.

42. Gallis HA, Berman RA, Cate TR, Hamilton JD, Gunnells JC, Stickel DL. Fungal Infection Following Renal-Transplantation. *Arch Intern Med.* 1975; 135:1163-1172.

43. Halvorsen RA, Duncan JD, Merten DF, Gallis HA, Putman CE. Pulmonary Blastomycosis: Radiologic Manifestations. *Radiology.* 1984; 150:1-5.

44. Khoury MB, Godwin JD, Ravin CE, Gallis HA, Halvorsen RA, Putman CE. Thoracic Cryptococcosis: Immunologic Competence and Radiologic Appearance. *AJR American Journal of Roentgenology.* 1984; 142:893-896.

45. Hansen JP, Falconer JA, Gallis HA, Hamilton JD. Inadequate Sensitivity of Tuberculin Tine Test for Screening Employee Populations. *J Occup Environ Med.* 1982; 24:602-604.

46. Kim JH, Langston AA, Gallis HA. Miliary Tuberculosis: Epidemiology, Clinical Manifestations, Diagnosis, and Outcome. *Rev Infect Dis.* 1990; 12:583-590.

47. Quitugua TN, Seaworth BJ, Weis SE, et al. Transmission of Drug-Resistant Tuberculosis in Texas and Mexico. *J Clin Microbiol.* 2002; 40:2716-2724.

48. Seaworth BJ. Multidrug-Resistant Tuberculosis. *Infect Dis Clin North Am.* 2002; 16:73-105.

49. Smith KC, Seaworth BJ. Drug-Resistant Tuberculosis: Controversies and Challenges in Pediatrics. *Expert Rev Anti Infect Ther.* 2005; 3:995-1010.

50. Dukes CS, Matthews TJ, Weinberg JB. Human Immunodeficiency Virus Type 1 Infection of Human Monocytes and Macrophages Does Not Alter Their Ability to Generate an Oxidative Burst. *J Infect Dis.* 1993; 168:459-462.

51. Dukes CS, Matthews TJ, Rivadeneira ED, Weinberg JB. Neopterin Production by HIV-1-Infected Mononuclear Phagocytes. *J Leukoc Biol.* 1994; 56:650-653.

52. Dukes CS, Yu Y, Rivadeneira ED, et al. Cellular CD44s as a Determinant of Human Immunodeficiency Virus Type 1 Infection and Cellular Tropism. *J Virol.* 1995; 69:4000-4005.

53. Dukes CS, Matthews TJ, Lambert DM, Dreyer GB, Petteway SR, Weinberg JB. Potent Inhibition of HIV Type 1 Infection of Mononuclear Phagocytes by Synthetic Peptide Analogs of HIV Type 1 Protease Substrates. *AIDS Res Hum Retroviruses.* 1996; 12:777-782.

54. Dukes CS, Sugarman J, Cegielski JP, Lallinger GJ, Mwakyusa DH. Severe Cutaneous Hypersensitivity Reactions During Treatment of Tuberculosis in Patients with HIV Infection in Tanzania. *Trop Geogr Med.* 1992; 44:308-311.

55. Cegielski JP, Lwakatare J, Dukes CS, et al. Tuberculous Pericarditis in Tanzanian Patients with and without HIV Infection. *Tuber Lung Dis.* 1994; 75:429-434.

56. Boggess KA, Myers ER, Hamilton CD. Antepartum or Postpartum Isoniazid Treatment of Latent Tuberculosis Infection. *Obstet Gynecol.* 2000; 96:757-762.

57. Vernon A, Burman W, Benator D, Khan A, Bozeman L. Acquired Rifamycin Monoresistance in Patients with HIV-Related Tuberculosis Treated with Once-Weekly Rifapentine and Isoniazid. Tuberculosis Trials Consortium. *Lancet.* 1999; 353:1843-1847.

58. Hamilton CD, Stout JE, Goodman PC, et al. The Value of End-of-Treatment Chest Radiograph in Predicting Pulmonary Tuberculosis Relapse. *Int J Tuberc Lung Dis.* 2008; 12:1059-1064.

59. Miller WC, Thielman NM, Swai N, et al. Delayed-Type Hypersensitivity Testing in Tanzanian Adults with HIV Infection. *J Acquir Immune Defic Syndr Hum Retrovirol.* 1996; 12:303-308.

60. Cegielski JP, Goetz MB, Jacobson JM, et al. Gender Differences in Early Suspicion of Tuberculosis in Hospitalized, High-Risk Patients During 4 Epidemic Years, 1987 to 1990. *Infect Control Hosp Epidemiol.* 1997; 18:237-243.

61. Clark PA, Cegielski JP, Hassell W. Tb or Not Tb? Increasing Door-to-Door Response to Screening. *Public Health Nurs.* 1997; 14:268-271.

62. Cegielski JP, Devlin BH, Morris AJ, et al. Comparison of PCR, Culture, and Histopathology for Diagnosis of Tuberculous Pericarditis. *J Clin Microbiol.* 1997; 35:3254-3257.

63. Wilson RW, Yang Z, Kelley M, et al. Evidence from Molecular Fingerprinting of Limited Spread of Drug-Resistant Tuberculosis in Texas. *J Clin Microbiol.* 1999; 37:3255-3259.

64. Robison VA, Griffith DE, Cegielski JP, Reed MR, Nash DR. A Tuberculosis Hotline in Tyler: A Texas Resource for Primary Health Care Providers for the Control of Infectious Diseases. *Tex Med.* 1998; 94:48-52.

65. Centers for Disease Control. Primary Multidrug-Resistant Tuberculosis--Ivanovo Oblast, Russia, 1999. *MMWR Morb Mortal Wkly Rep.* 1999; 48:661-663.

66. Perkins MD. Mycobacterial Diagnostics: The Role of the Laboratory. A Review. *Rev Soc Bras Med Trop.* 1994; 27.

67. Desjardin LE, Perkins MD, Teixeira L, Cave MD, Eisenach KD. Alkaline Decontamination of Sputum Specimens Adversely Effects Stability of Mycobacterial mRNA. *J Clin Microbiol.* 1996; 34:2435-2439.

68. Desjardin LE, Chen Y, Perkins MD, Teixeira L, Cave MD, Eisenach KD. Comparison of the Abi 7700 System (Taqman) and Competitive PCR for Quantification of Is6110 DNA in Sputum During Treatment of Tuberculosis. *J Clin Microbiol.* 1998; 36:1964-1968.

69. Hellyer TJ, DesJardin LE, Teixeira L, Perkins MD, Cave MD, Eisenach KD. Detection of Viable Mycobacterium Tuberculosis by Reverse Transcriptase-Strand Displacement Amplification of mRNA. *J Clin Microbiol.* 1999; 37:518-523.

70. Desjardin LE, Perkins MD, Wolski K, et al. Measurement of Sputum Mycobacterium Tuberculosis Messenger RNA as a Surrogate for Response to Chemotherapy. *Am J Respir Crit Care Med.* 1999; 160:203-210.

71. Wallis RS, Patil S, Cheon SH, et al. Drug Tolerance in Mycobacterium Tuberculosis. *Antimicrob Agents Chemother.* 1999; 43:2600-2606.

72. Wallis RS, Perkins M, Phillips M, et al. Induction of the Antigen 85 Complex of M. Tuberculosis in Sputum: A Determinant of Outcome in Pulmonary Tuberculosis Treatment. *J Infect Dis.* 1998; 178:1115-1121.

73. Perkins MD, Kritski AL. Diagnostic Testing in the Control of Tuberculosis. *Bull World Health Organ.* 2002; 80:512-513.

74. Frothingham R, Allen RL, Wilson KH. Rapid 16s Ribosomal DNA Sequencing from a Single Colony without DNA Extraction or Purification. *Biotechniques.* 1991; 11:40-44.

75. Frothingham R, Wilson KH. Sequence-Based Differentiation of Strains in the Mycobacterium Avium Complex. *J Bacteriol.* 1993; 175:2818-2825.

76. Frothingham R, Wilson KH. Molecular Phylogeny of the Mycobacterium Avium Complex Demonstrates Clinically Meaningful Divisions. *J Infect Dis.* 1994; 169:305-312.

77. Frothingham R, Hills HG, Wilson KH. Extensive DNA Sequence Conservation Throughout the Mycobacterium Tuberculosis Complex. *J Clin Microbiol.* 1994; 32:1639-1643.

78. Frothingham R, Meeker-O'Connell WA. Genetic Diversity in the Mycobacterium Tuberculosis Complex Based on Variable Numbers of Tandem DNA Repeats. *Microbiology.* 1998; 144:1189-1196.

79. Kremer K, Van Soolingen D, Frothingham R, et al. Comparison of Methods Based on Different Molecular Epidemiological Markers for Typing of Mycobacterium Tuberculosis Complex Strains: Interlaboratory Study of Discriminatory Power and Reproducibility. *J Clin Microbiol.* 1999; 37:2607-2618.

80. Talbot EA, Reller LB, Frothingham R. Bone Marrow Cultures for the Diagnosis of Mycobacterial and Fungal Infections in Patients Infected with the Human Immunodeficiency Virus. *Int J Tuberc Lung Dis.* 1999; 3:908-912.

81. Frothingham R, Meeker-O'Connell WA, Talbot EA, George JW, Kreuzer KN. Identification, Cloning, and Expression of the Escherichia Coli Pyrazinamidase and Nicotinamidase Gene, Pnca. *Antimicrob Agents Chemother.* 1996; 40:1426-1431.

82. Talbot EA, Williams DL, Frothingham R. PCR Identification of Mycobacterium Bovis Bcg. *J Clin Microbiol.* 1997; 35:566-569.

83. Velaz-Faircloth M, Cobb AJ, Horstman AL, Talbot EA, Henry SC, Frothingham R. Protection against Mycobacterium Avium by DNA Vaccines Expressing Mycobacterial Antigens as Fusion Proteins with Green Fluorescent Protein. *Infect Immun.* 1999; 67:4243-4250.

84. Teixeira L, Perkins MD, Johnson JL, et al. Infection and Disease among Household Contacts of Patients with Multidrug-Resistant Tuberculosis. *Int J Tuberc Lung Dis.* 2001; 5:321-328.

85. Talbot EA, Kenyon TA, Halabi S, Moeti TL, More K, Binkin NJ. Knowledge, Attitudes and Beliefs Regarding Tuberculosis Preventive Therapy for HIV-Infected Persons, Botswana, 1999. *Int J Tuberc Lung Dis.* 2000; 4:1156-1163.

86. Talbot EA, Moore M, McCray E, Binkin NJ. Tuberculosis among Foreign-Born Persons in the United States, 1993-1998. *JAMA.* 2000; 284:2894-2900.

1. Campbell WE, High M, Kennedy EJ. *Foundations for Excellence : 75 Years of Duke Medicine.* Durham, N.C.: Duke University Medical Center Library; 2006.

2. Bridges RM. Dr. Joseph Willis Beard: A Mighty Man, 1902-1983. *N C Med J.* 1985; 46:303-309.

3. Handler P, Wyngaarden JB. The Bio-Medical Research Training Program of Duke University. *J Med Educ.* 1961; 36:1587-1594.

4. Wyngaarden JB. The Clinical Investigator as an Endangered Species. *N Engl J Med.* 1979; 301:1254-1259.

5. D. Bernard Amos Interviewed by James F. Gifford. Digital Recording, 7 May, 1992. Durham, North Carolina: Duke University Medical Center Archives, Oral History Collection; 1992.

6. Wolfgang Karl Joklik Interviewed by Jessica Roseberry, May 15, 2007. Durham, NC: Duke University Medical Center Archives, Biographical Files Collection; 2007.

7. Hayes EC, Lee PW, Miller SE, Joklik WK. The Interaction of a Series of Hybridoma IgGs with Reovirus Particles. Demonstration That the Core Protein Lambda 2 Is Exposed on the Particle Surface. *Virology.* 1981; 108:147-155.

8. Joklik WK. The Molecular Basis of the Viral Eclipse Phase. *Prog Med Virol.* 1965; 7:44-96.

9. Joklik WK, Roner MR. Molecular Recognition in the Assembly of the Segmented Reovirus Genome. *Prog Nucleic Acid Res Mol Biol.* 1996; 53:249-281.

10. Lai MH, Joklik WK. The Induction of Interferon by Temperature-Sensitive Mutants of Reovirus, UV-Irradiated Reovirus, and Subviral Reovirus Particles. *Virology.* 1973; 51:191-204.

11. McCrae MA, Joklik WK. The Nature of the Polypeptide Encoded by Each of the 10 Double-Stranded RNA Segments of Reovirus Type 3. *Virology.* 1978; 89:578-593.

12. Roner MR, Lin PN, Nepluev I, Kong LJ, Joklik WK. Identification of Signals Required for the Insertion of Heterologous Genome Segments into the Reovirus Genome. *Proc Natl Acad Sci U S A.* 1995; 92:12362-12366.

13. Smith RE, Zweerink HJ, Joklik WK. Polypeptide Components of Virions, Top Component and Cores of Reovirus Type 3. *Virology.* 1969; 39:791-810.

14. Starnes MC, Joklik WK. Reovirus Protein Lambda 3 Is a Poly(C)-Dependent Poly(G) Polymerase. *Virology.* 1993; 193:356-366.

15. Becker Y, Joklik WK. Messenger RNA in Cells Infected with Vaccinia Virus. *Proc Natl Acad Sci U S A.* 1964; 51:577-585.

16. Pickup DJ, Ink BS, Hu W, Ray CA, Joklik WK. Hemorrhage in Lesions Caused by Cowpox Virus Is Induced by a Viral Protein That Is Related to Plasma Protein Inhibitors of Serine Proteases. *Proc Natl Acad Sci U S A*. 1986; 83:7698-7702.

17. Joklik WK, Moss B, Fields BN, Bishop DH, Sandakhchiev LS. Why the Small-pox Virus Stocks Should Not Be Destroyed. *Science*. 1993; 262:1225-1226.

18. Stone KR, Smith RE, Joklik WK. Changes in Membrane Polypeptides That Occur When Chick Embryo Fibroblasts and NRK Cells Are Transformed with Avian Sarcoma Viruses. *Virology*. 1974; 58:86-100.

19. Hizi A, Joklik WK. The Beta Subunit of the DNA Polymerase of Avian Sar-coma Virus Strain B77 Is a Phosphoprotein. *Virology*. 1977; 78:571-575.

20. Hizi A, Joklik WK. RNA-Dependent DNA Polymerase of Avian Sarcoma Virus B77. I. Isolation and Partial Characterization of the Alpha, Beta2, and Alphabeta Forms of the Enzyme. *J Biol Chem*. 1977; 252:2281-2289.

21. Hizi A, Leis JP, Joklik WK. The RNA-Dependent DNA Polymerase of Avian Sarcoma Virus B77. Binding of Viral and Nonviral Ribonucleic Acids to the Alpha, Beta2, and Alphabeta Forms of the Enzyme. *J Biol Chem*. 1977; 252:6878-6884.

22. Hizi A, Leis JP, Joklik WK. RNA-Dependent DNA Polymerase of Avian Sarcoma Virus B77. II. Comparison of the Catalytic Properties of the Alpha, Beta2, and Alphabeta Enzyme Forms. *J Biol Chem*. 1977; 252:2290-2295.

23. Hizi A, McCrae MA, Joklik WK. Studies on the Amino Acid Sequence Con-tent of Proteins Specified by the Gag and Pol Genes of Avian Sarcoma Virus B77. *Virology*. 1978; 89:272-284.

24. Bonar RA, Sverak L, Bolognesi DP, Langlois AJ, Beard D, Beard JW. Ribonu-cleic Acid Components of BAI Strain a (Myeloblastosis) Avian Tumor Virus. *Cancer Res*. 1967; 27:1138-1157.

25. Bauer H, Bolognesi DP. Polypeptides of Avian RNA Tumor Viruses. II. Sero-logical Characterization. *Virology*. 1970; 42:1113-1126.

26. Bolognesi DP, Bauer H, Gelderblom H, Huper G. Polypeptides of Avian RNA Tumor Viruses. IV. Components of the Viral Envelope. *Virology*. 1972; 47:551-566.

27. Bolognesi DP, Gelderblom H, Bauer H, Molling K, Huper G. Polypeptides of Avian RNA Tumor Viruses. V. Analysis of the Virus Core. *Virology*. 1972; 47:567-578.

28. Bolognesi DP, Montelaro RC, Frank H, Schafer W. Assembly of Type C On-cornaviruses: A Model. *Science*. 1978; 199:183-186.

29. Schafer W, Schwarz H, Thiel HJ, Wecker E, Bolognesi DP. Properties of Mouse Leukemia Viruses. XIII. Serum Therapy of Virus-Induced Murine Leukemias. *Virology*. 1976; 75:401-418.

30. Schafer W, Bolognesi DP, de Noronha F, et al. Immuno-Prophylaxis and -Therapy of C-Type Oncorna Viral Diseases in Mice and Cats. *Med Microbiol Immunol.* 1977; 164:217-229.

31. Schafer W, Schwarz H, Thiel HJ, Fischinger PJ, Bolognesi DP. Properties of Mouse Leukemia Viruses. Xiv. Prevention of Spontaneous AKR Leukemia by Treatment with Group-Specific Antibody against the Major Virus Gp71 Glycoprotein. *Virology.* 1977; 83:207-210.

32. Poiesz BJ, Ruscetti FW, Gazdar AF, Bunn PA, Minna JD, Gallo RC. Detection and Isolation of Type C Retrovirus Particles from Fresh and Cultured Lymphocytes of a Patient with Cutaneous T-Cell Lymphoma. *Proc Natl Acad Sci U S A.* 1980; 77:7415-7419.

33. Haynes BF, Miller SE, Palker TJ, et al. Identification of Human T Cell Leukemia Virus in a Japanese Patient with Adult T Cell Leukemia and Cutaneous Lymphomatous Vasculitis. *Proc Natl Acad Sci U S A.* 1983; 80:2054-2058.

34. Markert ML, Boeck A, Hale LP, et al. Transplantation of Thymus Tissue in Complete Digeorge Syndrome. *N Engl J Med.* 1999; 341:1180-1189.

35. Markert ML, Kostyu DD, Ward FE, et al. Successful Formation of a Chimeric Human Thymus Allograft Following Transplantation of Cultured Postnatal Human Thymus. *J Immunol.* 1997; 158:998-1005.

36. Lapierre D. Beyond Love. Warner Books Inc; 1991:277-380.

37. O'Reilly B, Field NE. The inside Story of the AIDS Drug. *Fortune.* Vol 122 1990:112-129.

38. Noland T. David W. Barry. *Yale Medicine* 1997:14-19.

39. Ranii D. David W. Barry Obituary. *The News & Observer* January 30, 2002.

40. Durack DT, Lukes AS, Bright DK. New Criteria for Diagnosis of Infective Endocarditis: Utilization of Specific Echocardiographic Findings. Duke Endocarditis Service. *Am J Med.* 1994; 96:200-209.

41. Durack DT, Ackerman SJ, Loegering DA, Gleich GJ. Purification of Human Eosinophil-Derived Neurotoxin. *Proc Natl Acad Sci U S A.* 1981; 78:5165-5169.

42. Perfect JR, Lang SD, Durack DT. Chronic Cryptococcal Meningitis: A New Experimental Model in Rabbits. *Am J Pathol.* 1980; 101:177-194.

43. Snyderman R, Durack DT, McCarty GA, Ward FE, Meadows L. Deficiency of the Fifth Component of Complement in Human Subjects. Clinical, Genetic and Immunologic Studies in a Large Kindred. *Am J Med.* 1979; 67:638-645.

44. Durack DT, Sumi SM, Klebanoff SJ. Neurotoxicity of Human Eosinophils. *Proc Natl Acad Sci U S A.* 1979; 76:1443-1447.

45. Durack DT, Petersdorf RG, Beeson PB. Penicillin Prophylaxis of Experimental S. Viridans Endocarditis. *Trans Assoc Am Physicians.* 1972; 85:222-230.

46. Durack DT, Beeson PB. Protective Role of Complement in Experimental Escherichia Coli Endocarditis. *Infect Immun.* 1977; 16:213-217.

47. Gottlieb MS, Schroff R, Schanker HM, et al. Pneumocystis Carinii Pneumonia and Mucosal Candidiasis in Previously Healthy Homosexual Men: Evidence of a New Acquired Cellular Immunodeficiency. *N Engl J Med*. 1981; 305:1425-1431.

48. Friedman-Kien. Pneumocystis Pneumonia. *MMWR Morb Mortal Wkly Rep*. 1981; 30:250-252.

49. Masur H, Michelis MA, Greene JB, et al. An Outbreak of Community-Acquired Pneumocystis Carinii Pneumonia: Initial Manifestation of Cellular Immune Dysfunction. *N Engl J Med*. 1981; 305:1431-1438.

50. Siegal FP, Lopez C, Hammer GS, et al. Severe Acquired Immunodeficiency in Male Homosexuals, Manifested by Chronic Perianal Ulcerative Herpes Simplex Lesions. *N Engl J Med*. 1981; 305:1439-1444.

51. Durack DT. Opportunistic Infections and Kaposi's Sarcoma in Homosexual Men. *N Engl J Med*. 1981; 305:1465-1467.

52. Kilpatrick JJ. Grants for Safe Sodomy. *Chicago Sun-Times*. November 14, 1987.

53. Peterson KM. *Sing Me to Heaven*. Grand Rapids, Michigan: Brazos Press, A Division of Baker Book House Company; 2003.

54. Lekus IK. `Health Care, the AIDS Crisis, and Politics of Community: The North Carolina Lesbian and Gay Health Project, 1982-1996. In: Black A, ed. *Modern American Queer History*. Philadelphia, PA: Temple University Press; 2001.

55. Barre-Sinoussi F, Chermann JC, Rey F, et al. Isolation of a T-Lymphotropic Retrovirus from a Patient at Risk for Acquired Immune Deficiency Syndrome (AIDS). *Science*. 1983; 220:868-871.

56. Nobelprize.org. The Nobel Prize in Physiology or Medicine. 2008; http:www.nobelprize.org/nobel_prizes/medicine/laureates/2008. Accessed July 14, 2014.

57. Gallo RC, Salahuddin SZ, Popovic M, et al. Frequent Detection and Isolation of Cytopathic Retroviruses (HTLV-III) from Patients with AIDS and at Risk for AIDS. *Science*. 1984; 224:500-503.

58. Popovic M, Sarngadharan MG, Read E, Gallo RC. Detection, Isolation, and Continuous Production of Cytopathic Retroviruses (HTLV-III) from Patients with AIDS and Pre-AIDS. *Science*. 1984; 224:497-500.

59. White GC, 2nd, Matthews TJ, Weinhold KJ, et al. HTLV-III Seroconversion Associated with Heat-Treated Factor VIII Concentrate. *Lancet*. 1986; 1:611-612.

60. Haynes B, White G, Palker T, et al. Clinical and Virologic Correlates of Serum Antibodies to Human T-Cell Lymphotropic Virus (HTLV) Type-1, Type-II, and Type-III in Hemophilia. *Clin Res*. 1985; 33:A342-A342.

61. Duesberg P. HIV Is Not the Cause of AIDS. *Science*. 1988; 241:514, 517.

62. Mitsuya H, Weinhold KJ, Furman PA, et al. 3'-Azido-3'-Deoxythymidine (Bw A509u): An Antiviral Agent That Inhibits the Infectivity and Cytopathic Effect of Human T-Lymphotropic Virus Type III/Lymphadenopathy-Associated Virus in Vitro. *Proc Natl Acad Sci U S A*. 1985; 82:7096-7100.

63. Yarchoan R, Klecker RW, Weinhold KJ, et al. Administration of 3'-Azido-3'-Deoxythymidine, an Inhibitor of HTLV-III/LAV Replication, to Patients with AIDS or AIDS-Related Complex. *Lancet*. 1986; 1:575-580.

64. Fischl MA, Richman DD, Grieco MH, et al. The Efficacy of Azidothymidine (AZT) in the Treatment of Patients with AIDS and AIDS-Related Complex. A Double-Blind, Placebo-Controlled Trial. *N Engl J Med*. 1987; 317:185-191.

65. Brook I. Approval of Zidovudine (AZT) for Acquired Immunodeficiency Syndrome. A Challenge to the Medical and Pharmaceutical Communities. *JAMA*. 1987; 258:1517.

66. Fischl MA, Parker CB, Pettinelli C, et al. A Randomized Controlled Trial of a Reduced Daily Dose of Zidovudine in Patients with the Acquired Immunodeficiency Syndrome. The AIDS Clinical Trials Group. *N Engl J Med*. 1990; 323:1009-1014.

67. Volberding PA, Lagakos SW, Koch MA, et al. Zidovudine in Asymptomatic Human Immunodeficiency Virus Infection. A Controlled Trial in Persons with Fewer Than 500 CD4-Positive Cells Per Cubic Millimeter. The AIDS Clinical Trials Group of the National Institute of Allergy and Infectious Diseases. *N Engl J Med*. 1990; 322:941-949.

68. Cooper E. New Drug Application 19-655 Supplement S-011. In: Food and Drug Administration, ed. Vol New Drug Application 19-655 Supplement S-011. Washington, DC: Food and Drug Administration; 1990:1-2.

69. Hamilton JD, Hartigan PM, Simberkoff MS, et al. A Controlled Trial of Early Versus Late Treatment with Zidovudine in Symptomatic Human Immunodeficiency Virus Infection. Results of the Veterans Affairs Cooperative Study. *N Engl J Med*. 1992; 326:437-443.

70. Aboulker JP, Swart AM. Preliminary Analysis of the Concorde Trial. Concorde Coordinating Committee. *Lancet*. 1993; 341:889-890.

71. St Clair MH, Hartigan PM, Andrews JC, Vavro CL, Simberkoff MS, Hamilton JD. Zidovudine Resistance, Syncytium-Inducing Phenotype, and HIV Disease Progression in a Case-Control Study. The VA Cooperative Study Group. *J Acquir Immune Defic Syndr*. 1993; 6:891-897.

72. O'Brien WA, Hartigan PM, Martin D, et al. Changes in Plasma HIV-1 RNA and CD4+ Lymphocyte Counts and the Risk of Progression to AIDS. Veterans Affairs Cooperative Study Group on AIDS. *N Engl J Med*. 1996; 334:426-431.

73. O'Brien WA, Hartigan PM, Daar ES, Simberkoff MS, Hamilton JD. Changes in Plasma HIV RNA Levels and CD4+ Lymphocyte Counts Predict Both

Response to Antiretroviral Therapy and Therapeutic Failure. VA Cooperative Study Group on AIDS. *Ann Intern Med.* 1997; 126:939-945.

74. McKinney RE, Jr., Wilfert C. Growth as a Prognostic Indicator in Children with Human Immunodeficiency Virus Infection Treated with Zidovudine. AIDS Clinical Trials Group Protocol 043 Study Group. *J Pediatr.* 1994; 125:728-733.

75. McKinney RE, Jr., Wilfert CM. Lymphocyte Subsets in Children Younger Than 2 Years Old: Normal Values in a Population at Risk for Human Immunodeficiency Virus Infection and Diagnostic and Prognostic Application to Infected Children. *Pediatr Infect Dis J.* 1992; 11:639-644.

76. Valentine ME, Jackson CR, Vavro C, et al. Evaluation of Surrogate Markers and Clinical Outcomes in Two-Year Follow-up of Eighty-Six Human Immunodeficiency Virus-Infected Pediatric Patients. *Pediatr Infect Dis J.* 1998; 17:18-23.

77. McKinney RE, Jr. Antiviral Therapy for Human Immunodeficiency Virus Infection in Children. *Pediatr Clin North Am.* 1991; 38:133-151.

78. McKinney RE, Jr., Pizzo PA, Scott GB, et al. Safety and Tolerance of Intermittent Intravenous and Oral Zidovudine Therapy in Human Immunodeficiency Virus-Infected Pediatric Patients. Pediatric Zidovudine Phase I Study Group. *J Pediatr.* 1990; 116:640-647.

79. Tudor-Williams G, St Clair MH, McKinney RE, et al. HIV-1 Sensitivity to Zidovudine and Clinical Outcome in Children. *Lancet.* 1992; 339:15-19.

80. Brady MT, McGrath N, Brouwers P, et al. Randomized Study of the Tolerance and Efficacy of High- Versus Low-Dose Zidovudine in Human Immunodeficiency Virus-Infected Children with Mild to Moderate Symptoms (AIDS Clinical Trials Group 128). Pediatric AIDS Clinical Trials Group. *J Infect Dis.* 1996; 173:1097-1106.

81. Pizzo PA, Wilfert CM. Treatment Considerations for Children with Human Immunodeficiency Virus Infection. *Pediatr Infect Dis J.* 1990; 9:690-699.

82. Connor EM, Sperling RS, Gelber R, et al. Reduction of Maternal-Infant Transmission of Human Immunodeficiency Virus Type 1 with Zidovudine Treatment. Pediatric AIDS Clinical Trials Group Protocol 076 Study Group. *N Engl J Med.* 1994; 331:1173-1180.

83. Sperling RS, Shapiro DE, Coombs RW, et al. Maternal Viral Load, Zidovudine Treatment, and the Risk of Transmission of Human Immunodeficiency Virus Type 1 from Mother to Infant. Pediatric AIDS Clinical Trials Group Protocol 076 Study Group. *N Engl J Med.* 1996; 335:1621-1629.

84. Shapiro DE, Sperling RS, Coombs RW. Effect of Zidovudine on Perinatal HIV-1 Transmission and Maternal Viral Load. Pediatric AIDS Clinical Trials Group 076 Study Group. *Lancet.* 1999; 354:156; author reply 157-158.

85. Furman PA, Fyfe JA, St Clair MH, et al. Phosphorylation of 3'-Azido-3'-Deoxythymidine and Selective Interaction of the 5'-Triphosphate with Human Immunodeficiency Virus Reverse Transcriptase. *Proc Natl Acad Sci U S A.* 1986; 83:8333-8337.

86. Surbone A, Yarchoan R, McAtee N, et al. Treatment of the Acquired Immunodeficiency Syndrome (AIDS) and AIDS-Related Complex with a Regimen of 3'-Azido-2',3'-Dideoxythymidine (Azidothymidine or Zidovudine) and Acyclovir. A Pilot Study. *Ann Intern Med.* 1988; 108:534-540.

87. McKinney RE, Jr., Maha MA, Connor EM, et al. A Multicenter Trial of Oral Zidovudine in Children with Advanced Human Immunodeficiency Virus Disease. The Protocol 043 Study Group. *N Engl J Med.* 1991; 324:1018-1025.

88. McKinney RE, Jr., Johnson GM, Stanley K, et al. A Randomized Study of Combined Zidovudine-Lamivudine Versus Didanosine Monotherapy in Children with Symptomatic Therapy-Naive HIV-1 Infection. The Pediatric AIDS Clinical Trials Group Protocol 300 Study Team. *J Pediatr.* 1998; 133:500-508.

89. Englund JA, Baker CJ, Raskino C, et al. Zidovudine, Didanosine, or Both as the Initial Treatment for Symptomatic HIV-Infected Children. AIDS Clinical Trials Group (ACTG) Study 152 Team. *N Engl J Med.* 1997; 336:1704-1712.

90. Palker TJ, Scearce RM, Miller SE, et al. Monoclonal Antibodies against Human T Cell Leukemia-Lymphoma Virus (HTLV) P24 Internal Core Protein. Use as Diagnostic Probes and Cellular Localization of HTLV. *J Exp Med.* 1984; 159:1117-1131.

91. Haynes BF, Robert-Guroff M, Metzgar RS, et al. Monoclonal Antibody against Human T Cell Leukemia Virus P19 Defines a Human Thymic Epithelial Antigen Acquired During Ontogeny. *J Exp Med.* 1983; 157:907-920.

92. Palker TJ, Bolognesi DP, Haynes BF. Human T-Cell Leukemia/Lymphoma Virus: Studies of Host-Virus Interaction. *Curr Top Microbiol Immunol.* 1985; 115:247-266.

93. Palker TJ, Matthews TJ, Langlois A, et al. Polyvalent Human Immunodeficiency Virus Synthetic Immunogen Comprised of Envelope Gp120 T Helper Cell Sites and B Cell Neutralization Epitopes. *J Immunol.* 1989; 142:3612-3619.

94. Palker TJ, Clark ME, Langlois AJ, et al. Type-Specific Neutralization of the Human Immunodeficiency Virus with Antibodies to Env-Encoded Synthetic Peptides. *Proc Natl Acad Sci U S A.* 1988; 85:1932-1936.

95. Weinhold KJ, Lyerly HK, Stanley SD, Austin AA, Matthews TJ, Bolognesi DP. HIV-1 Gp120-Mediated Immune Suppression and Lymphocyte Destruction in the Absence of Viral Infection. *J Immunol.* 1989; 142:3091-3097.

96. Ferrari G, Humphrey W, McElrath MJ, et al. Clade B-Based HIV-1 Vaccines Elicit Cross-Clade Cytotoxic T Lymphocyte Reactivities in Uninfected Volunteers. *Proc Natl Acad Sci U S A.* 1997; 94:1396-1401.

97. Hart MK, Palker TJ, Matthews TJ, et al. Synthetic Peptides Containing T and B Cell Epitopes from Human Immunodeficiency Virus Envelope Gp120 Induce Anti-HIV Proliferative Responses and High Titers of Neutralizing Antibodies in Rhesus Monkeys. *J Immunol*. 1990; 145:2677-2685.

98. Bartlett JA, Wasserman SS, Hicks CB, et al. Safety and Immunogenicity of an Hla-Based HIV Envelope Polyvalent Synthetic Peptide Immunogen. DATRI 010 Study Group. Division of AIDS Treatment Research Initiative. *AIDS*. 1998; 12:1291-1300.

99. Staats HF, Ennis FA, Jr. Il-1 Is an Effective Adjuvant for Mucosal and Systemic Immune Responses When Coadministered with Protein Immunogens. *J Immunol*. 1999; 162:6141-6147.

100. Staats HF, Montgomery SP, Palker TJ. Intranasal Immunization Is Superior to Vaginal, Gastric, or Rectal Immunization for the Induction of Systemic and Mucosal Anti-HIV Antibody Responses. *AIDS Res Hum Retroviruses*. 1997; 13:945-952.

101. Staats HF, Nichols WG, Palker TJ. Mucosal Immunity to HIV-1: Systemic and Vaginal Antibody Responses after Intranasal Immunization with the HIV-1 C4/V3 Peptide T1SP10 MN(A). *J Immunol*. 1996; 157:462-472.

102. Weinberg JB, Spiegel RA, Blazey DL, et al. Human T-Cell Lymphotropic Virus I and Adult T-Cell Leukemia: Report of a Cluster in North Carolina. *Am J Med*. 1988; 85:51-58.

103. Gao F, Robertson DL, Morrison SG, et al. The Heterosexual Human Immunodeficiency Virus Type 1 Epidemic in Thailand Is Caused by an Intersubtype (A/E) Recombinant of African Origin. *J Virol*. 1996; 70:7013-7029.

104. Gao F, Bailes E, Robertson DL, et al. Origin of HIV-1 in the Chimpanzee Pan Troglodytes Troglodytes. *Nature*. 1999; 397:436-441.

105. Wild C, Oas T, McDanal C, Bolognesi D, Matthews T. A Synthetic Peptide Inhibitor of Human Immunodeficiency Virus Replication: Correlation between Solution Structure and Viral Inhibition. *Proc Natl Acad Sci U S A*. 1992; 89:10537-10541.

106. Kilby JM, Hopkins S, Venetta TM, et al. Potent Suppression of HIV-1 Replication in Humans by T-20, a Peptide Inhibitor of Gp41-Mediated Virus Entry. *Nat Med*. 1998; 4:1302-1307.

107. Haynes BF, Markert ML, Sempowski GD, Patel DD, Hale LP. The Role of the Thymus in Immune Reconstitution in Aging, Bone Marrow Transplantation, and HIV-1 Infection. *Annu Rev Immunol*. 2000; 18:529-560.

108. Markert ML, Hicks CB, Bartlett JA, et al. Effect of Highly Active Antiretroviral Therapy and Thymic Transplantation on Immunoreconstitution in HIV Infection. *AIDS Res Hum Retroviruses*. 2000; 16:403-413.

109. Colvin RA, Garcia-Blanco MA. Unusual Structure of the Human Immuno-deficiency Virus Type 1 Trans-Activation Response Element. *J Virol.* 1992; 66:930-935.

110. Colvin RA, White SW, Garcia-Blanco MA, Hoffman DW. Structural Features of an RNA Containing the CUGGGA Loop of the Human Immunodefi-ciency Virus Type 1 Trans-Activation Response Element. *Biochemistry.* 1993; 32:1105-1112.

111. Hoffman DW, Colvin RA, Garcia-Blanco MA, White SW. Structural Features of the Trans-Activation Response RNA Element of Equine Infectious Anemia Virus. *Biochemistry.* 1993; 32:1096-1104.

112. Sune C, Hayashi T, Liu Y, Lane WS, Young RA, Garcia-Blanco MA. CA150, a Nuclear Protein Associated with the RNA Polymerase II Holoenzyme, Is Involved in Tat-Activated Human Immunodeficiency Virus Type 1 Tran-scription. *Mol Cell Biol.* 1997; 17:6029-6039.

113. Sune C, Garcia-Blanco MA. Transcriptional Cofactor CA150 Regulates RNA Polymerase II Elongation in a Tata-Box-Dependent Manner. *Mol Cell Biol.* 1999; 19:4719-4728.

114. Carty SM, Goldstrohm AC, Sune C, Garcia-Blanco MA, Greenleaf AL. Pro-tein-Interaction Modules That Organize Nuclear Function: FF Domains of CA150 Bind the Phosphoctd of RNA Polymerase II. *Proc Natl Acad Sci U S A.* 2000; 97:9015-9020.

115. Goldstrohm AC, Albrecht TR, Sune C, Bedford MT, Garcia-Blanco MA. The Transcription Elongation Factor CA150 Interacts with RNA Polymerase II and the Pre-mRNA Splicing Factor SF1. *Mol Cell Biol.* 2001; 21:7617-7628.

116. Bohjanen PR, Colvin RA, Puttaraju M, Been MD, Garcia-Blanco MA. A Small Circular Tar RNA Decoy Specifically Inhibits Tat-Activated HIV-1 Transcription. *Nucleic Acids Res.* 1996; 24:3733-3738.

117. Bohjanen PR, Liu Y, Garcia-Blanco MA. Tar RNA Decoys Inhibit Tat-Acti-vated HIV-1 Transcription after Preinitiation Complex Formation. *Nucleic Acids Res.* 1997; 25:4481-4486.

118. Garcia-Blanco MA, Cullen BR. Molecular Basis of Latency in Pathogenic Human Viruses. *Science.* 1991; 254:815-820.

119. Montefiori DC, Baba TW, Li A, Bilska M, Ruprecht RM. Neutralizing and In-fection-Enhancing Antibody Responses Do Not Correlate with the Differen-tial Pathogenicity of SIVmac239delta3 in Adult and Infant Rhesus Monkeys. *J Immunol.* 1996; 157:5528-5535.

120. Montefiori DC, Cornell RJ, Zhou JY, Zhou JT, Hirsch VM, Johnson PR. Complement Control Proteins, CD46, CD55, and CD59, as Common Surface Constituents of Human and Simian Immunodeficiency Viruses and Possible Targets for Vaccine Protection. *Virology.* 1994; 205:82-92.

121. Pilgrim AK, Pantaleo G, Cohen OJ, et al. Neutralizing Antibody Responses to Human Immunodeficiency Virus Type 1 in Primary Infection and Long-Term-Nonprogressive Infection. *J Infect Dis.* 1997; 176:924-932.

122. Bradney AP, Scheer S, Crawford JM, Buchbinder SP, Montefiori DC. Neutralization Escape in Human Immunodeficiency Virus Type 1-Infected Long-Term Nonprogressors. *J Infect Dis.* 1999; 179:1264-1267.

123. Langlois AJ, Desrosiers RC, Lewis MG, et al. Neutralizing Antibodies in Sera from Macaques Immunized with Attenuated Simian Immunodeficiency Virus. *J Virol.* 1998; 72:6950-6955.

124. Montefiori DC, Reimann KA, Wyand MS, et al. Neutralizing Antibodies in Sera from Macaques Infected with Chimeric Simian-Human Immunodeficiency Virus Containing the Envelope Glycoproteins of Either a Laboratory-Adapted Variant or a Primary Isolate of Human Immunodeficiency Virus Type 1. *J Virol.* 1998; 72:3427-3431.

125. Crawford JM, Earl PL, Moss B, et al. Characterization of Primary Isolate-Like Variants of Simian-Human Immunodeficiency Virus. *J Virol.* 1999; 73:10199-10207.

126. Staats HF, Bradney CP, Gwinn WM, et al. Cytokine Requirements for Induction of Systemic and Mucosal Ctl after Nasal Immunization. *J Immunol.* 2001; 167:5386-5394.

127. Bradney CP, Sempowski GD, Liao HX, Haynes BF, Staats HF. Cytokines as Adjuvants for the Induction of Anti-Human Immunodeficiency Virus Peptide Immunoglobulin G (IgG) and IgA Antibodies in Serum and Mucosal Secretions after Nasal Immunization. *J Virol.* 2002; 76:517-524.

128. Egan MA, Chong SY, Hagen M, et al. A Comparative Evaluation of Nasal and Parenteral Vaccine Adjuvants to Elicit Systemic and Mucosal HIV-1 Peptide-Specific Humoral Immune Responses in Cynomolgus Macaques. *Vaccine.* 2004; 22:3774-3788.

129. Cullen BR, Greene WC. Regulatory Pathways Governing HIV-1 Replication. *Cell.* 1989; 58:423-426.

130. Bohnlein E, Lowenthal JW, Siekevitz M, Ballard DW, Franza BR, Greene WC. The Same Inducible Nuclear Proteins Regulates Mitogen Activation of Both the Interleukin-2 Receptor-Alpha Gene and Type 1 HIV. *Cell.* 1988; 53:827-836.

131. Ballard DW, Bohnlein E, Lowenthal JW, Wano Y, Franza BR, Greene WC. HTLV-I Tax Induces Cellular Proteins That Activate the Kappa B Element in the Il-2 Receptor Alpha Gene. *Science.* 1988; 241:1652-1655.

132. Rimsky L, Hauber J, Dukovich M, et al. Functional Replacement of the HIV-1 Rev Protein by the HTLV-1 Rex Protein. *Nature.* 1988; 335:738-740.

133. Rimsky L, Dodon MD, Dixon EP, Greene WC. Trans-Dominant Inactivation of HTLV-I and HIV-1 Gene Expression by Mutation of the HTLV-I Rex Transactivator. *Nature*. 1989; 341:453-456.

134. Hammes SR, Dixon EP, Malim MH, Cullen BR, Greene WC. Nef Protein of Human Immunodeficiency Virus Type 1: Evidence against Its Role as a Transcriptional Inhibitor. *Proc Natl Acad Sci U S A*. 1989; 86:9549-9553.

135. Ballard DW, Walker WH, Doerre S, et al. The V-Rel Oncogene Encodes a Kappa B Enhancer Binding Protein That Inhibits NF-Kappa B Function. *Cell*. 1990; 63:803-814.

136. Ahmed YF, Gilmartin GM, Hanly SM, Nevins JR, Greene WC. The HTLV-I Rex Response Element Mediates a Novel Form of mRNA Polyadenylation. *Cell*. 1991; 64:727-737.

137. Malim MH, Hauber J, Le SY, Maizel JV, Cullen BR. The HIV-1 Rev Trans-Activator Acts through a Structured Target Sequence to Activate Nuclear Export of Unspliced Viral mRNA. *Nature*. 1989; 338:254-257.

138. Malim MH, Bohnlein S, Hauber J, Cullen BR. Functional Dissection of the HIV-1 Rev Trans-Activator--Derivation of a Trans-Dominant Repressor of Rev Function. *Cell*. 1989; 58:205-214.

139. Malim MH, Tiley LS, McCarn DF, Rusche JR, Hauber J, Cullen BR. HIV-1 Structural Gene Expression Requires Binding of the Rev Trans-Activator to Its RNA Target Sequence. *Cell*. 1990; 60:675-683.

140. Hwang SS, Boyle TJ, Lyerly HK, Cullen BR. Identification of the Envelope V3 Loop as the Primary Determinant of Cell Tropism in HIV-1. *Science*. 1991; 253:71-74.

141. Weinberg JB, Matthews TJ, Cullen BR, Malim MH. Productive Human Immunodeficiency Virus Type 1 (HIV-1) Infection of Nonproliferating Human Monocytes. *J Exp Med*. 1991; 174:1477-1482.

142. Madore SJ, Cullen BR. Genetic Analysis of the Cofactor Requirement for Human Immunodeficiency Virus Type 1 Tat Function. *J Virol*. 1993; 67:3703-3711.

143. Bieniasz PD, Grdina TA, Bogerd HP, Cullen BR. Recruitment of a Protein Complex Containing Tat and Cyclin T1 to Tar Governs the Species Specificity of HIV-1 Tat. *EMBO J*. 1998; 17:7056-7065.

144. Bieniasz PD, Grdina TA, Bogerd HP, Cullen BR. Recruitment of Cyclin T1/P-TEFb to an HIV Type 1 Long Terminal Repeat Promoter Proximal RNA Target Is Both Necessary and Sufficient for Full Activation of Transcription. *Proc Natl Acad Sci U S A*. 1999; 96:7791-7796.

145. Ferrari G, Berend C, Ottinger J, et al. Replication-Defective Canarypox (ALVAC) Vectors Effectively Activate Anti-Human Immunodeficiency Virus-1 Cytotoxic T Lymphocytes Present in Infected Patients: Implications for Antigen-Specific Immunotherapy. *Blood*. 1997; 90:2406-2416.

146. Ferrari G, King K, Rathbun K, et al. Il-7 Enhancement of Antigen-Driven Activation/Expansion of HIV-1-Specific Cytotoxic T Lymphocyte Precursors (CTLp). *Clin Exp Immunol.* 1995; 101:239-248.

147. Ferrari G, Ottinger J, Place C, Nigida SM, Jr., Arthur LO, Weinhold KJ. The Impact of HIV-1 Infection on Phenotypic and Functional Parameters of Cellular Immunity in Chimpanzees. *AIDS Res Hum Retroviruses.* 1993; 9:647-656.

148. Ferrari G, Place CA, Ahearne PM, et al. Comparison of Anti-HIV-1 ADCC Reactivities in Infected Humans and Chimpanzees. *J Acquir Immune Defic Syndr.* 1994; 7:325-331.

149. Toso JF, Chen CH, Mohr JR, et al. Oligoclonal CD8 Lymphocytes from Persons with Asymptomatic Human Immunodeficiency Virus (HIV) Type 1 Infection Inhibit HIV-1 Replication. *J Infect Dis.* 1995; 172:964-973.

150. Hamilton JD, Fitzwilliam JF, Cheung KS, Lang DJ. Eeffects of Murine Cytomegalovirus-Infection on the Immune-Response to a Tumor Allograft. *Rev Infect Dis.* 1979; 1:976-987.

151. Hamilton JD, Fitzwilliam JF, Cheung KS, Shelburne J, Lang DJ, Amos DB. Viral Infection-Homograft Interaction in a Murine Model. *J Clin Invest.* 1978; 62:1303-1312.

152. Hamilton JD, Seaworth BJ. Transmission of Latent Cytomegalo-Virus in a Murine Kidney Tissue-Transplantation Model. *Transplantation.* 1985; 39:290-296.

153. Henry SC, Hamilton JD. Detection of Murine Cytomegalovirus Immediate Early-1 Transcripts in the Spleens of Latently Infected Mice. *J Infect Dis.* 1993; 167:950-954.

154. Klotman ME, Henry SC, Greene RC, Brazy PC, Klotman PE, Hamilton JD. Detection of Mouse Cytomegalovirus Nucleic-Acid in Latently Infected Mice by in Vitro Enzymatic Amplification. *J Infect Dis.* 1990; 161:220-225.

155. Klotman ME, Starnes D, Hamilton JD. Tthe Source of Murine Cytomegalo-Virus in Mice Receiving Kidney Allografts. *J Infect Dis.* 1985; 152:1192-1196.

156. Szmuness W, Stevens CE, Harley EJ, et al. Hepatitis B Vaccine in Medical Staff of Hemodialysis Units: Efficacy and Subtype Cross-Protection. *N Engl J Med.* 1982; 307:1481-1486.

157. Stevens CE, Alter HJ, Taylor PE, Zang EA, Harley EJ, Szmuness W. Hepatitis B Vaccine in Patients Receiving Hemodialysis. Immunogenicity and Efficacy. *N Engl J Med.* 1984; 311:496-501.

158. Hamilton JD. Hepatitis-B Virus-Vaccine - an Analysis of Its Potential Use in Medical Workers. *J Am Med Assoc.* 1983; 250:2145-2150.

159. Seeff LB, Wright EC, Zimmerman HJ, et al. Type-B Hepatitis after Needle-Stick Exposure - Prevention with Hepatitis-B Immune Globulin - Final

Report of Veterans-Administration Cooperative Study. *Ann Intern Med.* 1978; 88:285-293.

160. O'Brien WA, Hartigan P, Hamilton JD. Viral Load and Response to Treatment of HIV - Reply. *N Engl J Med.* 1996; 334:1672-1673.

161. Babiker A, Bartlett J, Breckenridge A, et al. Human Immunodeficiency Virus Type 1 RNA Level and CD4 Count as Prognostic Markers and Surrogate End Points: A Meta-Analysis. *AIDS Res Hum Retroviruses.* 2000; 16:1123-1133.

162. Gordin FM, Hartigan PM, Klimas NG, Zolla-Pazner SB, Simberkoff MS, Hamilton JD. Delayed-Type Hypersensitivity Skin Tests Are an Independent Predictor of Human Immunodeficiency Virus Disease Progression. Department of Veterans Affairs Cooperative Study Group. *J Infect Dis.* 1994; 169:893-897.

163. Klotman ME, Kim S, Buchbinder A, DeRossi A, Baltimore D, Wong-Staal F. Kinetics of Expression of Multiply Spliced RNA in Early Human Immunodeficiency Virus Type 1 Infection of Lymphocytes and Monocytes. *Proc Natl Acad Sci U S A.* 1991; 88:5011-5015.

164. Berneman ZN, Gartenhaus RB, Reitz MS, Jr., et al. Expression of Alternatively Spliced Human T-Lymphotropic Virus Type I Px mRNA in Infected Cell Lines and in Primary Uncultured Cells from Patients with Adult T-Cell Leukemia/Lymphoma and Healthy Carriers. *Proc Natl Acad Sci U S A.* 1992; 89:3005-3009.

165. Bruggeman LA, Thomson MM, Nelson PJ, et al. Patterns of HIV-1 mRNA Expression in Transgenic Mice Are Tissue-Dependent. *Virology.* 1994; 202:940-948.

166. Winston JA, Bruggeman LA, Ross MD, et al. Nephropathy and Establishment of a Renal Reservoir of HIV Type 1 During Primary Infection. *N Engl J Med.* 2001; 344:1979-1984.

167. Chang TL, Mosoian A, Pine R, Klotman ME, Moore JP. A Soluble Factor(S) Secreted from CD8(+) T Lymphocytes Inhibits Human Immunodeficiency Virus Type 1 Replication through Stat1 Activation. *J Virol.* 2002; 76:569-581.

168. Chang TL, Francois F, Mosoian A, Klotman ME. CAF-Mediated Human Immunodeficiency Virus (HIV) Type 1 Transcriptional Inhibition Is Distinct from Alpha-Defensin-1 HIV Inhibition. *J Virol.* 2003; 77:6777-6784.

169. Marras D, Bruggeman LA, Gao F, et al. Replication and Compartmentalization of HIV-1 in Kidney Epithelium of Patients with HIV-Associated Nephropathy. *Nat Med.* 2002; 8:522-526.

170. Francois F, Klotman ME. Phosphatidylinositol 3-Kinase Regulates Human Immunodeficiency Virus Type 1 Replication Following Viral Entry in Primary CD4+ T Lymphocytes and Macrophages. *J Virol.* 2003; 77:2539-2549.

171. Herold BC, Scordi-Bello I, Cheshenko N, et al. Mandelic Acid Condensation Polymer: Novel Candidate Microbicide for Prevention of Human

Immunodeficiency Virus and Herpes Simplex Virus Entry. *J Virol.* 2002; 76:11236-11244.

172. Chang TL, Vargas J, Jr., DelPortillo A, Klotman ME. Dual Role of Alpha-Defensin-1 in Anti-HIV-1 Innate Immunity. *J Clin Invest.* 2005; 115:765-773.

173. Dukes CS, Matthews TJ, Lambert DM, Dreyer GB, Petteway SR, Weinberg JB. Potent Inhibition of HIV Type 1 Infection of Mononuclear Phagocytes by Synthetic Peptide Analogs of HIV Type 1 Protease Substrates. *AIDS Res Hum Retroviruses.* 1996; 12:777-782.

174. Dukes CS, Matthews TJ, Weinberg JB. Human Immunodeficiency Virus Type 1 Infection of Human Monocytes and Macrophages Does Not Alter Their Ability to Generate an Oxidative Burst. *J Infect Dis.* 1993; 168:459-462.

175. Dukes CS, Matthews TJ, Rivadeneira ED, Weinberg JB. Neopterin Production by HIV-1-Infected Mononuclear Phagocytes. *J Leukoc Biol.* 1994; 56:650-653.

176. Dukes CS, Yu Y, Rivadeneira ED, et al. Cellular CD44s as a Determinant of Human Immunodeficiency Virus Type 1 Infection and Cellular Tropism. *J Virol.* 1995; 69:4000-4005.

177. Cameron ML, Granger DL, Matthews TJ, Weinberg JB. Human Immunodeficiency Virus (HIV)-Infected Human Blood Monocytes and Peritoneal Macrophages Have Reduced Anticryptococcal Activity Whereas HIV-Infected Alveolar Macrophages Retain Normal Activity. *J Infect Dis.* 1994; 170:60-67.

178. Rivadeneira ED, Sauls DL, Yu Y, Haynes BF, Weinberg JB. Inhibition of HIV Type 1 Infection of Mononuclear Phagocytes by Anti-CD44 Antibodies. *AIDS Res Hum Retroviruses.* 1995; 11:541-546.

179. Weinberg JB, Sauls DL, Misukonis MA, Shugars DC. Inhibition of Productive Human Immunodeficiency Virus-1 Infection by Cobalamins. *Blood.* 1995; 86:1281-1287.

180. Weinberg JB, Shugars DC, Sherman PA, Sauls DL, Fyfe JA. Cobalamin Inhibition of HIV-1 Integrase and Integration of HIV-1 DNA into Cellular DNA. *Biochem Biophys Res Commun.* 1998; 246:393-397.

181. Hanly SM, Rimsky LT, Malim MH, et al. Comparative Analysis of the HTLV-I Rex and HIV-1 Rev Trans-Regulatory Proteins and Their RNA Response Elements. *Genes Dev.* 1989; 3:1534-1544.

182. Kim JH, Kaufman PA, Hanly SM, Rimsky LT, Greene WC. Rex Transregulation of Human T-Cell Leukemia Virus Type II Gene Expression. *J Virol.* 1991; 65:405-414.

183. Arima N, Molitor JA, Smith MR, Kim JH, Daitoku Y, Greene WC. Human T-Cell Leukemia Virus Type I Tax Induces Expression of the Rel-Related Family of Kappa B Enhancer-Binding Proteins: Evidence for a Pretranslational Component of Regulation. *J Virol.* 1991; 65:6892-6899.

184. Kim JH, Ratto S, Sitz KV, et al. Consequences of Stable Transduction and Antigen-Inducible Expression of the Human Interleukin-7 Gene on Tetanus-Toxoid-Specific T Cells. *Hum Gene Ther.* 1994; 5:1457-1466.

185. Kim JH, Mosca JD, Vahey MT, McLinden RJ, Burke DS, Redfield RR. Consequences of Human Immunodeficiency Virus Type 1 Superinfection of Chronically Infected Cells. *AIDS Res Hum Retroviruses.* 1993; 9:875-882.

186. Kim JH, McLinden RJ, Mosca JD, et al. Transcriptional Effects of Superinfection in HIV Chronically Infected T Cells: Studies in Dually Infected Clones. *J Acquir Immune Defic Syndr Hum Retrovirol.* 1996; 12:329-342.

187. Kim JH, Loveland JE, Sitz KV, et al. Expansion of Restricted Cellular Immune Responses to HIV-1 Envelope by Vaccination: Il-7 and Il-12 Differentially Augment Cellular Proliferative Responses to HIV-1. *Clin Exp Immunol.* 1997; 108:243-250.

188. Michael NL, Chang G, Kim JH, Birx DL. Dynamics of Cell-Free Viral Burden in HIV-1-Infected Patients. *J Acquir Immune Defic Syndr Hum Retrovirol.* 1997; 14:237-242.

189. Bohjanen PR, Johnson MD, Szczech LA, et al. Steady-State Pharmacokinetics of Lamivudine in Human Immunodeficiency Virus-Infected Patients with End-Stage Renal Disease Receiving Chronic Dialysis. *Antimicrob Agents Chemother.* 2002; 46:2387-2392.

190. Gottfredsson M, Bohjanen PR. Human Immunodeficiency Virus Type I as a Target for Gene Therapy. *Front Biosci.* 1997; 2:d619-634.

191. Jacobson MA, De Gruttola V, Reddy M, et al. The Predictive Value of Changes in Serologic and Cell Markers of HIV Activity for Subsequent Clinical Outcome in Patients with Asymptomatic HIV Disease Treated with Zidovudine. *AIDS.* 1995; 9:727-734.

192. Volberding PA, Lagakos SW, Grimes JM, et al. The Duration of Zidovudine Benefit in Persons with Asymptomatic HIV Infection. Prolonged Evaluation of Protocol 019 of the AIDS Clinical Trials Group. *JAMA.* 1994; 272:437-442.

193. Staszewski S, Hill AM, Bartlett J, et al. Reductions in HIV-1 Disease Progression for Zidovudine/Lamivudine Relative to Control Treatments: A Meta-Analysis of Controlled Trials. *AIDS.* 1997; 11:477-483.

194. Shadduck PP, Weinberg JB, Haney AF, et al. Lack of Enhancing Effect of Human Anti-Human Immunodeficiency Virus Type 1 (HIV-1) Antibody on HIV-1 Infection of Human Blood Monocytes and Peritoneal Macrophages. *J Virol.* 1991; 65:4309-4316.

195. Stine KC, Tyler DS, Stanley SD, Bartlett JA, Bolognesi DP, Weinhold KJ. The Effect of AZT on in Vitro Lymphokine-Activated Killer (LAK) Activity in Human Immunodeficiency Virus Type-1 (HIV-1) Infected Individuals. *Cell Immunol.* 1991; 136:165-172.

196. Chen CH, Weinhold KJ, Bartlett JA, Bolognesi DP, Greenberg ML. CD8+ T Lymphocyte-Mediated Inhibition of HIV-1 Long Terminal Repeat Transcription: A Novel Antiviral Mechanism. *AIDS Res Hum Retroviruses.* 1993; 9:1079-1086.

197. Cohen OJ, Pantaleo G, Holodniy M, et al. Decreased Human Immunodeficiency Virus Type 1 Plasma Viremia During Antiretroviral Therapy Reflects Downregulation of Viral Replication in Lymphoid Tissue. *Proc Natl Acad Sci U S A.* 1995; 92:6017-6021.

198. Hicks CB, Benson PM, Lupton GP, Tramont EC. Seronegative Secondary Syphilis in a Patient Infected with the Human Immunodeficiency Virus (HIV) with Kaposi Sarcoma. A Diagnostic Dilemma. *Ann Intern Med.* 1987; 107:492-495.

199. Hicks CB, Myers SA, Giner J. Resolution of Intractable Molluscum Contagiosum in a Human Immunodeficiency Virus-Infected Patient after Institution of Antiretroviral Therapy with Ritonavir. *Clin Infect Dis.* 1997; 24:1023-1025.

200. Fowler VG, Jr., Hicks CB, Kirkland KB. The Name Game: Lamivudine-Lamotrigine Dispensing Error Presenting as Human Immunodeficiency Virus-Associated Fever of Unknown Origin. *Int J STD AIDS.* 1999; 10:685-686.

201. Miller WC, Thielman NM, Swai N, et al. Diagnosis and Screening of HIV/ AIDS Using Clinical Criteria in Tanzanian Adults. *J Acquir Immune Defic Syndr Hum Retrovirol.* 1995; 9:408-414.

202. Miller WC, Thielman NM, Swai N, et al. Delayed-Type Hypersensitivity Testing in Tanzanian Adults with HIV Infection. *J Acquir Immune Defic Syndr Hum Retrovirol.* 1996; 12:303-308.

CITATIONS FOR CHAPTER 5

1. Ashhurst AP. The Centenary of Lister (1827-1927). A Tale of Sepsis and Antisepsis. *Ann Med Hist.* 1927; 9:205-221.

2. Selwyn S. Hospital Infection: The First 2500 Years. *J Hosp Infect.* 1991; 18 Suppl A:5-64.

3. Caton R. Two Lectures on the Temples and Ritual of Asklepios at Epidauros and Athens. Royal Institution of Great Britain; 1899; London, England.

4. Pringle J. *Observations on the Diseases of the Army, in Camp and Garrison.* London, England: Miller, Wilson and Payne; 1752:102-294.

5. Pringle J. *Observations on the Nature and Cure of Hospital and Jayl-Fevers. In a Letter to Dr. Mead.* London, England: A. Millar; 1750.

6. Pringle J. *Observations on the Diseases of the Army, in Camp and Garrison.* 4th ed. London: A. Millar and D. Wilson; 1764:263-267.

7. White C. *A Treatise on the Management of Pregnant and Lying-in Women.* London, England: Dilly; 1773.

8. Gordon A. *A Treatise on the Epidemic Puerperal Fever of Aberdeen.* London, England: Robinson; 1795.

9. Thomas RV. *The Modern Practice of Physic.* 6th ed. London, England: Longman, Hurst, Rees, Orme and Brown; 1819.

10. Holmes OW. The Contagiousness of Puerperal Fever (First Published in 1843). *Medical Classics.* 1936-1937; 1:211-243.

11. Semmelweiss IF. *The Etiology, the Concept and the Prophylaxis of Childbed Fever.* Birmingham, England: Classics of Medicine Library; 1861.

12. Brocklesby R. *Oeconomical and Medical Observations in Two Parts, from the Year 1758 to the Year 1763, Inclusive, Tending to the Improvement of Military Hospitals and to the Cure of Camp Diseases Incident to Soldiers.* London: Becket and De Hondt; 1764:53-91.

13. Simpson JY. On the Analogy between Puerperal and Surgical Fever. *Edinb Mon J Med Sci.* 1850; 11:414-429.

14. Simpson JY. Clinical Lectures. *Med Times Gaz.* 1859; 39:411-491.

15. Simpson JY. On the Communicability and Propagation of Puerperal Fever. *Edinb Mon J Med Sci.* 1851; 13:72-81.

16. Simpson JY. Our Existing System of Hospitalism and Its Effects. *Edinb Med J.* 1869; 14:1084-1115.

17. Simpson JY. Some Propositions on Hospitalism, by the Late Sir Jy Simpson. *Lancet.* 1870; 2:698-701.

18. Nightingale F. Notes on the Sanitary Condition of Hospitals, and on Defects in the Construction of Hospital Wards. *Trans Natl Assoc Promot Soc Sci.* 1858:462-482.

19. Lister J. On a New Method of Treating Compound Fracture, Abscess, Etc. With Observarions on the Conditions of Suppuration and on the Antiseptic Principle in the Practice of Surgery. *Lancet.* 1867; 1:326-509, 357–359, 387–389, 509–511.

20. Lister J. An Address on the Effect of the Antiseptic Treatment Upon the General Salubrity of Surgical Hospitals. *Br Med J.* 1875; 2:769-772.

21. Ritchie J. *History of the Laboratory of the Royal College of Physicians of Edinburgh.* Edinburgh, Scotland: Royal College of Physicians; 1953.

22. Williams R, Blowers R, Garrod LP, Shooter RA. *Hospital Infection.* London: Lloyd-Luke (Medical Books); 1960.

23. Gardner A, Stamp M, Bowgen JA, Moore B. The Infection Control Sister. *Lancet.* 1962; 2:710-712.

24. Lowbury EJL, Ayliffe GAJ, Geddes AM, Williams JD. *Control of Hospital Infection: A Practical Handbook.* 2nd ed. London: Chapman and Hall; 1981.

25. Meers PD, Ayliffe GAJ, Emmerson AM, et al. Report on the National Survey of Infection in Hospitals. *J Hosp Infect.* 1980; 2:1-51.

26. Selwyn S. Less Surgical and More Medical Infections in Hospital? *Br Med J.* 1982; 284:1895-1896.

27. Selwyn S. Hard Facts About Hospital Infections(a Review of Hosptial-Acquired Infection: Principles and Preventionby G. A. F. Ayliffe, B. J. Collins, and L. J. Taylor. Wright Psg, 1982) *Br Med J.* 1982; 285:1109.

28. Meers PD. The Organization of Infection Control in Hospitals. *J Hosp Infect.* 1980; 1:187-191.

29. Medical Research Council. *The Prevention of 'Hospital Infection' of Wounds. Mrc War Memorandum No. 6.* London HMSO; 1941.

30. Medical Research Council. *The Control of Cross-Infection in Hospitals.* London, England: Medical Research Council;1944.

31. Haley RW, Culver DH, White JW, et al. The Efficacy of Infection Surveillance and Control Programs in Preventing Nosocomial Infections in Us Hospitals. *Am J Epidemiol.* 1985; 121:182-205.

32. Estes H. Department of Family Medicine. 2010.

33. Jackson G. Employee Health. 2011.

34. Blumberg BS, Melartin L, Guint RA, Werner B. Family Studies of a Human Serum Isoantigen System (Australia Antigen). *Am J Hum Genet.* 1966; 18:594-608.

35. Hamilton JD, Hatch MH, Gutman RA. Serological Evidence of Cross Infection in a Dialysis Unit Hepatitis-B Epidemic. *Kidney Int.* 1974; 6:118-122.

36. Seeff LB, Wright EC, Zimmerman HJ, et al. Type-B Hepatitis after Needle-Stick Exposure - Prevention with Hepatitis-B Immune Globulin - Final Report of Veterans-Administration Cooperative Study. *Ann Intern Med.* 1978; 88:285-293.

37. Szmuness W, Stevens CE, Harley EJ, et al. Hepatitis B Vaccine in Medical Staff of Hemodialysis Units: Efficacy and Subtype Cross-Protection. *N Engl J Med.* 1982; 307:1481-1486.

38. Stevens CE, Alter HJ, Taylor PE, Zang EA, Harley EJ, Szmuness W. Hepatitis B Vaccine in Patients Receiving Hemodialysis. Immunogenicity and Efficacy. *N Engl J Med.* 1984; 311:496-501.

39. Hamilton JD. Hepatitis-B Virus-Vaccine - an Analysis of Its Potential Use in Medical Workers. *J Am Med Assoc.* 1983; 250:2145-2150.

40. Centers for Disease Control. Recommendations for Prevention of HIV Transmission in Healthcare Settings. *MMWR Morb Mortal Wkly Rep.* 1987; 36:3S-18S.

41. OSHA. Occupational Exposure to Blood-Borne Pathogens; Final Rule. Federal Register, vol. 56: US Dept. of Labor; 1991:64175-64182.

42. Sexton DJ. Letter to John D. Hamilton. Durham, NC: Duke University; 2011.

43. Milano CA, Kesler K, Archibald N, Sexton DJ, Jones RH. Mediastinitis after Coronary Artery Bypass Graft Surgery. Risk Factors and Long-Term Survival. *Circulation*. 1995; 92:2245-2251.

44. Marr KA, Sexton DJ, Conlon PJ, Corey GR, Schwab SJ, Kirkland KB. Catheter-Related Bacteremia and Outcome of Attempted Catheter Salvage in Patients Undergoing Hemodialysis. *Ann Intern Med*. 1997; 127:275-280.

45. Fowler VG, Sanders LL, Sexton DJ, et al. Outcome of Staphylococcus Aureus Bacteremia According to Compliance with Infectious Diseases Specialist Recommendations: Experience with 244 Patients. *Clin Infect Dis*. 1998; 27:478-486.

46. Abramson MA, Sexton DJ. Nosocomial Methicillin-Resistant and Methicillin Susceptible Staphylococcus Aureus Primary Bacteremia: And What Costs? *Infect Control Hosp Epidemiol*. 1999; 20:4.

47. Kirkland KB, Briggs JP, Trivette SL, Wilkinson WE, Sexton DJ. The Impact of Surgical-Site Infections in the 1990s: Attributable Mortality, Excess Length of Hospitalization, and Extra Costs. *Infect Control Hosp Epidemiol*. 1999; 20:725-731.

CITATIONS FOR CHAPTER 6

1. D. Bernard Amos Interviewed by James F. Gifford. Digital Recording, 7 May, 1992. Durham, North Carolina: Duke University Medical Center Archives, Oral History Collection; 1992.

2. Abernathy RS. David Tillerson Smith. *N C Med J*. 1990; 51:473-481.

3. Conant NF, Smith DT, Callaway JL, eds. *Manual of Clinical Mycology*. 3rd ed. Philadelphia: W. B. Saunders Co; 1971.

4. Smith DT. Fungous Infections in the United States. *J Am Med Assoc*. 1949; 141:1223-1226.

5. Smith DT. Immunologic Types of Blastomycosis; a Report on 40 Cases. *Ann Intern Med*. 1949; 31:463-469.

6. Smith DT. Progressive Primary Tuberculosis in the Adult and Its Differentiation from Lymphomas and Mycotic Infections. *N Engl J Med*. 1949; 241:198-202.

7. Smith DT, Conant NF. Diagnosis of Mycotic Infections. II. Deep Mycoses of the Lungs and Other Internal Organs. *The Merck report*. 1956; 65:19-24.

8. Smith DT. Therapy for Pulmonary and Systemic Fungus Diseases. *Postgrad Med*. 1956; 20:18-25.

9. Mitchell TG. Origin, Foundation and Scope of Medical Mycology at Duke University Medical Center. Received by John Hamilton. Durham, NC 2011:1-11.

10. Conant NF. *Conant Curriculum Vitae, Bibliography*. Duke University;1974.

11. Conant NF. Laboratory Diagnosis of Pulmonary Mycoses. *Am Rev Tuberc.* 1950; 61:690-704.

12. Friedman L, Conant NF. Immunologic Studies on the Etiologic Agents of North and South American Blastomycosis. I. Comparison of Hypersensitivity Reactions. *Mycopathol Mycol Appl.* 1953; 6:310-316.

13. Friedman L, Conant NF. Immunologic Studies on the Etiologic Agents of North and South American Blastomycosis. II. Comparison of Serologic Reactions. *Mycopathol Mycol Appl.* 1953; 6:317-324.

14. Conant NF, Smith DT. Diagnosis of Mycotic Infections. I. Superficial Mycoses. *The Merck report.* 1955; 64:19-24.

15. Conant NF. The Role of the Mycologist in Medicine. *Mycologia.* 1967; 59:30-38.

16. Gehlbach SH, Hamilton JD, Conant NF. Coccidioidomycosis. An Occupational Disease in Cotton Mill Workers. *Arch Intern Med.* 1973; 131:254-255.

17. Smith DT, Conant NF. Diagnosis of Mycotic Infections. III. Deep Mycoses of Mucous Membranes, Skin, and Subcutaneous Tissues. *The Merck report.* 1956; 65:22-24.

18. Smith JG, Jr., Harris JS, Conant NF, Smith DT. An Epidemic of North American Blastomycosis. *J Am Med Assoc.* 1955; 158:641-646.

19. Downing JG, Conant NF. Medical Progress: Mycotic Infections. *N Engl J Med.* 1945; 233:153-161, 181-188.

20. Mazumdar M. History of Immunology. In: Paul W, ed. *Fundamental Immunology.* Philadelphia, PA: Lippincott Williams & Wilkins; 2003:23-46.

21. Tedder TF, Dawson JR. In Memoriam: D. Bernard Amos, April 16, 1923-May 15, 2003. *J Immunol.* 2003; 171:6316-6317.

22. Amos DB, Gorer PA, Mikulska ZB. An Analysis of an Antigenic System in the Mouse (the H-2 System). *Proc R Soc London, Ser B.* 1955; 144:369-380.

23. Amos DB. The Use of Simpliried Systems as an Aid to the Interpretation of Mechanisms of Graft Rejection. *Prog Allergy.* 1962; 6:468-538.

24. Amos DB, Peacocke N. Leucoagglutination. A Modified Technique and Preliminary Results of Absorption with Tissues. *Proceedings of the 9th Congress of the European Society of Haematology* Lisbon, Portugal: S. Karger; 1963:1132-1140.

25. Amos DB, Peacocke N. Leukoagglutination Technique. *Histocompatibility Testing 1965.* Copenhagen, Denmark: Munksgaard; 1965.

26. Stickel DL, Amos DB, Robinson RR, et al. Renal Transplantation with Donor Recipient Tissue-Matching: Preliminary Report of First Case in North Carolina. *North Caroline Medical Journal.* 1965; 26:379-383.

27. Bach FH, Amos DB. Hu-1 Major Histocompatibility Locus in Man. *Science.* 1967; 156:1506-1508.

28. Amos DB. Human Histocompatibility Locus Hl-A. *Science.* 1968; 159:659-660.

29. Amos DB. Genetic and Antigenetic Aspects of Human Histocompatibility Systems. *Adv Immunol.* 1969; 10:251-297.

30. Yunis EJ, Amos DB. Three Closely Linked Genetic Systems Relevant to Transplantation. *Proc Natl Acad Sci U S A.* 1971; 68:3031-3035.

31. Berke G, Amos DB. Mechanism of Lymphocyte-Mediated Cytolysis-Lmc Cycle and Its Role in Transplantation Immunity. *Transplant Rev.* 1973; 17:71-107.

32. Blumenthal MN, Amos DB, Noreen H. Genetic Mapping of Ir Locus in Man: Linkage to Second Locus of Hl-A. *Science.* 1974; 184:1301-1303.

33. Dasgupta JD, Cemach K, Dubey DP, Yunis EJ, Amos DB. The Role of Class I Histocompatibility Antigens in the Regulation of T-Cell Activation. *Proc Natl Acad Sci U S A.* 1987; 84:1094-1098.

34. Stickel DL, Amos DB, Zmijewski CM, Glenn JF, Robinson RR. Human Renal Transplantation with Donor Selection by Leukocyte Typing. *Transplantation.* 1967; 5:Suppl:1024-1029.

35. Stickel DL, Seigler HF, Amos DB, et al. Immunogenetics of Consanguineous Allografts in Man. II. Correlation of Renal Allografting with Hl-a Genotyping. *Ann Surg.* 1970; 172:160-179.

36. Seigler HF, Stickel DL, Ward FE, et al. Correlations of Test Skin Grafts and Renal Allograft Function in Human Subjects Genotyped for Hl-A. *Surgery.* 1970; 68:86-91.

37. Chen D-F. *Clinical Transplantation Immunology Testing.* Durham, NC: Duke University; 2012.

38. Tinckam K. Histocompatibility Methods. *Transplant Rev.* 2009; 23:80-93.

39. Buckley RH, Schiff SE, Schiff RI, et al. Hematopoietic Stem-Cell Transplantation for the Treatment of Severe Combined Immunodeficiency. *N Engl J Med.* 1999; 340:508-516.

40. Buckley RH, Schiff RI, Schiff SE, et al. Human Severe Combined Immunodeficiency: Genetic, Phenotypic, and Functional Diversity in One Hundred Eight Infants. *J Pediatr.* 1997; 130:378-387.

41. Buckley RH. Primary Immunodeficiency Diseases Due to Defects in Lymphocytes. *N Engl J Med.* 2000; 343:1313-1324.

42. Buckley RH, Wray BB, Belmaker EZ. Extreme Hyperimmunoglobulinemia E and Undue Susceptibility to Infection. *Pediatrics.* 1972; 49:59-70.

43. Buckley RH, Schiff SE, Sampson HA, et al. Development of Immunity in Human Severe Primary T Cell Deficiency Following Haploidentical Bone Marrow Stem Cell Transplantation. *J Immunol.* 1986; 136:2398-2407.

44. Puel A, Ziegler SF, Buckley RH, Leonard WJ. Defective Il7r Expression in T(-)B(+)Nk(+) Severe Combined Immunodeficiency. *Nat Genet.* 1998; 20:394-397.

45. Buckley RH, Dees SC, O'Fallon WM. Serum Immunoglobulins. I. Levels in Normal Children and in Uncomplicated Childhood Allergy. *Pediatrics.* 1968; 41:600-611.

46. Buckley RH, Dees SC, O'Fallon WM. Serum Immunoglobulins. II. Levels in Children Subject to Recurrent Infection. *Pediatrics.* 1968; 42:50-60.

47. Buckley RH, Fiscus SA. Serum Igd and Ige Concentrations in Immunodeficiency Diseases. *J Clin Invest.* 1975; 55:157-165.

48. Schiff SE, Kurtzberg J, Buckley RH. Studies of Human Bone Marrow Treated with Soybean Lectin and Sheep Erythrocytes: Stepwise Analysis of Cell Morphology, Phenotype and Function. *Clin Exp Immunol.* 1987; 68:685-693.

49. Ghory P, Schiff S, Buckley R. Appearance of Multiple Benign Paraproteins During Early Engraftment of Soy Lectin T Cell-Depleted Haploidentical Bone Marrow Cells in Severe Combined Immunodeficiency. *J Clin Immunol.* 1986; 6:161-169.

50. Buckley RH. Bone Marrow Transplantation in Primary Immunodeficiency. In: Rich RR, ed. *Clinical Immunology: Principles and Practice.* St Louis: Mosby-Year Book; 1996:1813-1830.

51. *Rebecca H. Buckley Interviewed by Jessica Roseberry, February 19, 2007.* Durham, NC: Duke University Medical Center Archives, Biographical Files Collection; 2007.

52. Ali H, Richardson RM, Haribabu B, Snyderman R. Chemoattractant Receptor Cross-Desensitization. *J Biol Chem.* 1999; 274:6027-6030.

53. Smith CD, Cox CC, Snyderman R. Receptor-Coupled Activation of Phosphoinositide-Specific Phospholipase C by an N Protein. *Science.* 1986; 232:97-100.

54. Verghese MW, Smith CD, Snyderman R. Potential Role for a Guanine Nucleotide Regulatory Protein in Chemoattractant Receptor Mediated Polyphosphoinositide Metabolism, Ca++ Mobilization and Cellular Responses by Leukocytes. *Biochem Biophys Res Commun.* 1985; 127:450-457.

55. Snyderman R, Goetzl EJ. Molecular and Cellular Mechanisms of Leukocyte Chemotaxis. *Science.* 1981; 213:830-837.

56. Williams LT, Snyderman R, Pike MC, Lefkowitz RJ. Specific Receptor Sites for Chemotactic Peptides on Human Polymorphonuclear Leukocytes. *Proc Natl Acad Sci U S A.* 1977; 74:1204-1208.

57. Snyderman R, Pike MC. An Inhibitor of Macrophage Chemotaxis Produced by Neoplasms. *Science.* 1976; 192:370-372.

58. Snyderman R, Pike MC, Altman LC. Abnormalities of Leukocyte Chemotaxis in Human Disease. *Ann N Y Acad Sci.* 1975; 256:386-401.

59. Polakis PG, Weber RF, Nevins B, Didsbury JR, Evans T, Snyderman R. Identification of the Ral and Racl Gene Products, Low Molecular Mass Gtp-Binding Proteins from Human Platelets. *J Biol Chem.* 1989; 264:16383-16389.

60. Snyderman R, Shin HS, Phillips JK, Gewurz H, Mergenhagen SE. A Neutrophil Chemotatic Factor Derived from C'5 Upon Interaction of Guinea Pig Serum with Endotoxin. *J Immunol.* 1969; 103:413-422.

61. Snyderman R, Gewurz H, Mergenhagen SE. Interactions of the Complement System with Endotoxic Lipopolysaccharide. Generation of a Factor Chemotactic for Polymorphonuclear Leukocytes. *J Exp Med.* 1968; 128:259-275.

62. Joklik WK. Adventures of a Biochemist in Virology. *J Biol Chem.* 2005; 280:40385-40397.

63. Roseberry J. Interview. In: Joklik WK, ed. Durham, North Carolina: Medical Center Archives; 2007.

64. Manning M, Mitchell TG. Morphogenesis of Candida Albicans and Cytoplasmic Proteins Associated with Differences in Morphology, Strain, or Temperature. *J Bacteriol.* 1980; 144:258-273.

65. Small JM, Mitchell TG. Binding of Purified and Radioiodinated Capsular Polysaccharides from Cryptococcus Neoformans Serotype a Strains to Capsule-Free Mutants. *Infect Immun.* 1986; 54:742-750.

66. Small JM, Mitchell TG, Wheat RW. Strain Variation in Composition and Molecular Size of the Capsular Polysaccharide of Cryptococcus Neoformans Serotype A. *Infect Immun.* 1986; 54:735-741.

67. Miller MF, Mitchell TG. Killing of Cryptococcus Neoformans Strains by Human Neutrophils and Monocytes. *Infect Immun.* 1991; 59:24-28.

68. Mitchell TG, Perfect JR. Cryptococcosis in the Era of AIDS--100 Years after the Discovery of Cryptococcus Neoformans. *Clin Microbiol Rev.* 1995; 8:515-548.

69. Graser Y, Volovsek M, Arrington J, et al. Molecular Markers Reveal That Population Structure of the Human Pathogen Candida Albicans Exhibits Both Clonality and Recombination. *Proc Natl Acad Sci U S A.* 1996; 93:12473-12477.

70. Xu J, Mitchell TG, Vilgalys R. PCR-Restriction Fragment Length Polymorphism (Rflp) Analyses Reveal Both Extensive Clonality and Local Genetic Differences in Candida Albicans. *Mol Ecol.* 1999; 8:59-73.

71. Xu J, Vilgalys R, Mitchell TG. Lack of Genetic Differentiation between Two Geographically Diverse Samples of Candida Albicans Isolated from Patients Infected with Human Immunodeficiency Virus. *J Bacteriol.* 1999; 181:1369-1373.

72. Xu J, Ali RY, Gregory DA, et al. Uniparental Mitochondrial Transmission in Sexual Crosses in Cryptococcus Neoformans. *Curr Microbiol.* 2000; 40:269-273.

73. Xu J, Vilgalys R, Mitchell TG. Multiple Gene Genealogies Reveal Recent Dispersion and Hybridization in the Human Pathogenic Fungus Cryptococcus Neoformans. *Mol Ecol.* 2000; 9:1471-1481.

74. Durack DT. Opportunistic Infections and Kaposi's Sarcoma in Homosexual Men. *N Engl J Med.* 1981; 305:1465-1467.

75. Durack DT, Ackerman SJ, Loegering DA, Gleich GJ. Purification of Human Eosinophil-Derived Neurotoxin. *Proc Natl Acad Sci U S A.* 1981; 78:5165-5169.

76. Durack DT, Sumi SM, Klebanoff SJ. Neurotoxicity of Human Eosinophils. *Proc Natl Acad Sci U S A.* 1979; 76:1443-1447.

77. Perfect JR, Durack DT. Treatment of Experimental Cryptococcal Meningitis with Amphotericin B, 5-Fluorocytosine, and Ketoconazole. *J Infect Dis.* 1982; 146:429-435.

78. Perfect JR, Durack DT, Hamilton JD, Gallis HA. Failure of Ketoconazole in Cryptococcal Meningitis. *Jama-Journal of the American Medical Association.* 1982; 247:3349-3351.

79. Perfect JR, Lang SD, Durack DT. Chronic Cryptococcal Meningitis: A New Experimental Model in Rabbits. *Am J Pathol.* 1980; 101:177-194.

80. Snyderman R, Durack DT, McCarty GA, Ward FE, Meadows L. Deficiency of the Fifth Component of Complement in Human Subjects. Clinical, Genetic and Immunologic Studies in a Large Kindred. *Am J Med.* 1979; 67:638-645.

81. Toffaletti DL, Rude TH, Johnston SA, Durack DT, Perfect JR. Gene Transfer in Cryptococcus Neoformans by Use of Biolistic Delivery of DNA. *J Bacteriol.* 1993; 175:1405-1411.

82. Weinhold KJ, Lyerly HK, Matthews TJ, et al. Cellular Anti-Gp120 Cytolytic Reactivities in HIV-1 Seropositive Individuals. *Lancet.* 1988; 1:902-905.

83. Yarchoan R, Klecker RW, Weinhold KJ, et al. Administration of 3'-Azi-do-3'-Deoxythymidine, an Inhibitor of HTLV-III/LAV Replication, to Patients with AIDS or AIDS-Related Complex. *Lancet.* 1986; 1:575-580.

84. Lodge JK, Jackson-Machelski E, Toffaletti DL, Perfect JR, Gordon JI. Targeted Gene Replacement Demonstrates That Myristoyl-Coa: Protein N-Myristoyl-transferase Is Essential for Viability of Cryptococcus Neoformans. *Proc Natl Acad Sci U S A.* 1994; 91:12008-12012.

85. Odom A, Muir S, Lim E, Toffaletti DL, Perfect J, Heitman J. Calcineurin Is Required for Virulence of Cryptococcus Neoformans. *EMBO J.* 1997; 16:2576-2589.

86. Keene JD. Ribonucleoprotein Infrastructure Regulating the Flow of Genetic Information between the Genome and the Proteome. *Proc Natl Acad Sci U S A.* 2001; 98:7018-7024.

87. Keene JD, Tenenbaum SA. Eukaryotic mRNPs May Represent Posttranscriptional Operons. *Mol Cell.* 2002; 9:1161-1167.

88. Keene JD, Lager PJ. Post-Transcriptional Operons and Regulons Co-Ordinating Gene Expression. *Chromosome Res.* 2005; 13:327-337.

89. Keene JD. RNA Regulons: Coordination of Post-Transcriptional Events. *Nat Rev Genet.* 2007; 8:533-543.

90. Mukherjee N, Lager PJ, Friedersdorf MB, Thompson MA, Keene JD. Coordinated Posttranscriptional mRNA Population Dynamics During T-Cell Activation. *Mol Syst Biol.* 2009; 5:288.

91. Mukherjee N, Corcoran DL, Nusbaum JD, et al. Integrative Regulatory Mapping Indicates That the RNA-Binding Protein Hur Couples Pre-mRNA Processing and mRNA Stability. *Mol Cell.* 2011; 43:327-339.

92. Chambers JC, Keene JD. Isolation and Analysis of Cdna Clones Expressing Human Lupus La Antigen. *Proc Natl Acad Sci U S A.* 1985; 82:2115-2119.

93. Chambers JC, Kenan D, Martin BJ, Keene JD. Genomic Structure and Amino Acid Sequence Domains of the Human La Autoantigen. *J Biol Chem.* 1988; 263:18043-18051.

94. Query CC, Bentley RC, Keene JD. A Common RNA Recognition Motif Identified within a Defined U1 RNA Binding Domain of the 70k U1 Snrnp Protein. *Cell.* 1989; 57:89-101.

95. St Clair EW, Pisetsky DS, Reich CF, Keene JD. Analysis of Autoantibody Binding to Different Regions of the Human La Antigen Expressed in Recombinant Fusion Proteins. *J Immunol.* 1988; 141:4173-4180.

96. Query CC, Keene JD. A Human Autoimmune Protein Associated with U1 RNA Contains a Region of Homology That Is Cross-Reactive with Retroviral P30gag Antigen. *Cell.* 1987; 51:211-220.

97. Levine TD, Gao F, King PH, Andrews LG, Keene JD. Hel-N1: An Autoimmune RNA-Binding Protein with Specificity for 3' Uridylate-Rich Untranslated Regions of Growth Factor Mrnas. *Mol Cell Biol.* 1993; 13:3494-3504.

98. Jain RG, Andrews LG, McGowan KM, Pekala PH, Keene JD. Ectopic Expression of Hel-N1, an RNA-Binding Protein, Increases Glucose Transporter (Glut1) Expression in 3t3-L1 Adipocytes. *Mol Cell Biol.* 1997; 17:954-962.

99. Keene JD. Why Is Hu Where? Shuttling of Early-Response-Gene Messenger RNA Subsets. *Proc Natl Acad Sci U S A.* 1999; 96:5-7.

100. Antic D, Lu N, Keene JD. Elav Tumor Antigen, Hel-N1, Increases Translation of Neurofilament M mRNA and Induces Formation of Neurites in Human Teratocarcinoma Cells. *Genes Dev.* 1999; 13:449-461.

101. Gao FB, Carson CC, Levine T, Keene JD. Selection of a Subset of Mrnas from Combinatorial 3' Untranslated Region Libraries Using Neuronal RNA-Binding Protein Hel-N1. *Proc Natl Acad Sci U S A.* 1994; 91:11207-11211.

102. Tenenbaum SA, Carson CC, Lager PJ, Keene JD. Identifying mRNA Subsets in Messenger Ribonucleoprotein Complexes by Using Cdna Arrays. *Proc Natl Acad Sci U S A.* 2000; 97:14085-14090.

103. Liu K, Howell DN, Perfect JR, Schell WA. Morphologic Criteria for the Preliminary Identification of Fusarium, Paecilomyces, and Acremonium Species by Histopathology. *Am J Clin Pathol.* 1998; 109:45-54.

104. Schell WA, Perfect JR. Fatal, Disseminated Acremonium Strictum Infection in a Neutropenic Host. *J Clin Microbiol.* 1996; 34:1333-1336.

105. McGinnis MR, Schell WA, Carson J. Phaeoannellomyces and the Phaeococ-comycetaceae, New Dematiaceous Blastomycete Taxa. *Sabouraudia*. 1985; 23:179-188.

106. Perfect JR, Cox GM, Dodge RK, Schell WA. In Vitro and in Vivo Efficacies of the Azole Sch56592 against Cryptococcus Neoformans. *Antimicrob Agents Chemother*. 1996; 40:1910-1913.

107. del Poeta M, Schell WA, Perfect JR. In Vitro Antifungal Activity of Pneumocandin L-743,872 against a Variety of Clinically Important Molds. *Antimicrob Agents Chemother*. 1997; 41:1835-1836.

108. Cameron ML, Schell WA, Bruch S, Bartlett JA, Waskin HA, Perfect JR. Correlation of in Vitro Fluconazole Resistance of Candida Isolates in Relation to Therapy and Symptoms of Individuals Seropositive for Human Immunodeficiency Virus Type 1. *Antimicrob Agents Chemother*. 1993; 37:2449-2453.

109. Granger DL, Perfect JR, Durack DT. Virulence of Cryptococcus Neoformans. Regulation of Capsule Synthesis by Carbon Dioxide. *J Clin Invest*. 1985; 76:508-516.

110. Granger DL, Hibbs JB, Jr., Perfect JR, Durack DT. Specific Amino Acid (L-Arginine) Requirement for the Microbiostatic Activity of Murine Macrophages. *J Clin Invest*. 1988; 81:1129-1136.

111. Anstey NM, Weinberg JB, Hassanali MY, et al. Nitric Oxide in Tanzanian Children with Malaria: Inverse Relationship between Malaria Severity and Nitric Oxide Production/Nitric Oxide Synthase Type 2 Expression. *J Exp Med*. 1996; 184:557-567.

112. Granger DL, Hibbs JB, Jr., Perfect JR, Durack DT. Metabolic Fate of L-Arginine in Relation to Microbiostatic Capability of Murine Macrophages. *J Clin Invest*. 1990; 85:264-273.

113. Granger DL, Perfect JR, Durack DT. Macrophage-Mediated Fungistasis: Requirement for a Macromolecular Component in Serum. *J Immunol*. 1986; 137:693-701.

114. Granger DL, Perfect JR, Durack DT. Macrophage-Mediated Fungistasis in Vitro: Requirements for Intracellular and Extracellular Cytotoxicity. *J Immunol*. 1986; 136:672-680.

115. Boockvar KS, Granger DL, Poston RM, et al. Nitric Oxide Produced During Murine Listeriosis Is Protective. *Infect Immun*. 1994; 62:1089-1100.

116. Alspaugh JA, Granger DL. Inhibition of Cryptococcus Neoformans Replication by Nitrogen Oxides Supports the Role of These Molecules as Effectors of Macrophage-Mediated Cytostasis. *Infect Immun*. 1991; 59:2291-2296.

117. Naslund PK, Miller WC, Granger DL. Cryptococcus Neoformans Fails to Induce Nitric Oxide Synthase in Primed Murine Macrophage-Like Cells. *Infect Immun*. 1995; 63:1298-1304.

118. Granger DL, Hibbs JB, Jr., Broadnax LM. Urinary Nitrate Excretion in Relation to Murine Macrophage Activation. Influence of Dietary L-Arginine and Oral Ng-Monomethyl-L-Arginine. *J Immunol.* 1991; 146:1294-1302.

119. Boockvar KS, Maybodi M, Poston RM, Kurlander RL, Granger DL. Nitric Oxide in Listeriosis. In: Fang FC, ed. *Nitric Oxide in Infection.* New York: Kluwer Academic/Plenum Publishers; 1999:447-471.

120. Anstey N, al. E. Nitric Oxide in Tanzanian Children with Malaria. *J Exp Med.* 1996; 184:557-567.

121. Weinstein MP, Mirrett S, Wilson ML, Reimer LG, Reller LB. Controlled Evaluation of 5 Versus 10 Milliliters of Blood Cultured in Aerobic Bact/Alert Blood Culture Bottles. *J Clin Microbiol.* 1994; 32:2103-2106.

122. Morris AJ, Wilson SJ, Marx CE, Wilson ML, Mirrett S, Reller LB. Clinical Impact of Bacteria and Fungi Recovered Only from Broth Cultures. *J Clin Microbiol.* 1995; 33:161-165.

123. Perkins MD, Mirrett S, Reller LB. Rapid Bacterial Antigen Detection Is Not Clinically Useful. *J Clin Microbiol.* 1995; 33:1486-1491.

124. Zaidi AK, Knaut AL, Mirrett S, Reller LB. Value of Routine Anaerobic Blood Cultures for Pediatric Patients. *J Pediatr.* 1995; 127:263-268.

125. Zaidi AK, Harrell LJ, Rost JR, Reller LB. Assessment of Similarity among Coagulase-Negative Staphylococci from Sequential Blood Cultures of Neonates and Children by Pulsed-Field Gel Electrophoresis. *J Infect Dis.* 1996; 174:1010-1014.

126. Weinstein MP, Towns ML, Quartey SM, et al. The Clinical Significance of Positive Blood Cultures in the 1990s: A Prospective Comprehensive Evaluation of the Microbiology, Epidemiology, and Outcome of Bacteremia and Fungemia in Adults. *Clin Infect Dis.* 1997; 24:584-602.

127. Archibald LK, den Dulk MO, Pallangyo KJ, Reller LB. Fatal Mycobacterium Tuberculosis Bloodstream Infections in Febrile Hospitalized Adults in Dar Es Salaam, Tanzania. *Clin Infect Dis.* 1998; 26:290-296.

128. Morris AJ, Wilson ML, Reller LB. Application of Rejection Criteria for Stool Ovum and Parasite Examinations. *J Clin Microbiol.* 1992; 30:3213-3216.

129. Morris AJ, Tanner DC, Reller LB. Rejection Criteria for Endotracheal Aspirates from Adults. *J Clin Microbiol.* 1993; 31:1027-1029.

130. Fan K, Morris AJ, Reller LB. Application of Rejection Criteria for Stool Cultures for Bacterial Enteric Pathogens. *J Clin Microbiol.* 1993; 31:2233-2235.

131. Heitman J, Movva NR, Hall MN. Targets for Cell Cycle Arrest by the Immunosuppressant Rapamycin in Yeast. *Science.* 1991; 253:905-909.

132. Alspaugh JA, Perfect JR, Heitman J. Cryptococcus Neoformans Mating and Virulence Are Regulated by the G-Protein Alpha Subunit Gpa1 and Camp. *Genes Dev.* 1997; 11:3206-3217.

133. Lorenz MC, Heitman J. The Mep2 Ammonium Permease Regulates Pseudohyphal Differentiation in Saccharomyces Cerevisiae. *EMBO J.* 1998; 17:1236-1247.

134. Cardenas ME, Cutler NS, Lorenz MC, Di Como CJ, Heitman J. The Tor Signaling Cascade Regulates Gene Expression in Response to Nutrients. *Genes Dev.* 1999; 13:3271-3279.

135. Cox GM, Mukherjee J, Cole GT, Casadevall A, Perfect JR. Urease as a Virulence Factor in Experimental Cryptococcosis. *Infect Immun.* 2000; 68:443-448.

136. Yue C, Cavallo LM, Alspaugh JA, et al. The Ste12alpha Homolog Is Required for Haploid Filamentation but Largely Dispensable for Mating and Virulence in Cryptococcus Neoformans. *Genetics.* 1999; 153:1601-1615.

137. Cruz MC, Cavallo LM, Gorlach JM, et al. Rapamycin Antifungal Action Is Mediated Via Conserved Complexes with Fkbp12 and Tor Kinase Homologs in Cryptococcus Neoformans. *Mol Cell Biol.* 1999; 19:4101-4112.

138. Lengeler KB, Wang P, Cox GM, Perfect JR, Heitman J. Identification of the Mata Mating-Type Locus of Cryptococcus Neoformans Reveals a Serotype a Mata Strain Thought to Have Been Extinct. *Proc Natl Acad Sci U S A.* 2000; 97:14455-14460.

139. Gorlach J, Fox DS, Cutler NS, Cox GM, Perfect JR, Heitman J. Identification and Characterization of a Highly Conserved Calcineurin Binding Protein, Cbp1/Calcipressin, in Cryptococcus Neoformans. *EMBO J.* 2000; 19:3618-3629.

140. Cox GM, Toffaletti DL, Perfect JR. Dominant Selection System for Use in Cryptococcus Neoformans. *J Med Vet Mycol.* 1996; 34:385-391.

141. Cruz MC, Sia RA, Olson M, Cox GM, Heitman J. Comparison of the Roles of Calcineurin in Physiology and Virulence in Serotype D and Serotype a Strains of Cryptococcus Neoformans. *Infect Immun.* 2000; 68:982-985.

142. Cox GM, Rude TH, Dykstra CC, Perfect JR. The Actin Gene from Cryptococcus Neoformans: Structure and Phylogenetic Analysis. *J Med Vet Mycol.* 1995; 33:261-266.

143. Skrzynia C, Binninger DM, Alspaugh JA, 2nd, Pukkila PJ. Molecular Characterization of Trp1, a Gene Coding for Tryptophan Synthetase in the Basidiomycete Coprinus Cinereus. *Gene.* 1989; 81:73-82.

144. Alspaugh JA, Perfect JR, Heitman J. Cryptococcus Neoformans Mating and Virulence Are Regulated by the G-Protein Alpha-Subunit Gpa1 and Camp. *Genes Dev.* 1997; 11:3206-3217.

145. Alspaugh JA, Perfect JR. Infections Due to Zygomycetes and Other Rare Opportunistic Fungi. *Semin Respir Crit Care Med.* 1997; 18:265-279.

146. Lyon GM, Alspaugh JA, Meredith FT, et al. Mycoplasma Hominis Pneumonia Complicating Bilateral Lung Transplantation: Case Report and Review of the Literature. *Chest.* 1997; 112:1428-1432.

147. Alspaugh JA, Perfect JR, Heitman J. Signal Transduction Pathways Regulating Differentiation and Pathogenicity of Cryptococcus Neoformans. *Fungal Genet Biol.* 1998; 25:1-14.

148. Heitman J, Allen B, Alspaugh JA, Kwon-Chung KJ. On the Origins of Congenic Matalpha and Mata Strains of the Pathogenic Yeast Cryptococcus Neoformans. *Fungal Genet Biol.* 1999; 28:1-5.

149. Sudarshan S, Davidson RC, Heitman J, Alspaugh JA. Molecular Analysis of the Cryptococcus Neoformans Ade2 Gene, a Selectable Marker for Transformation and Gene Disruption. *Fungal Genet Biol.* 1999; 27:36-48.

150. Hamilton JD, Seaworth BJ. Transmission of Latent Murine Cytomegalovirus by Kidney Tissue-Transplantation. *Clin Res.* 1983; 31:A364-A364.

151. Hamilton JD, Seaworth BJ. Transmission of Latent Cytomegalo-Virus in a Murine Kidney Tissue-Transplantation Model. *Transplantation.* 1985; 39:290-296.

152. Rubin RH, Tolkoffrubin NE, Oliver D, et al. Multicenter Seroepidemiologic Study of the Impact of Cytomegalo-Virus Infection on Renal-Transplantation. *Transplantation.* 1985; 40:243-249.

153. Hamilton JD, Fitzwilliam JF, Cheung KS, Lang DJ. Eeffects of Murine Cytomegalovirus-Infection on the Immune-Response to a Tumor Allograft. *Rev Infect Dis.* 1979; 1:976-987.

154. Hamilton JD, Fitzwilliam J, Shelburne J, Cheung KS. Depressed Cytotoxicity (Lmc) in Cytomegalovirus Infected Mice. *Clin Res.* 1977; 25:A359-A359.

155. Henry SC, Klotman ME, Hamilton JD. Detection of Murine Cytomegalo-Virus RNA Transcripts by in Vitro Enzymatic Amplication of Cdna. *Clin Res.* 1989; 37:A431-A431.

156. Henry SC, Hamilton JD. Detection of Murine Cytomegalovirus Immediate Early-1 Transcripts in the Spleens of Latently Infected Mice. *J Infect Dis.* 1993; 167:950-954.

157. Henry SC, Schmader K, Brown TT, et al. Enhanced Green Fluorescent Protein as a Marker for Localizing Murine Cytomegalovirus in Acute and Latent Infection. *J Virol Methods.* 2000; 89:61-73.

158. Kanj S, Hamilton J. The 10 Most Common Questions About Cytomegalovirus Infection in Solid Organ Transplant Recipients. *Infectious Diseases in Clinical Practice.* 1997; 6:29-32.

159. Kanj SS, Sharara AI, Clavien PA, Hamilton JD. Cytomegalovirus Infection Following Liver Transplantation: Review of the Literature. *Clin Infect Dis.* 1996; 22:537-549.

160. Klotman ME, Henry SC, Hamilton JD. Determinants of the Source of Cytomegalovirus in Murine Renal-Allograft Recipients. *Transplantation.* 1987; 44:636-639.

161. Klotman ME, Porter KP, Best CF, Hamilton JD, Klotman PE. Morphologic and Functional-Effects of Acute Cytomegalo-Virus (Cmv) Infection on Rat Kidneysi. *Clin Res.* 1988; 36:A459-A459.

162. Porter KR, Starnes DM, Hamilton JD. Reactivation of Latent Murine Cytomegalo-Virus from Kidney. *Kidney Int.* 1985; 28:922-925.

163. Yu YH, Henry SC, Xu FJ, Daley GG, Hamilton JD. The Suppressive Effect of Cytomegalovirus on Bone-Marrow - a Murine Model. *Cancer Research Therapy & Control.* 1993; 3:259-267.

164. Yu YH, Henry SC, Xu FJ, Hamilton JD. Expression of a Murine Cytomegalovirus Early-Late Protein in Latently Infected Mice. *J Infect Dis.* 1995; 172:371-379.

165. Schmader KE, Rahija RJ, Porter KR, Hamilton JD. Murine Cytomegalovirus Gene Amplification and Culture after Submaxillary Salivary Gland Biopsy. *Lab Anim Sci.* 1991; 41:396-400.

166. Schmader KE, Rahija R, Porter KR, Daley G, Hamilton JD. Aging and Reactivation of Latent Murine Cytomegalovirus. *J Infect Dis.* 1992; 166:1403-1407.

167. Schmader KE, Henry SC, Rahija RJ, Yu Y, Daley GG, Hamilton JD. Mouse Cytomegalovirus Reactivation in Severe Combined Immune Deficient Mice after Implantation of Latently Infected Salivary Gland. *J Infect Dis.* 1995; 172:531-534.

168. Schmader KE, Studenski S, MacMillan J, Grufferman S, Cohen HJ. Are Stressful Life Events Risk Factors for Herpes Zoster? *J Am Geriatr Soc.* 1990; 38:1188-1194.

169. Schmader KE, Gerorge LK, Burchett BM, Hamilton JD, Pieper CF. Race and Stress in the Incidence of Herpes Zoster in Older Adults. *J Am Geriatr Soc.* 1998; 46:973-977.

170. Schmader KE, George LK, Newton R, Hamilton JD. The Accuracy of Self-Reports of Herpes Zoster. *J Clin Epidemiol.* 1994; 47:1271-1276.

171. Oxman MN, Levin MJ, Johnson GR, et al. A Vaccine to Prevent Herpes Zoster and Postherpetic Neuralgia in Older Adults. *N Engl J Med.* 2005; 352:2271-2278.

172. Markert ML. Purine Nucleoside Phosphorylase Deficiency. *Immunodefic Rev.* 1991; 3:45-81.

173. Markert ML, Hutton JJ, Wiginton DA, States JC, Kaufman RE. Adenosine Deaminase (Ada) Deficiency Due to Deletion of the Ada Gene Promoter and First Exon by Homologous Recombination between Two Alu Elements. *J Clin Invest.* 1988; 81:1323-1327.

174. Markert ML, Norby-Slycord C, Ward FE. A High Proportion of Ada Point Mutations Associated with a Specific Alanine-to-Valine Substitution. *Am J Hum Genet.* 1989; 45:354-361.

175. Markert ML, Finkel BD, McLaughlin TM, et al. Mutations in Purine Nucleoside Phosphorylase Deficiency. *Hum Mutat.* 1997; 9:118-121.

176. Markert ML, Watson TJ, Kaplan I, Hale LP, Haynes BF. The Human Thymic Microenvironment During Organ Culture. *Clin Immunol Immunopathol.* 1997; 82:26-36.

177. Markert ML, Kostyu DD, Ward FE, et al. Successful Formation of a Chimeric Human Thymus Allograft Following Transplantation of Cultured Postnatal Human Thymus. *J Immunol.* 1997; 158:998-1005.

178. Davis CM, McLaughlin TM, Watson TJ, et al. Normalization of the Peripheral Blood T Cell Receptor V Beta Repertoire after Cultured Postnatal Human Thymic Transplantation in Digeorge Syndrome. *J Clin Immunol.* 1997; 17:167-175.

179. Markert ML, Hummell DS, Rosenblatt HM, et al. Complete Digeorge Syndrome: Persistence of Profound Immunodeficiency. *J Pediatr.* 1998; 132:15-21.

180. Collard HR, Boeck A, Mc Laughlin TM, et al. Possible Extrathymic Development of Nonfunctional T Cells in a Patient with Complete Digeorge Syndrome. *Clin Immunol.* 1999; 91:156-162.

181. Markert ML, Boeck A, Hale LP, et al. Transplantation of Thymus Tissue in Complete Digeorge Syndrome. *N Engl J Med.* 1999; 341:1180-1189.

182. Denning SM, Kurtzberg J, Le PT, Tuck DT, Singer KH, Haynes BF. Human Thymic Epithelial Cells Directly Induce Activation of Autologous Immature Thymocytes. *Proc Natl Acad Sci U S A.* 1988; 85:3125-3129.

183. Le PT, Kurtzberg J, Brandt SJ, Niedel JE, Haynes BF, Singer KH. Human Thymic Epithelial Cells Produce Granulocyte and Macrophage Colony-Stimulating Factors. *J Immunol.* 1988; 141:1211-1217.

184. Haynes BF, Martin ME, Kay HH, Kurtzberg J. Early Events in Human T Cell Ontogeny. Phenotypic Characterization and Immunohistologic Localization of T Cell Precursors in Early Human Fetal Tissues. *J Exp Med.* 1988; 168:1061-1080.

185. Kurtzberg J, Denning SM, Nycum LM, Singer KH, Haynes BF. Immature Human Thymocytes Can Be Driven to Differentiate into Nonlymphoid Lineages by Cytokines from Thymic Epithelial Cells. *Proc Natl Acad Sci U S A.* 1989; 86:7575-7579.

186. Peters WP, Rosner G, Ross M, et al. Comparative Effects of Granulocyte-Macrophage Colony-Stimulating Factor (Gm-Csf) and Granulocyte Colony-Stimulating Factor (G-Csf) on Priming Peripheral Blood Progenitor Cells for Use with Autologous Bone Marrow after High-Dose Chemotherapy. *Blood.* 1993; 81:1709-1719.

187. Kurtzberg J, Graham M, Casey J, Olson J, Stevens CE, Rubinstein P. The Use of Umbilical Cord Blood in Mismatched Related and Unrelated Hemopoietic Stem Cell Transplantation. *Blood Cells*. 1994; 20:275-283; discussion 284.

188. McCowage GB, Kurtzberg J, Rubinstein P. Transplantation of Cord-Blood Cells. *N Engl J Med*. 1995; 333:67; author reply 68-69.

189. Kurtzberg J. Umbilical Cord Blood: A Novel Alternative Source of Hematopoietic Stem Cells for Bone Marrow Transplantation. *J Hematother*. 1996; 5:95-96.

190. Kurtzberg J, Laughlin M, Graham ML, et al. Placental Blood as a Source of Hematopoietic Stem Cells for Transplantation into Unrelated Recipients. *N Engl J Med*. 1996; 335:157-166.

191. Sugarman J, Reisner EG, Kurtzberg J. Ethical Aspects of Banking Placental Blood for Transplantation. *JAMA*. 1995; 274:1783-1785.

192. Laughlin MJ, Rizzieri DA, Smith CA, et al. Hematologic Engraftment and Reconstitution of Immune Function Post Unrelated Placental Cord Blood Transplant in an Adult with Acute Lymphocytic Leukemia. *Leuk Res*. 1998; 22:215-219.

193. Rubinstein P, Carrier C, Scaradavou A, et al. Outcomes among 562 Recipients of Placental-Blood Transplants from Unrelated Donors. *N Engl J Med*. 1998; 339:1565-1577.

194. Schlegel PG, Aharoni R, Chen Y, et al. A Synthetic Random Basic Copolymer with Promiscuous Binding to Class II Major Histocompatibility Complex Molecules Inhibits T-Cell Proliferative Responses to Major and Minor Histocompatibility Antigens in Vitro and Confers the Capacity to Prevent Murine Graft-Versus-Host Disease in Vivo. *Proc Natl Acad Sci U S A*. 1996; 93:5061-5066.

195. Schlegel PG, Chao NJ. Immunomodulatory Peptides with High Binding Affinity for Class II Mhc Molecules for the Prevention of Graft-Versus-Host Disease. *Leuk Lymphoma*. 1996; 23:11-16.

196. Behar E, Chao NJ, Hiraki DD, et al. Polymorphism of Adhesion Molecule Cd31 and Its Role in Acute Graft-Versus-Host Disease. *N Engl J Med*. 1996; 334:286-291.

197. Chao NJ, Parker PM, Niland JC, et al. Paradoxical Effect of Thalidomide Prophylaxis on Chronic Graft-Vs.-Host Disease. *Biol Blood Marrow Transplant*. 1996; 2:86-92.

198. Chen Y, Zeng D, Schlegel PG, Fidler J, Chao NJ. Pg27, an Extract of Tripterygium Wilfordii Hook F, Induces Antigen-Specific Tolerance in Bone Marrow Transplantation in Mice. *Blood*. 2000; 95:705-710.

199. Moore J, Silberman HR, Cohen HJ, Carpenter C. *The History of Cancer Care, Research, and Education at Duke*. Durham, NC: Creative Services, Duke University; 2012.

200. Kanj SS, Tapson V, Davis RD, Madden J, Browning I. Infections in Patients with Cystic Fibrosis Following Lung Transplantation. *Chest.* 1997; 112:924-930.

201. Kanj SS, Welty-Wolf K, Madden J, et al. Fungal Infections in Lung and Heart-Lung Transplant Recipients. Report of 9 Cases and Review of the Literature. *Medicine.* 1996; 75:142-156.

202. Palmer SM, Jr., Kanj SS, Davis RD, Tapson VF. A Case of Disseminated Infection with Nocardia Brasiliensis in a Lung Transplant Recipient. *Transplantation.* 1997; 63:1189-1190.

203. Alexander BD, Perfect JR. Antifungal Resistance Trends Towards the Year 2000. Implications for Therapy and New Approaches. *Drugs.* 1997; 54:657-678.

204. Palmer SM, Alexander BD, Sanders LL, et al. Significance of Blood Stream Infection after Lung Transplantation: Analysis in 176 Consecutive Patients. *Transplantation.* 2000; 69:2360-2366.

CITATIONS FOR CHAPTER 7

1. Petersdorf RG. The Doctors' Dilemma. *N Engl J Med.* 1978; 299:628-634.

2. Tucker WB. The Evolution of the Cooperative Studies in the Chemotherapy of Tuberculosis of the Veterans Administration and Armed Forces of the U.S.A. An Account of the Evolving Education of the Physician in Clinical Pharmacology. *Bibl Tuberc.* 1960; 15:1-68.

3. Harris HW. Chemotherapy of Tuberculosis: The Beginning. In: Rom WN, Garay SM, eds. *Tuberculosis.* Boston, MA: Little, Brown and Company; 1996:745–749.

4. Hays MT. *VA Research 1925-1980.* Bloomington, IN: Authorhouse; 2010.

5. Conant NF, Smith DT, Callaway JL, eds. *Manual of Clinical Mycology.* 3rd ed. Philadelphia: W. B. Saunders Co; 1971.

6. Callaway JL. Dermatophytes as Opportunistic Organisms. *Lab Invest.* 1962; 11:1132-1133.

7. Callaway JL. Management of Common Dermatoses. *GP.* 1953; 8:75-81.

8. Callaway JL, Noojin RO, et al. The Treatment of Early Syphilis with Penicillin. *South Med J.* 1946; 39:718-726.

9. Callaway JL, Noojin RO, et al. The Use of Penicillin in the Treatment of Syphilis of the Central Nervous System; a Report of 100 Persons. *Am J Syph Gonorrhea Vener Dis.* 1946; 30:110-124.

10. Callaway JL, Riley KA. The Present Status of Penicillin in the Treatment of Syphilis. *South Med Surg.* 1948; 110:204-206.

11. Callaway JL, Olansky S. Superficial Mycotic Infections of the Skin: Diagnosis and Therapy. *J Chronic Dis.* 1957; 5:518-527.

12. Heald PW, Burton CS, Callaway JL. Shingles. *N C Med J.* 1983; 44:500-501.

13. Reque PG, Callaway JL. The Present Status of Penicillin in the Treatment of Syphilis. *N C Med J.* 1946; 7:469-473.

14. Buckley CE, 3rd, Nagaya H, Sieker HO. Altered Immunologic Activity in Sarcoidosis. *Ann Intern Med.* 1966; 64:508-520.

15. Dowell AR, Sieker HO, Schwartzman R. Atypical Periodic Respiration in an Obese Patient. *Arch Intern Med.* 1967; 120:591-598.

16. Gauer OH, Henry JP, Sieker HO. Changes in Central Venous Pressure after Moderate Hemorrhage and Transfusion in Man. *Circ Res.* 1956; 4:79-84.

17. Gauer OH, Henry JP, Sieker HO. Cardiac Receptors and Fluid Volume Control. *Prog Cardiovasc Dis.* 1961; 4:1-26.

18. Manfredi F, Sieker HO. The Effect of Carbon Dioxide on the Pulmonary Circulation. *J Clin Invest.* 1960; 39:295-301.

19. Nagaya H, Sieker HO. Allograft Survival: Effect of Antiserums to Thymus Glands and Lymphocytes. *Science.* 1965; 150:1181-1182.

20. Nagaya H, Sieker HO. Effects of Anti-Thymus and Anti-Lymphocyte Sera on Allograft Survival. *Trans Assoc Am Physicians.* 1966; 79:205-220.

21. Nagaya H, Sieker HO. Effects of Antithymus Serum and Antilymphocyte Serum on the Incidence of Lymphoid Leukemia. *Proc Soc Exp Biol Med.* 1969; 131:891-895.

22. Nagaya H, Sieker HO. Pathogenetic Mechanisms of Interstitial Pulmonary Fibrosis in Patients with Serum Antinuclear Factor. A Histologic and Clinical Correlation. *Am J Med.* 1972; 52:51-62.

23. Sieker HO, Gauer OH, Henry JP. The Effect of Continuous Negative Pressure Breathing on Water and Electrolyte Excretion by the Human Kidney. *J Clin Invest.* 1954; 33:572-577.

24. Sieker HO, Hickam JB. Carbon Dioxide Intoxication: The Clinical Syndrome, Its Etiology and Management with Particular Reference to the Use of Mechanical Respirators. *Medicine.* 1956; 35:389-423.

25. Sieker HO, Saltzman HA. Medical Considerations and Applications of Hyperbaric Oxygenation. *JAMA.* 1965; 193:31-36.

26. Behnke AR, Saltzman HA. Hyperbaric Oxygenation. *N Engl J Med.* 1967; 276:1423-1429 contd.

27. Dowell AR, Kylstra JA, Saltzman HA. Effects of a Hyperbaric Environment on Gas Exchange and Ventilation in Patients with Chronic Bronchitis and Emphysema. *Am Rev Respir Dis.* 1967; 96:389-399.

28. Fulkerson WJ, Coleman RE, Ravin CE, Saltzman HA. Diagnosis of Pulmonary Embolism. *Arch Intern Med.* 1986; 146:961-967.

29. Hempel FG, Jobsis FF, LaManna JL, Rosenthal MR, Saltzman HA. Oxidation of Cerebral Cytochrome Aa3 by Oxygen Plus Carbon Dioxide at Hyperbaric Pressures. *J Appl Physiol.* 1977; 43:873-879.

30. Heyman A, Saltzman HA, Whalen RE. The Use of Hyperbaric Oxygenation in the Treatment of Cerebral Ischemia and Infarction. *Circulation.* 1966; 33:II20-27.

31. Holland JA, Hill GB, Wolfe WG, Osterhout S, Saltzman HA, Brown IW, Jr. Experimental and Clinical Experience with Hyperbaric Oxygen in the Treatment of Clostridial Myonecrosis. *Surgery.* 1975; 77:75-85.

32. Johnston WW, Saltzman HA, Bufkin JH, Smith DT. The Tuberculin Test and the Diagnosis of Clinical Tuberculosis. *Am Rev Respir Dis.* 1960; 81:189-195.

33. Kariman K, Hempel FG, Jobsis FF, Burns SR, Saltzman HA. In Vivo Comparison of Cerebral Tissue Po2 and Cytochrome Aa3 Reduction-Oxidation State in Cats During Hemorrhagic Shock. *J Clin Invest.* 1981; 68:21-27.

34. Manfredi F, Sieker HO, Spoto AP, Saltzman HA. Severe Carbon Dioxide Intoxication. Treatment with Organic Buffer (Trihvdroxymethvlaminomethane). *JAMA.* 1960; 173:999-1003.

35. Saltzman HA. Rational Normobaric and Hyperbaric Oxygen Therapy. *Ann Intern Med.* 1967; 67:843-852.

36. Saltzman HA, Chick EW, Conant NF. Nocardiosis as a Complication of Other Diseases. *Lab Invest.* 1962; 11:1110-1117.

37. Stein PD, Terrin ML, Hales CA, et al. Clinical, Laboratory, Roentgenographic, and Electrocardiographic Findings in Patients with Acute Pulmonary Embolism and No Pre-Existing Cardiac or Pulmonary Disease. *Chest.* 1991; 100:598-603.

38. Choppin PW, Osterhout S, Tamm I. Immunological Characteristics of N.Y. Strains of Influenza a Virus from the 1957 Pandemic. *Proc Soc Exp Biol Med.* 1958; 98:513-520.

39. Osterhout S, Tamm I. Measurement of Herpes Simplex Virus by the Plaque Technique in Human Amnion Cells. *J Immunol.* 1959; 83:442-447.

40. Fuson RL, Saltzman HA, Smith WW, Whalen RE, Osterhout S, Parker RT. Clinical Hyperbaric Oxygenation with Severe Oxygen Toxicity: Report of a Case. *N Engl J Med.* 1965; 273:415-419.

41. Wallace AG, Young WG, Jr., Osterhout S. Treatment of Acute Bacterial Endocarditis by Valve Excision and Replacement. *Circulation.* 1965; 31:450-453.

42. Georgiade NG, Lucas MC, O'Fallon WM, Osterhout S. A Comparison of Methods for the Quantitation of Bacteria in Burn Wounds. I. Experimental Evaluation. *Am J Clin Pathol.* 1970; 53:35-39.

43. Georgiade NG, Lucas MC, Osterhout S. A Comparison of Methods for the Quantitation of Bacteria in Burn Wounds. II. Clinical Evaluation. *Am J Clin Pathol.* 1970; 53:40-42.

44. Hill GB, Osterhout S. Experimental Effects of Hyperbaric Oxgen on Selected Clostridial Species. I. In-Vitro Studies. *J Infect Dis.* 1972; 125:17-25.

45. Hill GB, Osterhout S. Experimental Effects of Hyperbaric Oxygen on Selected Clostridial Species. II. In-Vitro Studies in Mice. *J Infect Dis.* 1972; 125:26-35.

46. Castle M, Wilfert CM, Cate TR, Osterhout S. Antibiotic Use at Duke University Medical Center. *JAMA.* 1977; 237:2819-2822.

47. Greenfield JC. *Duke Chief Medical Residents.* Durham, North Carolina: Carolina Academic Press; 2005.

48. Kass EH, Hayes KM. A History of the Idsa. *Rev Infect Dis.* 1988; 10:1-158.

49. Hayes EC, Lee PW, Miller SE, Joklik WK. The Interaction of a Series of Hybridoma IgGs with Reovirus Particles. Demonstration That the Core Protein Lambda 2 Is Exposed on the Particle Surface. *Virology.* 1981; 108:147-155.

50. Joklik WK. The Molecular Basis of the Viral Eclipse Phase. *Prog Med Virol.* 1965; 7:44-96.

51. Joklik WK, Moss B, Fields BN, Bishop DH, Sandakhchiev LS. Why the Smallpox Virus Stocks Should Not Be Destroyed. *Science.* 1993; 262:1225-1226.

52. Joklik WK, Roner MR. Molecular Recognition in the Assembly of the Segmented Reovirus Genome. *Prog Nucleic Acid Res Mol Biol.* 1996; 53:249-281.

53. Lai MH, Joklik WK. The Induction of Interferon by Temperature-Sensitive Mutants of Reovirus, UV-Irradiated Reovirus, and Subviral Reovirus Particles. *Virology.* 1973; 51:191-204.

54. McCrae MA, Joklik WK. The Nature of the Polypeptide Encoded by Each of the 10 Double-Stranded RNA Segments of Reovirus Type 3. *Virology.* 1978; 89:578-593.

55. Pickup DJ, Ink BS, Hu W, Ray CA, Joklik WK. Hemorrhage in Lesions Caused by Cowpox Virus Is Induced by a Viral Protein That Is Related to Plasma Protein Inhibitors of Serine Proteases. *Proc Natl Acad Sci U S A.* 1986; 83:7698-7702.

56. Roner MR, Lin PN, Nepluev I, Kong LJ, Joklik WK. Identification of Signals Required for the Insertion of Heterologous Genome Segments into the Reovirus Genome. *Proc Natl Acad Sci U S A.* 1995; 92:12362-12366.

57. Smith RE, Zweerink HJ, Joklik WK. Polypeptide Components of Virions, Top Component and Cores of Reovirus Type 3. *Virology.* 1969; 39:791-810.

58. Starnes MC, Joklik WK. Reovirus Protein Lambda 3 Is a Poly(C)-Dependent Poly(G) Polymerase. *Virology.* 1993; 193:356-366.

59. Chambers JC, Keene JD. Isolation and Analysis of Cdna Clones Expressing Human Lupus La Antigen. *Proc Natl Acad Sci U S A.* 1985; 82:2115-2119.

60. Chambers JC, Kenan D, Martin BJ, Keene JD. Genomic Structure and Amino Acid Sequence Domains of the Human La Autoantigen. *J Biol Chem.* 1988; 263:18043-18051.

61. Query CC, Bentley RC, Keene JD. A Common RNA Recognition Motif Identified within a Defined U1 RNA Binding Domain of the 70k U1 Snrnp Protein. *Cell.* 1989; 57:89-101.

62. Query CC, Keene JD. A Human Autoimmune Protein Associated with U1 RNA Contains a Region of Homology That Is Cross-Reactive with Retroviral P30gag Antigen. *Cell.* 1987; 51:211-220.
63. St Clair EW, Pisetsky DS, Reich CF, Keene JD. Analysis of Autoantibody Binding to Different Regions of the Human La Antigen Expressed in Recombinant Fusion Proteins. *J Immunol.* 1988; 141:4173-4180.
64. St Clair EW, Pisetsky DS, Reich CF, Chambers JC, Keene JD. Quantitative Immunoassay of Anti-La Antibodies Using Purified Recombinant La Antigen. *Arthritis Rheum.* 1988; 31:506-514.
65. Levine TD, Gao F, King PH, Andrews LG, Keene JD. Hel-N1: An Autoimmune RNA-Binding Protein with Specificity for 3' Uridylate-Rich Untranslated Regions of Growth Factor Mrnas. *Mol Cell Biol.* 1993; 13:3494-3504.
66. Jain RG, Andrews LG, McGowan KM, Pekala PH, Keene JD. Ectopic Expression of Hel-N1, an RNA-Binding Protein, Increases Glucose Transporter (Glut1) Expression in 3t3-L1 Adipocytes. *Mol Cell Biol.* 1997; 17:954-962.
67. Keene JD. Why Is Hu Where? Shuttling of Early-Response-Gene Messenger RNA Subsets. *Proc Natl Acad Sci U S A.* 1999; 96:5-7.
68. Antic D, Lu N, Keene JD. Elav Tumor Antigen, Hel-N1, Increases Translation of Neurofilament M mRNA and Induces Formation of Neurites in Human Teratocarcinoma Cells. *Genes Dev.* 1999; 13:449-461.
69. Keene JD. Ribonucleoprotein Infrastructure Regulating the Flow of Genetic Information between the Genome and the Proteome. *Proc Natl Acad Sci U S A.* 2001; 98:7018-7024.
70. Tenenbaum SA, Carson CC, Lager PJ, Keene JD. Identifying mRNA Subsets in Messenger Ribonucleoprotein Complexes by Using Cdna Arrays. *Proc Natl Acad Sci U S A.* 2000; 97:14085-14090.
71. Gao FB, Carson CC, Levine T, Keene JD. Selection of a Subset of Mrnas from Combinatorial 3' Untranslated Region Libraries Using Neuronal RNA-Binding Protein Hel-N1. *Proc Natl Acad Sci U S A.* 1994; 91:11207-11211.
72. Ray CA, Black RA, Kronheim SR, et al. Viral Inhibition of Inflammation: Cowpox Virus Encodes an Inhibitor of the Interleukin-1 Beta Converting Enzyme. *Cell.* 1992; 69:597-604.
73. Spriggs MK, Hruby DE, Maliszewski CR, et al. Vaccinia and Cowpox Viruses Encode a Novel Secreted Interleukin-1-Binding Protein. *Cell.* 1992; 71:145-152.
74. Hu FQ, Smith CA, Pickup DJ. Cowpox Virus Contains Two Copies of an Early Gene Encoding a Soluble Secreted Form of the Type II Tnf Receptor. *Virology.* 1994; 204:343-356.
75. Smith CA, Hu FQ, Smith TD, et al. Cowpox Virus Genome Encodes a Second Soluble Homologue of Cellular Tnf Receptors, Distinct from Crmb, That Binds Tnf but Not Lt Alpha. *Virology.* 1996; 223:132-147.

76. Loparev VN, Parsons JM, Knight JC, et al. A Third Distinct Tumor Necrosis Factor Receptor of Orthopoxviruses. *Proc Natl Acad Sci U S A.* 1998; 95:3786-3791.

77. Panus JF, Smith CA, Ray CA, Smith TD, Patel DD, Pickup DJ. Cowpox Virus Encodes a Fifth Member of the Tumor Necrosis Factor Receptor Family: A Soluble, Secreted Cd30 Homologue. *Proc Natl Acad Sci U S A.* 2002; 99:8348-8353.

78. Oie KL, Pickup DJ. Cowpox Virus and Other Members of the Orthopoxvirus Genus Interfere with the Regulation of NF-Kappab Activation. *Virology.* 2001; 288:175-187.

79. Patel DD, Pickup DJ. Messenger Rnas of a Strongly-Expressed Late Gene of Cowpox Virus Contain 5'-Terminal Poly(a) Sequences. *EMBO J.* 1987; 6:3787-3794.

80. Nevins JR, Joklik WK. Poly (a) Sequences of Vaccinia Virus Messenger RNA: Nature, Mode of Addition and Function During Translation in Vitra and in Vivo. *Virology.* 1975; 63:1-14.

81. Nevins JR, Joklik WK. Isolation and Properties of the Vaccinia Virus DNA-Dependent RNA Polymerase. *J Biol Chem.* 1977; 252:6930-6938.

82. Nevins JR, Joklik WK. Isolation and Partial Characterization of the Poly(a) Polymerases from Hela Cells Infected with Vaccinia Virus. *J Biol Chem.* 1977; 252:6939-6947.

83. Nevins JR, Darnell JE, Jr. Steps in the Processing of Ad2 mRNA: Poly(a)+ Nuclear Sequences Are Conserved and Poly(a) Addition Precedes Splicing. *Cell.* 1978; 15:1477-1493.

84. Kovesdi I, Reichel R, Nevins JR. Identification of a Cellular Transcription Factor Involved in E1a Trans-Activation. *Cell.* 1986; 45:219-228.

85. Kovesdi I, Reichel R, Nevins JR. E1a Transcription Induction: Enhanced Binding of a Factor to Upstream Promoter Sequences. *Science.* 1986; 231:719-722.

86. Chellappan SP, Hiebert S, Mudryj M, Horowitz JM, Nevins JR. The E2f Transcription Factor Is a Cellular Target for the Rb Protein. *Cell.* 1991; 65:1053-1061.

87. Bild AH, Yao G, Chang JT, et al. Oncogenic Pathway Signatures in Human Cancers as a Guide to Targeted Therapies. *Nature.* 2006; 439:353-357.

88. Marciniak RA, Garcia-Blanco MA, Sharp PA. Identification and Characterization of a Hela Nuclear Protein That Specifically Binds to the Trans-Activation-Response (Tar) Element of Human Immunodeficiency Virus. *Proc Natl Acad Sci U S A.* 1990; 87:3624-3628.

89. Garcia-Blanco MA, Cullen BR. Molecular Basis of Latency in Pathogenic Human Viruses. *Science.* 1991; 254:815-820.

90. Colvin RA, Garcia-Blanco MA. Unusual Structure of the Human Immunodeficiency Virus Type 1 Trans-Activation Response Element. *J Virol.* 1992; 66:930-935.

91. Colvin RA, White SW, Garcia-Blanco MA, Hoffman DW. Structural Features of an RNA Containing the CUGGGA Loop of the Human Immunodeficiency Virus Type 1 Trans-Activation Response Element. *Biochemistry.* 1993; 32:1105-1112.

92. Hoffman DW, Colvin RA, Garcia-Blanco MA, White SW. Structural Features of the Trans-Activation Response RNA Element of Equine Infectious Anemia Virus. *Biochemistry.* 1993; 32:1096-1104.

93. Sune C, Garcia-Blanco MA. Transcriptional Trans Activation by Human Immunodeficiency Virus Type 1 Tat Requires Specific Coactivators That Are Not Basal Factors. *J Virol.* 1995; 69:3098-3107.

94. Sune C, Hayashi T, Liu Y, Lane WS, Young RA, Garcia-Blanco MA. CA150, a Nuclear Protein Associated with the RNA Polymerase II Holoenzyme, Is Involved in Tat-Activated Human Immunodeficiency Virus Type 1 Transcription. *Mol Cell Biol.* 1997; 17:6029-6039.

95. Sune C, Garcia-Blanco MA. Transcriptional Cofactor CA150 Regulates RNA Polymerase II Elongation in a Tata-Box-Dependent Manner. *Mol Cell Biol.* 1999; 19:4719-4728.

96. Carty SM, Goldstrohm AC, Sune C, Garcia-Blanco MA, Greenleaf AL. Protein-Interaction Modules That Organize Nuclear Function: FF Domains of CA150 Bind the Phosphoctd of RNA Polymerase II. *Proc Natl Acad Sci U S A.* 2000; 97:9015-9020.

97. Goldstrohm AC, Greenleaf AL, Garcia-Blanco MA. Co-Transcriptional Splicing of Pre-Messenger Rnas: Considerations for the Mechanism of Alternative Splicing. *Gene.* 2001; 277:31-47.

98. Bohjanen PR, Colvin RA, Puttaraju M, Been MD, Garcia-Blanco MA. A Small Circular Tar RNA Decoy Specifically Inhibits Tat-Activated HIV-1 Transcription. *Nucleic Acids Res.* 1996; 24:3733-3738.

99. Bohjanen PR, Liu Y, Garcia-Blanco MA. Tar RNA Decoys Inhibit Tat-Activated HIV-1 Transcription after Preinitiation Complex Formation. *Nucleic Acids Res.* 1997; 25:4481-4486.

100. Heitman J, Movva NR, Hall MN. Targets for Cell Cycle Arrest by the Immunosuppressant Rapamycin in Yeast. *Science.* 1991; 253:905-909.

101. Odom A, Muir S, Lim E, Toffaletti DL, Perfect J, Heitman J. Calcineurin Is Required for Virulence of Cryptococcus Neoformans. *EMBO J.* 1997; 16:2576-2589.

102. Cruz MC, Goldstein AL, Blankenship JR, et al. Calcineurin Is Essential for Survival During Membrane Stress in Candida Albicans. *EMBO J.* 2002; 21:546-559.

103. Lorenz MC, Heitman J. The Mep2 Ammonium Permease Regulates Pseudohyphal Differentiation in Saccharomyces Cerevisiae. *EMBO J.* 1998; 17:1236-1247.

104. Cardenas ME, Cutler NS, Lorenz MC, Di Como CJ, Heitman J. The Tor Signaling Cascade Regulates Gene Expression in Response to Nutrients. *Genes Dev.* 1999; 13:3271-3279.

105. Alspaugh JA, Perfect JR, Heitman J. Cryptococcus Neoformans Mating and Virulence Are Regulated by the G-Protein Alpha Subunit Gpa1 and Camp. *Genes Dev.* 1997; 11:3206-3217.

106. Katz SL, Milovanovic MV, Enders JF. Propagation of Measles Virus in Cultures of Chick Embryo Cells. *Proc Soc Exp Biol Med.* 1958; 97:23-29.

107. Katz SL, Morley DC, Krugman S. Attenuated Measles Vaccine in Nigerian Children. *Am J Dis Child.* 1962; 103:402-405.

108. Enders JF, Katz SL, Milovanovic MV, Holloway A. Studies on an Attenuated Measles-Virus Vaccine. I. Development and Preparations of the Vaccine: Technics for Assay of Effects of Vaccination. *N Engl J Med.* 1960; 263:153-159.

109. Katz SL, Smith AP. Studies of the Resistance and Susceptibility to Viral Superinfection of Adenovirus 12 Hamster Tumor Cells. *Arch Gesamte Virusforsch.* 1967; 22:143-158.

110. Nicholas SW, Sondheimer DL, Willoughby AD, Yaffe SJ, Katz SL. Human Immunodeficiency Virus Infection in Childhood, Adolescence, and Pregnancy: A Status Report and National Research Agenda. *Pediatrics.* 1989; 83:293-308.

111. Katz SL. Poliovaccine Policy--Time for a Change. *Pediatrics.* 1996; 98:116-117.

112. Alpert JJ, Katz SL, Lissauer T. International Exchanges: A Missed Opportunity in Pediatric Graduate Education. *Arch Pediatr Adolesc Med.* 2006; 160:570-571.

113. Cooper LZ, Larson HJ, Katz SL. Protecting Public Trust in Immunization. *Pediatrics.* 2008; 122:149-153.

114. Larson HJ, Cooper LZ, Eskola J, Katz SL, Ratzan S. Addressing the Vaccine Confidence Gap. *Lancet.* 2011; 378:526-535.

115. Cheung KS, Huang ES, Lang DJ. Murine Cytomegalovirus: Detection of Latent Infection by Nucleic Acid Hybridization Technique. *Infect Immun.* 1980; 27:851-854.

116. Cheung KS, Lang DJ. Transmission and Activation of Cytomegalovirus with Blood Transfusion: A Mouse Model. *J Infect Dis.* 1977; 135:841-845.

117. Cheung KS, Lang DJ. Detection of Latent Cytomegalovirus in Murine Salivary and Prostate Explant Cultures and Cells. *Infect Immun.* 1977; 15:568-575.

118. Cheung KS, Li JK, Falletta JM, Wagner JL, Lang DJ. Murine Cytomegalovirus Infection: Hematological, Morphological, and Functional Study of Lymphoid Cells. *Infect Immun.* 1981; 33:239-249.

119. Cheung KS, Roche JK, Capel WD, Lang DJ. An Evaluation under Code of New Techniques for the Detection of Cytomegalovirus Antibodies: Sensitivity of Assays and Importance of Immune Complexes. *J Clin Lab Immunol.* 1981; 6:269-274.

120. Lang DJ. Cytomegalovirus Immunization: Status, Prospects, and Problems. *Rev Infect Dis.* 1980; 2:449-458.

121. Lang DJ. Cytomegalovirus Infections in Organ Transplantation and Post Transfusion. An Hypothesis. *Arch Gesamte Virusforsch.* 1972; 37:365-377.

122. Lang DJ. Cytomegalovirus Infections in Pregnancy and the Newborn. *Clin Obstet Gynecol.* 1970; 13:348-359.

123. Lang DJ, Kummer JF. Cytomegalovirus in Semen: Observations in Selected Populations. *J Infect Dis.* 1975; 132:472-473.

124. Lang DJ, Hanshaw JB. Cytomegalovirus Infection and the Postperfusion Syndrome. Recognition of Primary Infections in Four Patients. *N Engl J Med.* 1969; 280:1145-1149.

125. Wilfert CM, Buckley RH, Mohanakumar T, et al. Persistent and Fatal Central-Nervous-System Echovirus Infections in Patients with Agammaglobulinemia. *N Engl J Med.* 1977; 296:1485-1489.

126. McKinney RE, Jr., Katz SL, Wilfert CM. Chronic Enteroviral Meningoencephalitis in Agammaglobulinemic Patients. *Rev Infect Dis.* 1987; 9:334-356.

127. Wilfert CM, MacCormack JN, Kleeman K, et al. The Prevalence of Antibodies to Rickettsia Rickettsii in an Area Endemic for Rocky Mountain Spotted Fever. *J Infect Dis.* 1985; 151:823-831.

128. McKinney RE, Jr., Maha MA, Connor EM, et al. A Multicenter Trial of Oral Zidovudine in Children with Advanced Human Immunodeficiency Virus Disease. The Protocol 043 Study Group. *N Engl J Med.* 1991; 324:1018-1025.

129. Coutfoudis A, Coovadia HM, Wilfert CM. HIV, Infant Feeding and More Perils for Poor People: New Who Guidelines Encourage Review of Formula Milk Policies. *World Hosp Health Serv.* 2008; 44:45-48.

130. Stringer EM, Ekouevi DK, Coetzee D, et al. Coverage of Nevirapine-Based Services to Prevent Mother-to-Child HIV Transmission in 4 African Countries. *JAMA.* 2010; 304:293-302.

131. Kieffer MP, Nhlabatsi B, Mahdi M, Hoffman HJ, Kudiabor K, Wilfert CM. Improved Detection of Incident HIV Infection and Uptake of Pmtct Services in Labor and Delivery in a High HIV Prevalence Setting. *J Acquir Immune Defic Syndr.* 2011; 57:e85-91.

132. Fiscus SA, Adimora AA, Schoenbach VJ, et al. Perinatal HIV Infection and the Effect of Zidovudine Therapy on Transmission in Rural and Urban Counties. *JAMA.* 1996; 275:1483-1488.

133. Grundmann N, Iliff P, Stringer J, Wilfert C. Presumptive Diagnosis of Severe HIV Infection to Determine the Need for Antiretroviral Therapy in Children Less Than 18 Months of Age. *Bull World Health Organ.* 2011; 89:513-520.

134. Griffith JF, Katz SL. Subacute Sclerosing Panencephalitis, Laboratory Findings in 6 Cases. *Neurology.* 1968; 18:98-100.

135. Parker JC, Jr., Klintworth GK, Graham DG, Griffith JF. Uncommon Morphologic Features in Subacute Sclerosing Panencephalitis (Sspe). Report of Two Cases with Virus Recovery from One Autopsy Brain Specimen. *Am J Pathol.* 1970; 61:275-292.

136. Gutman LT, Moye J, Zimmer B, Tian C. Tuberculosis in Human Immunodeficiency Virus-Exposed or -Infected United States Children. *Pediatr Infect Dis J.* 1994; 13:963-968.

137. Gutman LT, Herman-Giddens ME, Phelps WC. Transmission of Human Genital Papillomavirus Disease: Comparison of Data from Adults and Children. *Pediatrics.* 1993; 91:31-38.

138. Gutman LT, St Claire KK, Weedy C, et al. Human Immunodeficiency Virus Transmission by Child Sexual Abuse. *Am J Dis Child.* 1991; 145:137-141.

139. Gutman LT, St Claire KK, Everett VD, et al. Cervical-Vaginal and Intraanal Human Papillomavirus Infection of Young Girls with External Genital Warts. *J Infect Dis.* 1994; 170:339-344.

140. Gutman LT, Ottesen EA, Quan TJ, Noce PS, Katz SL. An Inter-Familial Outbreak of Yersinia Enterocolitica Enteritis. *N Engl J Med.* 1973; 288:1372-1377.

141. Gutman LT, Wilfert CM, Eppes S. Herpes Simplex Virus Encephalitis in Children: Analysis of Cerebrospinal Fluid and Progressive Neurodevelopmental Deterioration. *J Infect Dis.* 1986; 154:415-421.

142. Rudd C, Rivadeneira ED, Gutman LT. Dosing Considerations for Oral Acyclovir Following Neonatal Herpes Disease. *Acta Paediatr.* 1994; 83:1237-1243.

143. Tiffany KF, Benjamin DK, Jr., Palasanthiran P, O'Donnell K, Gutman LT. Improved Neurodevelopmental Outcomes Following Long-Term High-Dose Oral Acyclovir Therapy in Infants with Central Nervous System and Disseminated Herpes Simplex Disease. *J Perinatol.* 2005; 25:156-161.

144. McKinney RE, Jr., Pizzo PA, Scott GB, et al. Safety and Tolerance of Intermittent Intravenous and Oral Zidovudine Therapy in Human Immunodeficiency Virus-Infected Pediatric Patients. Pediatric Zidovudine Phase I Study Group. *J Pediatr.* 1990; 116:640-647.

145. Tudor-Williams G, St Clair MH, McKinney RE, et al. HIV-1 Sensitivity to Zidovudine and Clinical Outcome in Children. *Lancet.* 1992; 339:15-19.

146. McKinney RE, Jr., Wilfert C. Growth as a Prognostic Indicator in Children with Human Immunodeficiency Virus Infection Treated with Zidovudine. AIDS Clinical Trials Group Protocol 043 Study Group. *J Pediatr.* 1994; 125:728-733.

147. Englund JA, Baker CJ, Raskino C, et al. Zidovudine, Didanosine, or Both as the Initial Treatment for Symptomatic HIV-Infected Children. AIDS Clinical Trials Group (ACTG) Study 152 Team. *N Engl J Med*. 1997; 336:1704-1712.

148. McKinney RE, Jr., Johnson GM, Stanley K, et al. A Randomized Study of Combined Zidovudine-Lamivudine Versus Didanosine Monotherapy in Children with Symptomatic Therapy-Naive HIV-1 Infection. The Pediatric AIDS Clinical Trials Group Protocol 300 Study Team. *J Pediatr*. 1998; 133:500-508.

149. Connor EM, Sperling RS, Gelber R, et al. Reduction of Maternal-Infant Transmission of Human Immunodeficiency Virus Type 1 with Zidovudine Treatment. Pediatric AIDS Clinical Trials Group Protocol 076 Study Group. *N Engl J Med*. 1994; 331:1173-1180.

150. LaFon SW, Lehrman SN, Barry DW. Prophylactically Administered Retrovir in Health Care Workers Potentially Exposed to the Human Immunodeficiency Virus. *J Infect Dis*. 1988; 158:503.

151. Surbone A, Yarchoan R, McAtee N, et al. Treatment of the Acquired Immunodeficiency Syndrome (AIDS) and AIDS-Related Complex with a Regimen of 3'-Azido-2',3'-Dideoxythymidine (Azidothymidine or Zidovudine) and Acyclovir. A Pilot Study. *Ann Intern Med*. 1988; 108:534-540.

152. Parks WP, Parks ES, Fischl MA, et al. HIV-1 Inhibition by Azidothymidine in a Concurrently Randomized Placebo-Controlled Trail. *J Aquired Immune Deficienct Syndrome*. 1988; 1:125-130.

153. Pizzo PA, Eddy J, Falloon J, et al. Effect of Continuous Intravenous Infusion of Zidovudine (AZT) in Children with Symptomatic HIV Infection. *N Engl J Med*. 1988; 319:889-896.

154. Creagh-Kirk T, Doi P, Andrews E, et al. Survival Experience among Patients with AIDS Receiving Zidovudine. Follow-up of Patients in a Compassionate Plea Program. *JAMA*. 1988; 260:3009-3015.

155. Hollander H, Lifson AR, Maha M, Blum R, Rutherford GW, Nusinoff-Lehrman S. Phase I Study of Low-Dose Zidovudine and Acyclovir in Asymptomatic Human Immunodeficiency Virus Seropositive Individuals. *Am J Med*. 1989; 87:628-632.

156. Patel DD, Ray CA, Drucker RP, Pickup DJ. A Poxvirus-Derived Vector That Directs High Levels of Expression of Cloned Genes in Mammalian Cells. *Proc Natl Acad Sci U S A*. 1988; 85:9431-9435.

157. Seaworth B, Drucker J, Starling J, Drucker R, Stevens C, Hamilton J. Hepatitis B Vaccines in Patients with Chronic Renal Failure before Dialysis. *J Infect Dis*. 1988; 157:332-337.

158. Walter EB, Drucker RP, Clements DA. A Major Barrier to Universal Hepatitis B Immunization of Infants. *Arch Pediatr Adolesc Med*. 1994; 148:538-539.

159. Combs SP, Walter EB, Drucker RP, Clements DA. Removing a Major Barrier to Universal Hepatitis B Immunization in Infants. *Arch Pediatr Adolesc Med.* 1996; 150:112-114.

160. Walter E, Sung J, Kahn Meine E, Drucker RP, Clements DA. Lack of Effectiveness of a Letter Reminder for Annual Influenza Immunization of Asthmatic Children. *Pediatr Infect Dis J.* 1997; 16:1187-1188.

161. Meine EK, Bailey SR, Drucker RP, Clements DA, Walter E. Varicella Vaccination in a Primary Care Pediatric Practice. *Arch Pediatr Adolesc Med.* 1998; 152:608-609.

162. Walter EB, Moggio MV, Drucker RP, Wilfert CM. Immunogenicity of Haemophilus B Conjugate Vaccine (Meningococcal Protein Conjugate) in Children with Prior Invasive Haemophilus Influenzae Type B Disease. *Pediatr Infect Dis J.* 1990; 9:632-635.

163. Walter EB, Drucker RP, McKinney RE, Wilfert CM. Myopathy in Human Immunodeficiency Virus-Infected Children Receiving Long-Term Zidovudine Therapy. *J Pediatr.* 1991; 119:152-155.

164. Walter EB, Elliott AJ, Regan AN, Drucker RP, Clements DA, Wilfert CM. Maternal Acceptance of Voluntary Human Immunodeficiency Virus Antibody Testing During the Newborn Period with the Guthrie Card. *Pediatr Infect Dis J.* 1995; 14:376-381.

165. Laskowitz DT, Drucker RP, Parsonnet J, Cross PC, Gesundheit N. Engaging Students in Dedicated Research and Scholarship During Medical School: The Long-Term Experiences at Duke and Stanford. *Acad Med.* 2010; 85:419-428.

166. Rennels MB, Hohenboken MJ, Reisinger KS, et al. Comparison of Acellular Pertussis-Diphtheria-Tetanus Toxoids and Haemophilus Influenzae Type B Vaccines Administered Separately Vs. Combined in Younger Vs. Older Toddlers. *Pediatr Infect Dis J.* 1998; 17:164-166.

167. Clements DA, Armstrong CB, Ursano AM, Moggio MM, Walter EB, Wilfert CM. Over Five-Year Follow-up of Oka/Merck Varicella Vaccine Recipients in 465 Infants and Adolescents. *Pediatr Infect Dis J.* 1995; 14:874-879.

168. Clements DA, Zaref JI, Bland CL, Walter EB, Coplan PM. Partial Uptake of Varicella Vaccine and the Epidemiological Effect on Varicella Disease in 11 Day-Care Centers in North Carolina. *Arch Pediatr Adolesc Med.* 2001; 155:455-461.

169. Clements DA, Langdon L, Bland C, Walter E. Influenza a Vaccine Decreases the Incidence of Otitis Media in 6- to 30-Month-Old Children in Day Care. *Arch Pediatr Adolesc Med.* 1995; 149:1113-1117.

170. Kuter B, Matthews H, Shinefield H, et al. Ten Year Follow-up of Healthy Children Who Received One or Two Injections of Varicella Vaccine. *Pediatr Infect Dis J.* 2004; 23:132-137.

171. Alexander KA, Phelps WC. A Fluorescence Anisotropy Study of DNA Binding by Hpv-11 E2c Protein: A Hierarchy of E2-Binding Sites. *Biochemistry.* 1996; 35:9864-9872.

172. Alexander KA, Phelps WC. Recent Advances in Diagnosis and Therapy of Human Papillomaviruses. *Expert opinion on investigational drugs.* 2000; 9:1753-1765.

173. Benjamin DK, Jr., Miller WC, Bayliff S, Martel L, Alexander KA, Martin PL. Infections Diagnosed in the First Year after Pediatric Stem Cell Transplantation. *Pediatr Infect Dis J.* 2002; 21:227-234.

174. Dixon EP, Pahel GL, Rocque WJ, et al. The E1 Helicase of Human Papillomavirus Type 11 Binds to the Origin of Replication with Low Sequence Specificity. *Virology.* 2000; 270:345-357.

175. Hartley KA, Alexander KA. Human Tata Binding Protein Inhibits Human Papillomavirus Type 11 DNA Replication by Antagonizing E1-E2 Protein Complex Formation on the Viral Origin of Replication. *J Virol.* 2002; 76:5014-5023.

176. Phelps WC, Alexander KA. Antiviral Therapy for Human Papillomaviruses: Rational and Prospects. *Ann Intern Med.* 1995; 123:368-382.

177. Englund JA, Walter EB, Fairchok MP, Monto AS, Neuzil KM. A Comparison of 2 Influenza Vaccine Schedules in 6- to 23-Month-Old Children. *Pediatrics.* 2005; 115:1039-1047.

178. Englund JA, Walter EB, Gbadebo A, Monto AS, Zhu Y, Neuzil KM. Immunization with Trivalent Inactivated Influenza Vaccine in Partially Immunized Toddlers. *Pediatrics.* 2006; 118:e579-585.

179. Walter EB, Neuzil KM, Zhu Y, et al. Influenza Vaccine Immunogenicity in 6- to 23-Month-Old Children: Are Identical Antigens Necessary for Priming? *Pediatrics.* 2006; 118:e570-578.

180. Walter EB, Allred N, Rowe-West B, Chmielewski K, Kretsinger K, Dolor RJ. Cocooning Infants: Tdap Immunization for New Parents in the Pediatric Office. *Acad Pediatr.* 2009; 9:344-347.

181. Walter EB, Englund JA, Blatter M, Nyberg J, Ruben FL, Decker MD. Trivalent Inactivated Influenza Virus Vaccine Given to Two-Month-Old Children: An Off-Season Pilot Study. *Pediatr Infect Dis J.* 2009; 28:1099-1104.

182. Walter EB, Allred NJ, Swamy GK, Hellkamp AS, Dolor RJ. Influenza Vaccination of Household Contacts of Newborns: A Hospital-Based Strategy to Increase Vaccination Rates. *Infect Control Hosp Epidemiol.* 2010; 31:1070-1073.

183. Tan W, Viera AJ, Rowe-West B, Grimshaw A, Quinn B, Walter EB. The Hpv Vaccine: Are Dosing Recommendations Being Followed? *Vaccine.* 2011; 29:2548-2554.

184. Chen WH, Winokur PL, Edwards KM, et al. Phase 2 Assessment of the Safety and Immunogenicity of Two Inactivated Pandemic Monovalent H1n1

Vaccines in Adults as a Component of the U.S. Pandemic Preparedness Plan in 2009. *Vaccine.* 2012; 30:4240-4248.

185. Cate TR, Couch RB, Fleet WF, Griffith WR, Gerone PJ, Knight V. Production of Tracheobronchitis in Volunteers with Rhinovirus in a Small-Particle Aerosol. *Am J Epidemiol.* 1965; 81:95-105.

186. Cate TR, Couch RB, Johnson KM. Studies with Rhinoviruses in Volunteers: Production of Illness, Effect of Naturally Acquired Antibody, and Demonstration of a Protective Effect Not Associated with Serum Antibody. *J Clin Invest.* 1964; 43:56-67.

187. Cate TR, Douglas RG, Jr., Couch RB. Interferon and Resistance to Upper Respiratory Virus Illness. *Proc Soc Exp Biol Med.* 1969; 131:631-636.

188. Cate TR, Douglas RG, Jr., Johnson KM, Couch RB, Knight V. Studies on the Inability of Rhinovirus to Survive and Replicate in the Intestinal Tract of Volunteers. *Proc Soc Exp Biol Med.* 1967; 124:1290-1295.

189. Cate TR, Kelly JR. Hong Kong Influenza Antigen Sensitivity and Decreased Interferon Response of Peripheral Lymphocytes. *Antimicrob Agents Chemother (Bethesda).* 1970; 10:156-160.

190. Cate TR, Mold NG. Increased Influenza Pneumonia Mortality of Mice Adoptively Immunized with Node and Spleen Cells Sensitized by Inactivated but Not Live Virus. *Infect Immun.* 1975; 11:908-914.

191. Hamilton JD, Elliott DM. Combined Activity of Amphotericin B and 5-Fluorocytosine against Cryptococcus Neoformans in Vitro and in Vivo in Mice. *J Infect Dis.* 1975; 131:129-137.

192. Hamilton JD, Bradley BA, Rood JJV. Leukemia-Specific Human Cytotoxic-T Cells. *Clin Res.* 1978; 26:A797-A797.

193. Goulmy E, Hamilton JD, Bradley BA. Anti-Self Hla May Be Clonally Expressed. *J Exp Med.* 1979; 149:545-550.

194. Hamilton JD, Bradley BA, Vanrood JJ. Monolayer Absorption of Human Cytotoxic T Cells: Evidence for Clonality. *Tissue Antigens.* 1979; 13:349-356.

195. Gilligan P. In Memorium. *Microbe Magazine.* 2009.

196. Durack DT, Beeson PB. Experimental Bacterial Endocarditis. I. Colonization of a Sterile Vegetation. *Br J Exp Pathol.* 1972; 53:44-49.

197. Durack DT, Beeson PB, Petersdorf RG. Experimental Bacterial Endocarditis. 3. Production and Progress of the Disease in Rabbits. *Br J Exp Pathol.* 1973; 54:142-151.

198. Durack DT, Petersdorf RG. Chemotherapy of Experimental Streptococcal Endocarditis. I. Comparison of Commonly Recommended Prophylactic Regimens. *J Clin Invest.* 1973; 52:592-598.

199. Durack DT, Lukes AS, Bright DK. New Criteria for Diagnosis of Infective Endocarditis: Utilization of Specific Echocardiographic Findings. Duke Endocarditis Service. *Am J Med.* 1994; 96:200-209.

200. Durack DT. Prevention of Infective Endocarditis. *N Engl J Med.* 1995; 332:38-44.

201. Durack DT. Letter: Chemoprophylaxis of Bacterial Endocarditis. *N Engl J Med.* 1975; 292:1080-1081.

202. Durack DT, Sumi SM, Klebanoff SJ. Neurotoxicity of Human Eosinophils. *Proc Natl Acad Sci U S A.* 1979; 76:1443-1447.

203. Durack DT, Ackerman SJ, Loegering DA, Gleich GJ. Purification of Human Eosinophil-Derived Neurotoxin. *Proc Natl Acad Sci U S A.* 1981; 78:5165-5169.

204. Durack DT. Opportunistic Infections and Kaposi's Sarcoma in Homosexual Men. *N Engl J Med.* 1981; 305:1465-1467.

205. Fischl MA, Richman DD, Grieco MH, et al. The Efficacy of Azidothymidine (AZT) in the Treatment of Patients with AIDS and AIDS-Related Complex. A Double-Blind, Placebo-Controlled Trial. *N Engl J Med.* 1987; 317:185-191.

206. Wyngaarden JB. *An Unimagined Life.* Durham, North Carolina: BW&A Books, Inc; 2009.

207. Bonner RE, Crevasse L, Ferrer MI, Greenfield JC, Jr. A New Computer Program for Comparative Analysis of Serial Scalar Electrocardiograms: Description and Performance of the 1976 Ibm Program. *Comput Biomed Res.* 1978; 11:103-118.

208. Chamis AL, Peterson GE, Cabell CH, et al. Staphylococcus Aureus Bacteremia in Patients with Permanent Pacemakers or Implantable Cardioverter-Defibrillators. *Circulation.* 2001; 104:1029-1033.

209. Greenfield JC, Jr., Cox RL. Instantaneous Pressure-Flow-Length Relationships in the Intact Human Heart. *Am J Med Sci.* 1968; 255:288-291.

210. Greenfield JC, Jr., Patel DJ. Relation between Pressure and Diameter in the Ascending Aorta of Man. *Circ Res.* 1962; 10:778-781.

211. Greenfield JC, Jr., Rembert JC, Young WG, Jr., Oldham HN, Jr., Alexander JA, Sabiston DC, Jr. Studies of Blood Flow in Aorta-to-Coronary Venous Bypass Grafts in Man. *J Clin Invest.* 1972; 51:2724-2735.

212. Greenfield JC, Jr., Tindall GT. Effect of Norepinephrine, Epinephrine, and Angiotensin on Blood Flow in the Internal Carotid Artery of Man. *J Clin Invest.* 1968; 47:1672-1684.

213. Greenfield JC, Jr., Tindall GT, Dillon ML, Mahaley MS. Mechanics of the Human Common Carotid Artery in Vivo. *Circ Res.* 1964; 15:240-246.

214. Hernandez RR, Greenfield JC, Jr., McCall BW. Pressure-Flow Studies in Hypertrophic Subaortic Stenosis. *J Clin Invest.* 1964; 43:401-407.

215. Moser KM, Hull R, Saltzman HA, Dantzker DR, Goldhaber SZ, Greenfield LJ. Recent Advances in Diagnosis of Pulmonary Embolism and Deep Venous Thrombosis. *Am Rev Respir Dis.* 1988; 138:1046-1047.

216. Rembert JC, Kleinman LH, Fedor JM, Wechsler AS, Greenfield JC, Jr. Myocardial Blood Flow Distribution in Concentric Left Ventricular Hypertrophy. *J Clin Invest.* 1978; 62:379-386.

217. Sadick N, Dube GP, McHale PA, Greenfield JC, Jr. Metabolic Mediation of Single Brief Diastolic Occlusion Reactive Hyperemic Responses. *Am J Physiol.* 1987; 253:H25-30.

218. Sexton DJ, Corey GR, Greenfield JC, Jr., Burton CS, Raoult D. Imported African Tick Bite Fever: A Case Report. *Am J Trop Med Hyg.* 1999; 60:865-867.

219. Cohen JI, Corey GR. Cytomegalovirus Infection in the Normal Host. *Medicine.* 1985; 64:100-114.

220. Sexton DJ, Corey GR. Rocky Mountain "Spotless" and "Almost Spotless" Fever: A Wolf in Sheep's Clothing. *Clin Infect Dis.* 1992; 15:439-448.

221. Corey GR, Campbell PT, Van Trigt P, et al. Etiology of Large Pericardial Effusions. *Am J Med.* 1993; 95:209-213.

222. Sexton DJ, Kanj SS, Wilson K, et al. The Use of a Polymerase Chain Reaction as a Diagnostic Test for Rocky Mountain Spotted Fever. *Am J Trop Med Hyg.* 1994; 50:59-63.

223. Miller WC, Corey GR, Lallinger GJ, Durack DT. International Health and Internal Medicine Residency Training: The Duke University Experience. *Am J Med.* 1995; 99:291-297.

224. Weinstein JW, Roe M, Towns M, et al. Resistant Enterococci: A Prospective Study of Prevalence, Incidence, and Factors Associated with Colonization in a University Hospital. *Infect Control Hosp Epidemiol.* 1996; 17:36-41.

225. Sexton DJ, Corey GR, Carpenter C, et al. Dual Infection with Ehrlichia Chaffeensis and a Spotted Fever Group Rickettsia: A Case Report. *Emerg Infect Dis.* 1998; 4:311-316.

226. Carpenter CF, Gandhi TK, Kong LK, et al. The Incidence of Ehrlichial and Rickettsial Infection in Patients with Unexplained Fever and Recent History of Tick Bite in Central North Carolina. *J Infect Dis.* 1999; 180:900-903.

227. Granger DL, Perfect JR, Durack DT. Virulence of Cryptococcus Neoformans. Regulation of Capsule Synthesis by Carbon Dioxide. *J Clin Invest.* 1985; 76:508-516.

228. Granger DL, Hibbs JB, Jr., Perfect JR, Durack DT. Specific Amino Acid (L-Arginine) Requirement for the Microbiostatic Activity of Murine Macrophages. *J Clin Invest.* 1988; 81:1129-1136.

229. Granger DL, Hibbs JB, Jr., Perfect JR, Durack DT. Metabolic Fate of L-Arginine in Relation to Microbiostatic Capability of Murine Macrophages. *J Clin Invest.* 1990; 85:264-273.

230. Granger DL, Perfect JR, Durack DT. Macrophage-Mediated Fungistasis: Requirement for a Macromolecular Component in Serum. *J Immunol.* 1986; 137:693-701.

231. Granger DL, Perfect JR, Durack DT. Macrophage-Mediated Fungistasis in Vitro: Requirements for Intracellular and Extracellular Cytotoxicity. *J Immunol.* 1986; 136:672-680.

232. Alspaugh JA, Granger DL. Inhibition of Cryptococcus Neoformans Replication by Nitrogen Oxides Supports the Role of These Molecules as Effectors of Macrophage-Mediated Cytostasis. *Infect Immun.* 1991; 59:2291-2296.

233. Granger DL, Hibbs JB, Jr., Broadnax LM. Urinary Nitrate Excretion in Relation to Murine Macrophage Activation. Influence of Dietary L-Arginine and Oral Ng-Monomethyl-L-Arginine. *J Immunol.* 1991; 146:1294-1302.

234. Naslund PK, Miller WC, Granger DL. Cryptococcus Neoformans Fails to Induce Nitric Oxide Synthase in Primed Murine Macrophage-Like Cells. *Infect Immun.* 1995; 63:1298-1304.

235. Ross JS, Keyhani S, Keenan PS, et al. Use of Recommended Ambulatory Care Services: Is the Veterans Affairs Quality Gap Narrowing? *Arch Intern Med.* 2008; 168:950-958.

236. Boockvar KS, Granger DL, Poston RM, et al. Nitric Oxide Produced During Murine Listeriosis Is Protective. *Infect Immun.* 1994; 62:1089-1100.

237. Anstey NM, Weinberg JB, Hassanali MY, et al. Nitric Oxide in Tanzanian Children with Malaria: Inverse Relationship between Malaria Severity and Nitric Oxide Production/Nitric Oxide Synthase Type 2 Expression. *J Exp Med.* 1996; 184:557-567.

238. Andersen GL, Simchock JM, Wilson KH. Identification of a Region of Genetic Variability among Bacillus Anthracis Strains and Related Species. *J Bacteriol.* 1996; 178:377-384.

239. Dawson JE, Anderson BE, Fishbein DB, et al. Isolation and Characterization of an Ehrlichia Sp. From a Patient Diagnosed with Human Ehrlichiosis. *J Clin Microbiol.* 1991; 29:2741-2745.

240. Wilson KH, Blitchington RB. Human Colonic Biota Studied by Ribosomal DNA Sequence Analysis. *Appl Environ Microbiol.* 1996; 62:2273-2278.

241. Wilson KH, Blitchington RB, Frothingham R, Wilson JA. Identification of the Whipple's Disease Bacillus. *N Engl J Med.* 1993; 328:62; author reply 63.

242. Wilson KH, Blitchington RB, Greene RC. Amplification of Bacterial 16s Ribosomal DNA with Polymerase Chain Reaction. *J Clin Microbiol.* 1990; 28:1942-1946.

243. Wilson KH, Freter R. Interaction of Clostridium Difficile and Escherichia Coli with Microfloras in Continuous-Flow Cultures and Gnotobiotic Mice. *Infect Immun.* 1986; 54:354-358.

244. Wilson KH, Perini F. Role of Competition for Nutrients in Suppression of Clostridium Difficile by the Colonic Microflora. *Infect Immun.* 1988; 56:2610-2614.

245. Wilson KH, Silva J, Fekety FR. Suppression of Clostridium Difficile by Normal Hamster Cecal Flora and Prevention of Antibiotic-Associated Cecitis. *Infect Immun.* 1981; 34:626-628.

246. Wilson KH, Wilson WJ, Radosevich JL, et al. High-Density Microarray of Small-Subunit Ribosomal DNA Probes. *Appl Environ Microbiol.* 2002; 68:2535-2541.

247. Agner RC, Gallis HA. Pericarditis: Differential Diagnostic Considerations. *Arch Intern Med.* 1979; 139:407-412.

248. Gallis HA, Berman RA, Cate TR, Hamilton JD, Gunnells JC, Stickel DL. Fungal Infection Following Renal Transplantation. *Arch Intern Med.* 1975; 135:1163-1172.

249. Smego RA, Jr., Moeller MB, Gallis HA. Trimethoprim-Sulfamethoxazole Therapy for Nocardia Infections. *Arch Intern Med.* 1983; 143:711-718.

250. Perfect JR, Durack DT, Gallis HA. Cryptococcemia. *Medicine.* 1983; 62:98-109.

251. Smego RA, Jr., Gallis HA. The Clinical Spectrum of Nocardia Brasiliensis Infection in the United States. *Rev Infect Dis.* 1984; 6:164-180.

252. Dismukes WE, Cloud G, Gallis HA, et al. Treatment of Cryptococcal Meningitis with Combination Amphotericin B and Flucytosine for Four as Compared with Six Weeks. *N Engl J Med.* 1987; 317:334-341.

253. Kim JH, Gallis HA. Observations on Spiraling Empiricism: Its Causes, Allure, and Perils, with Particular Reference to Antibiotic Therapy. *Am J Med.* 1989; 87:201-206.

254. Mitchell TG, Perfect JR. Cryptococcosis in the Era of AIDS--100 Years after the Discovery of Cryptococcus Neoformans. *Clin Microbiol Rev.* 1995; 8:515-548.

255. Toffaletti DL, Rude TH, Johnston SA, Durack DT, Perfect JR. Gene Transfer in Cryptococcus Neoformans by Use of Biolistic Delivery of DNA. *J Bacteriol.* 1993; 175:1405-1411.

256. Perfect JR, Lang SD, Durack DT. Chronic Cryptococcal Meningitis: A New Experimental Model in Rabbits. *Am J Pathol.* 1980; 101:177-194.

257. Perfect JR, Dismukes WE, Dromer F, et al. Clinical Practice Guidelines for the Management of Cryptococcal Disease: 2010 Update by the Infectious Diseases Society of America. *Clin Infect Dis.* 2010; 50:291-322.

258. Petzold EW, Himmelreich U, Mylonakis E, et al. Characterization and Regulation of the Trehalose Synthesis Pathway and Its Importance in the Pathogenicity of Cryptococcus Neoformans. *Infect Immun.* 2006; 74:5877-5887.

259. Kanj SS, Tapson V, Davis RD, Madden J, Browning I. Infections in Patients with Cystic Fibrosis Following Lung Transplantation. *Chest.* 1997; 112:924-930.

260. Kanj SS, Welty-Wolf K, Madden J, et al. Fungal Infections in Lung and Heart-Lung Transplant Recipients. Report of 9 Cases and Review of the Literature. *Medicine.* 1996; 75:142-156.

261. Kanj SS, Sharara AI, Clavien PA, Hamilton JD. Cytomegalovirus Infection Following Liver Transplantation: Review of the Literature. *Clin Infect Dis.* 1996; 22:537-549.

262. Kanj S, Hamilton J. The 10 Most Common Questions About Cytomegalovirus Infection in Solid Organ Transplant Recipients. *Infectious Diseases in Clinical Practice.* 1997; 6:29-32.

263. Dukes CS, Sugarman J, Cegielski JP, Lallinger GJ, Mwakyusa DH. Severe Cutaneous Hypersensitivity Reactions During Treatment of Tuberculosis in Patients with HIV Infection in Tanzania. *Trop Geogr Med.* 1992; 44:308-311.

264. Dukes CS, Matthews TJ, Weinberg JB. Human Immunodeficiency Virus Type 1 Infection of Human Monocytes and Macrophages Does Not Alter Their Ability to Generate an Oxidative Burst. *J Infect Dis.* 1993; 168:459-462.

265. Dukes CS, Yu Y, Rivadeneira ED, et al. Cellular CD44s as a Determinant of Human Immunodeficiency Virus Type 1 Infection and Cellular Tropism. *J Virol.* 1995; 69:4000-4005.

266. Glickman SW, Rasiel EB, Hamilton CD, Kubataev A, Schulman KA. Medicine. A Portfolio Model of Drug Development for Tuberculosis. *Science.* 2006; 311:1246-1247.

267. Stout JE, Saharia KK, Nageswaran S, Ahmed A, Hamilton CD. Racial and Ethnic Disparities in Pediatric Tuberculosis in North Carolina. *Arch Pediatr Adolesc Med.* 2006; 160:631-637.

268. Holland DP, Sanders GD, Hamilton CD, Stout JE. Costs and Cost-Effectiveness of Four Treatment Regimens for Latent Tuberculosis Infection. *Am J Respir Crit Care Med.* 2009; 179:1055-1060.

269. Velez DR, Hulme WF, Myers JL, et al. Nos2a, Tlr4, and Ifngr1 Interactions Influence Pulmonary Tuberculosis Susceptibility in African-Americans. *Hum Genet.* 2009; 126:643-653.

270. Sterling TR, Villarino ME, Borisov AS, et al. Three Months of Rifapentine and Isoniazid for Latent Tuberculosis Infection. *N Engl J Med.* 2011; 365:2155-2166.

271. Hamilton CD, Stout JE, Goodman PC, et al. The Value of End-of-Treatment Chest Radiograph in Predicting Pulmonary Tuberculosis Relapse. *Int J Tuberc Lung Dis.* 2008; 12:1059-1064.

272. Dukes Hamilton C, Sterling TR, Blumberg HM, et al. Extensively Drug-Resistant Tuberculosis: Are We Learning from History or Repeating It? *Clin Infect Dis.* 2007; 45:338-342.

273. Marr KA, Sexton DJ, Conlon PJ, Corey GR, Schwab SJ, Kirkland KB. Catheter-Related Bacteremia and Outcome of Attempted Catheter Salvage in Patients Undergoing Hemodialysis. *Ann Intern Med.* 1997; 127:275-280.

274. Kirkland KB, Briggs JP, Trivette SL, Wilkinson WE, Sexton DJ. The Impact of Surgical-Site Infections in the 1990s: Attributable Mortality, Excess Length of Hospitalization, and Extra Costs. *Infect Control Hosp Epidemiol.* 1999; 20:725-731.

275. Sexton DJ, Rollin PE, Breitschwerdt EB, et al. Life-Threatening Cache Valley Virus Infection. *N Engl J Med.* 1997; 336:547-549.

276. Fowler VG, Sanders LL, Sexton DJ, et al. Outcome of Staphylococcus Aureus Bacteremia According to Compliance with Infectious Diseases Specialist Recommendations: Experience with 244 Patients. *Clin Infect Dis.* 1998; 27:478-486.

277. Sexton DJ, Tenenbaum MJ, Wilson WR, et al. Ceftriaxone Once Daily for Four Weeks Compared with Ceftriaxone Plus Gentamicin Once Daily for Two Weeks for Treatment of Endocarditis Due to Penicillin-Susceptible Streptococci. Endocarditis Treatment Consortium Group. *Clin Infect Dis.* 1998; 27:1470-1474.

278. Friedman ND, Kaye KS, Stout JE, et al. Health Care--Associated Bloodstream Infections in Adults: A Reason to Change the Accepted Definition of Community-Acquired Infections. *Ann Intern Med.* 2002; 137:791-797.

279. McDonald JR, Friedman ND, Stout JE, Sexton DJ, Kaye KS. Risk Factors for Ineffective Therapy in Patients with Bloodstream Infection. *Arch Intern Med.* 2005; 165:308-313.

280. Anderson DJ, Richet H, Chen LF, et al. Seasonal Variation in Klebsiella Pneumoniae Bloodstream Infection on 4 Continents. *J Infect Dis.* 2008; 197:752-756.

281. Kaye KS, Anderson DJ, Choi Y, Link K, Thacker P, Sexton DJ. The Deadly Toll of Invasive Methicillin-Resistant Staphylococcus Aureus Infection in Community Hospitals. *Clin Infect Dis.* 2008; 46:1568-1577.

282. Cegielski JP, Ramiya K, Lallinger GJ, Mtulia IA, Mbaga IM. Pericardial Disease and Human Immunodeficiency Virus in Dar Es Salaam, Tanzania. *Lancet.* 1990; 335:209-212.

283. Miller WC, Shao JF, Weaver DJ, Shimokura GH, Paul DA, Lallinger GJ. Seroprevalence of Viral Hepatitis in Tanzanian Adults. *Trop Med Int Health.* 1998; 3:757-763.

284. Miller WC, Thielman NM, Swai N, et al. Diagnosis and Screening of HIV/AIDS Using Clinical Criteria in Tanzanian Adults. *J Acquir Immune Defic Syndr Hum Retrovirol.* 1995; 9:408-414.

285. Miller WC, Thielman NM, Swai N, et al. Delayed-Type Hypersensitivity Testing in Tanzanian Adults with HIV Infection. *J Acquir Immune Defic Syndr Hum Retrovirol*. 1996; 12:303-308.

286. Bartlett JA, Joklik WK. The Sequence of the Reovirus Serotype 3 L3 Genome Segment Which Encodes the Major Core Protein Lambda 1. *Virology*. 1988; 167:31-37.

287. Bartlett JA, DeMasi R, Dawson D, Hill A. Variability in Repeated Consecutive Measurements of Plasma Human Immunodeficiency Virus RNA in Persons Receiving Stable Nucleoside Reverse Transcriptase Inhibitor Therapy or No Treatment. *J Infect Dis*. 1998; 178:1803-1805.

288. Bartlett JA, Wasserman SS, Hicks CB, et al. Safety and Immunogenicity of an Hla-Based HIV Envelope Polyvalent Synthetic Peptide Immunogen. DATRI 010 Study Group. Division of AIDS Treatment Research Initiative. *AIDS*. 1998; 12:1291-1300.

289. Bartlett JA, DeMasi R, Quinn J, Moxham C, Rousseau F. Overview of the Effectiveness of Triple Combination Therapy in Antiretroviral-Naive HIV-1 Infected Adults. *AIDS*. 2001; 15:1369-1377.

290. Bartlett JA, Miralles GD, Sevin AD, et al. Addition of Cyclophosphamide to Antiretroviral Therapy Does Not Diminish the Cellular Reservoir in HIV-Infected Persons. *AIDS Res Hum Retroviruses*. 2002; 18:535-543.

291. Bartlett JA, Fath MJ, Demasi R, et al. An Updated Systematic Overview of Triple Combination Therapy in Antiretroviral-Naive HIV-Infected Adults. *AIDS*. 2006; 20:2051-2064.

292. Bartlett JA, Johnson J, Herrera G, et al. Long-Term Results of Initial Therapy with Abacavir and Lamivudine Combined with Efavirenz, Amprenavir/Ritonavir, or Stavudine. *J Acquir Immune Defic Syndr*. 2006; 43:284-292.

293. Bartlett JA, Ribaudo HJ, Wallis CL, et al. Lopinavir/Ritonavir Monotherapy after Virologic Failure of First-Line Antiretroviral Therapy in Resource-Limited Settings. *AIDS*. 2012; 26:1345-1354.

294. Crump JA, Ramadhani HO, Morrissey AB, et al. Bacteremic Disseminated Tuberculosis in Sub-Saharan Africa: A Prospective Cohort Study. *Clin Infect Dis*. 2012; 55:242-250.

295. Hicks CB, Benson PM, Lupton GP, Tramont EC. Seronegative Secondary Syphilis in a Patient Infected with the Human Immunodeficiency Virus (HIV) with Kaposi Sarcoma. A Diagnostic Dilemma. *Ann Intern Med*. 1987; 107:492-495.

296. Hicks CB, Myers SA, Giner J. Resolution of Intractable Molluscum Contagiosum in a Human Immunodeficiency Virus-Infected Patient after Institution of Antiretroviral Therapy with Ritonavir. *Clin Infect Dis*. 1997; 24:1023-1025.

297. Fowler VG, Jr., Hicks CB, Kirkland KB. The Name Game: Lamivudine-Lamo-trigine Dispensing Error Presenting as Human Immunodeficiency Virus-As-sociated Fever of Unknown Origin. *Int J STD AIDS.* 1999; 10:685-686.

298. Ole-Nguyaine S, Crump JA, Kibiki GS, et al. HIV-Associated Morbidity, Mor-tality and Diagnostic Testing Opportunities among Inpatients at a Referral Hospital in Northern Tanzania. *Ann Trop Med Parasitol.* 2004; 98:171-179.

299. Ramadhani HO, Thielman NM, Landman KZ, et al. Predictors of Incomplete Adherence, Virologic Failure, and Antiviral Drug Resistance among HIV-In-fected Adults Receiving Antiretroviral Therapy in Tanzania. *Clin Infect Dis.* 2007; 45:1492-1498.

300. Thielman NM, Chu HY, Ostermann J, et al. Cost-Effectiveness of Free HIV Voluntary Counseling and Testing through a Community-Based AIDS Service Organization in Northern Tanzania. *Am J Public Health.* 2006; 96:114-119.

301. Cox GM, Mukherjee J, Cole GT, Casadevall A, Perfect JR. Urease as a Virulence Factor in Experimental Cryptococcosis. *Infect Immun.* 2000; 68:443-448.

302. Yue C, Cavallo L, Alspaugh JA, et al. The Ste12alpha Homolog Is Required for Haploid Filamentation but Largely Dispensable for Mating and Virulence in Cryptococcus Neoformans. *Genetics.* 1999; 153:1601-1615.

303. Lengeler KB, Wang P, Cox GM, Perfect JR, Heitman J. Identification of the Mata Mating-Type Locus of Cryptococcus Neoformans Reveals a Serotype a Mata Strain Thought to Have Been Extinct. *Proc Natl Acad Sci U S A.* 2000; 97:14455-14460.

304. Gorlach J, Fox DS, Cutler NS, Cox GM, Perfect JR, Heitman J. Identifi-cation and Characterization of a Highly Conserved Calcineurin Binding Protein, Cbp1/Calcipressin, in Cryptococcus Neoformans. *EMBO J.* 2000; 19:3618-3629.

305. Cox GM, Toffaletti DL, Perfect JR. Dominant Selection System for Use in Cryptococcus Neoformans. *J Med Vet Mycol.* 1996; 34:385-391.

306. Cruz MC, Sia RA, Olson M, Cox GM, Heitman J. Comparison of the Roles of Calcineurin in Physiology and Virulence in Serotype D and Serotype a Strains of Cryptococcus Neoformans. *Infect Immun.* 2000; 68:982-985.

307. Cox GM, Rude TH, Dykstra CC, Perfect JR. The Actin Gene from Crypto-coccus Neoformans: Structure and Phylogenetic Analysis. *J Med Vet Mycol.* 1995; 33:261-266.

308. Amos DB, Hattler BG, Hutchin P, McCloskey R, Zmijewski CM. Skin Donor Selection by Leucocyte Typing. *Lancet.* 1966; 1:300-302.

309. Baron S, Buckler CE, McCloskey RV, Kirschstein RL. Role of Interferon During Viremia. I. Production of Circulating Interferon. *J Immunol.* 1966; 96:12-16.

310. Baron S, Buckler CE, Friedman RM, McCloskey RV. Role of Interferon During Viremia. II. Protective Action of Circulating Interferon. *J Immunol.* 1966; 96:17-24.

311. McCloskey RV. Diphtheria Antitoxin Titers in Hospital Workers after a Single Dose of Adult Type Diphtheria Tetanus Toxoid. *Am J Med Sci.* 1969; 258:209-213.

312. McCloskey RV, Eller JJ, Green M, Mauney CU, Richards SE. The 1970 Epidemic of Diphtheria in San Antonio. *Ann Intern Med.* 1971; 75:495-503.

313. McCloskey RV, Green MJ, Eller J, Smilack J. Treatment of Diphtheria Carriers: Benzathine Penicillin, Erythromycin, and Clindamycin. *Ann Intern Med.* 1974; 81:788-791.

314. McCloskey RV, Saragea A, Maximescu P. Phage Typing in Diphtheria Outbreaks in the Southwestern United States, 1968-1971. *J Infect Dis.* 1972; 126:196-199.

315. McCloskey RV, Straube RC, Sanders C, Smith SM, Smith CR. Treatment of Septic Shock with Human Monoclonal Antibody Ha-1a. A Randomized, Double-Blind, Placebo-Controlled Trial. Chess Trial Study Group. *Ann Intern Med.* 1994; 121:1-5.

316. Sanders WE, Jr., Morris JF, Alessi P, et al. Oral Ofloxacin for the Treatment of Acute Bacterial Pneumonia: Use of a Nontraditional Protocol to Compare Experimental Therapy with "Usual Care" in a Multicenter Clinical Trial. *Am J Med.* 1991; 91:261-266.

317. LeFrock JL, McCloskey RV. Cefotaxime Treatment of Skin and Skin Structure Infections: A Multicenter Study. *Clin Ther.* 1982; 5 Suppl A:19-25.

318. McCloskey RV. Clinical and Bacteriologic Efficacy of Ceftriaxone in the United States. *Am J Med.* 1984; 77:97-103.

319. Keitel WA, Couch RB, Quarles JM, Cate TR, Baxter B, Maassab HF. Trivalent Attenuated Cold-Adapted Influenza Virus Vaccine: Reduced Viral Shedding and Serum Antibody Responses in Susceptible Adults. *J Infect Dis.* 1993; 167:305-311.

320. Keitel WA, Cate TR, Couch RB, Huggins LL, Hess KR. Efficacy of Repeated Annual Immunization with Inactivated Influenza Virus Vaccines over a Five Year Period. *Vaccine.* 1997; 15:1114-1122.

321. Keitel WA, Atmar RL, Cate TR, et al. Safety of High Doses of Influenza Vaccine and Effect on Antibody Responses in Elderly Persons. *Arch Intern Med.* 2006; 166:1121-1127.

322. Keitel WA, Campbell JD, Treanor JJ, et al. Safety and Immunogenicity of an Inactivated Influenza a/H5n1 Vaccine Given with or without Aluminum Hydroxide to Healthy Adults: Results of a Phase I-II Randomized Clinical Trial. *J Infect Dis.* 2008; 198:1309-1316.

323. Keitel WA, Bond NL, Zahradnik JM, Cramton TA, Robbins JB. Clinical and Serological Responses Following Primary and Booster Immunization with Salmonella Typhi Vi Capsular Polysaccharide Vaccine. *Vaccine.* 1994; 12:195-199.

324. Keitel WA, Muenz LR, Decker MD, et al. A Randomized Clinical Trial of Acellular Pertussis Vaccines in Healthy Adults: Dose-Response Comparisons of 5 Vaccines and Implications for Booster Immunization. *J Infect Dis.* 1999; 180:397-403.

325. Eppes SC, Troutman JL, Gutman LT. Outcome of Treatment of Candidemia in Children Whose Central Catheters Were Removed or Retained. *Pediatr Infect Dis J.* 1989; 8:99-104.

326. Rankin JT, Jr., Eppes SB, Antczak JB, Joklik WK. Studies on the Mechanism of the Antiviral Activity of Ribavirin against Reovirus. *Virology.* 1989; 168:147-158.

327. Frothingham R, Meeker-O'Connell WA, Talbot EA, George JW, Kreuzer KN. Identification, Cloning, and Expression of the Escherichia Coli Pyrazinamidase and Nicotinamidase Gene, Pnca. *Antimicrob Agents Chemother.* 1996; 40:1426-1431.

328. Talbot EA, Williams DL, Frothingham R. PCR Identification of Mycobacterium Bovis Bcg. *J Clin Microbiol.* 1997; 35:566-569.

329. Talbot EA, Reller LB, Frothingham R. Bone Marrow Cultures for the Diagnosis of Mycobacterial and Fungal Infections in Patients Infected with the Human Immunodeficiency Virus. *Int J Tuberc Lung Dis.* 1999; 3:908-912.

330. Teixeira L, Perkins MD, Johnson JL, et al. Infection and Disease among Household Contacts of Patients with Multidrug-Resistant Tuberculosis. *Int J Tuberc Lung Dis.* 2001; 5:321-328.

331. Talbot EA, Frothingham R. Meningitis Due to Mycobacterium Bovis Bcg--Reactivation or Accidental Intrathecal Inoculation? *Clin Infect Dis.* 1996; 23:1335-1336.

332. Talbot EA, Halabi S, Manchanda R, Mwansa RA, Wells CD. Knowledge, Attitudes, and Beliefs About Directly-Administered Antiretroviral Therapy among Tuberculosis Patients, Botswana 2002. *Int J STD AIDS.* 2004; 15:282-283.

333. Talbot EA, Adams LV, von Reyn CF. The Importance of Culture for Diagnosing Tuberculosis. *Clin Infect Dis.* 2005; 41:1213-1214; author reply 1214-1215.

334. Talbot EA, Kenyon TA, Moeti TL, et al. HIV Risk Factors among Patients with Tuberculosis--Botswana 1999. *Int J STD AIDS.* 2002; 13:311-317.

335. Cheng AC, Murdoch DR, Harrell LJ, Barth Reller L. Clinical Profile and Strain Relatedness of Recurrent Enterococcal Bacteremia. *Scand J Infect Dis.* 2005; 37:642-646.

336. Vann RD, Pollock NW, Pieper CF, et al. Statistical Models of Acute Mountain Sickness. *High Alt Med Biol.* 2005; 6:32-42.

337. Murdoch DR, Woods CW, Zimmerman MD, et al. The Etiology of Febrile Illness in Adults Presenting to Patan Hospital in Kathmandu, Nepal. *Am J Trop Med Hyg.* 2004; 70:670-675.

338. Anderson DJ, Murdoch DR, Sexton DJ, et al. Risk Factors for Infective Endocarditis in Patients with Enterococcal Bacteremia: A Case-Control Study. *Infection.* 2004; 32:72-77.

339. Vann RD, Pollock NW, Pieper CF, et al. Epidemiological Modeling of Acute Mountain Sickness (Ams). A Prospective Data Collection Standard. *Adv Exp Med Biol.* 2003; 543:355-358.

340. Murdoch DR, Mirrett S, Harrell LJ, Donabedian SM, Zervos MJ, Reller LB. Comparison of Microscan Broth Microdilution, Synergy Quad Plate Agar Dilution, and Disk Diffusion Screening Methods for Detection of High-Level Aminoglycoside Resistance in Enterococcus Species. *J Clin Microbiol.* 2003; 41:2703-2705.

341. Murdoch DR, Mirrett S, Harrell LJ, Monahan JS, Reller LB. Sequential Emergence of Antibiotic Resistance in Enterococcal Bloodstream Isolates over 25 Years. *Antimicrob Agents Chemother.* 2002; 46:3676-3678.

342. Murdoch DR, Reller LB. Antimicrobial Susceptibilities of Group B Streptococci Isolated from Patients with Invasive Disease: 10-Year Perspective. *Antimicrob Agents Chemother.* 2001; 45:3623-3624.

343. Murdoch DR, Laing RT, Mills GD, et al. Evaluation of a Rapid Immunochromatographic Test for Detection of Streptococcus Pneumoniae Antigen in Urine Samples from Adults with Community-Acquired Pneumonia. *J Clin Microbiol.* 2001; 39:3495-3498.

344. Miller DL, Brazer S, Murdoch D, Reller LB, Corey GR. Significance of Clostridium Tertium Bacteremia in Neutropenic and Nonneutropenic Patients: Review of 32 Cases. *Clin Infect Dis.* 2001; 32:975-978.

345. Murdoch DR, Roberts SA, Fowler Jr VG, Jr., et al. Infection of Orthopedic Prostheses after Staphylococcus Aureus Bacteremia. *Clin Infect Dis.* 2001; 32:647-649.

346. Cegielski JP, Ortega YR, McKee S, et al. Cryptosporidium, Enterocytozoon, and Cyclospora Infections in Pediatric and Adult Patients with Diarrhea in Tanzania. *Clin Infect Dis.* 1999; 28:314-321.

347. Cegielski JP, Devlin BH, Morris AJ, et al. Comparison of PCR, Culture, and Histopathology for Diagnosis of Tuberculous Pericarditis. *J Clin Microbiol.* 1997; 35:3254-3257.

348. Daley CL, Mugusi F, Chen LL, et al. Pulmonary Complications of HIV Infection in Dar Es Salaam, Tanzania. Role of Bronchoscopy and Bronchoalveolar Lavage. *Am J Respir Crit Care Med.* 1996; 154:105-110.

349. Cegielski JP, Msengi AE, Miller SE. Enteric Viruses Associated with HIV Infection in Tanzanian Children with Chronic Diarrhea. *Pediatr AIDS HIV Infect.* 1994; 5:296-299.

350. Cegielski JP, Msengi AE, Dukes CS, et al. Intestinal Parasites and HIV Infection in Tanzanian Children with Chronic Diarrhea. *AIDS.* 1993; 7:213-221.

351. Jorgensen AF, Jensen VG, Shao JF, et al. Beta-2-Microglobulin as a Prognostic Marker for Patients with AIDS in Dar-Es-Salaam, Tanzania. *AIDS.* 1990; 4:11-12.

352. Hasler T, Handunnetti SM, Aguiar JC, et al. In Vitro Rosetting, Cytoadherence, and Microagglutination Properties of Plasmodium Falciparum-Infected Erythrocytes from Gambian and Tanzanian Patients. *Blood.* 1990; 76:1845-1852.

353. Aguiar JC, Albrecht GR, Cegielski P, et al. Agglutination of Plasmodium Falciparum-Infected Erythrocytes from East and West African Isolates by Human Sera from Distant Geographic Regions. *Am J Trop Med Hyg.* 1992; 47:621-632.

354. Clements DA, Weigle KA, Gilbert GL. A Case-Control Study Examining Risk Factors for Invasive Haemophilus Influenzae Type B Disease in Victoria, Australia 1988-90. *J Paediatr Child Health.* 1995; 31:513-518.

355. Clements DA, Guise IA, MacInnes SJ, Gilbert GL. Haemophilus Influenzae Type B Infections in Victoria, Australia, 1985-1989. *J Infect Dis.* 1992; 165 Suppl 1:S33-34.

356. Desai SA, Krogstad DJ, McCleskey EW. A Nutrient-Permeable Channel on the Intraerythrocytic Malaria Parasite. *Nature.* 1993; 362:643-646.

357. Desai SA, Rosenberg RL. Pore Size of the Malaria Parasite's Nutrient Channel. *Proc Natl Acad Sci U S A.* 1997; 94:2045-2049.

358. Eddins D, Lyford LK, Lee JW, Desai SA, Rosenberg RL. Permeant but Not Impermeant Divalent Cations Enhance Activation of Nondesensitizing Alpha(7) Nicotinic Receptors. *Am J Physiol Cell Physiol.* 2002; 282:C796-804.

359. Hanly SM, Rimsky LT, Malim MH, et al. Comparative Analysis of the HTLV-I Rex and HIV-1 Rev Trans-Regulatory Proteins and Their RNA Response Elements. *Genes Dev.* 1989; 3:1534-1544.

360. Kim JH, Kaufman PA, Hanly SM, Rimsky LT, Greene WC. Rex Transregulation of Human T-Cell Leukemia Virus Type II Gene Expression. *J Virol.* 1991; 65:405-414.

361. Arima N, Molitor JA, Smith MR, Kim JH, Daitoku Y, Greene WC. Human T-Cell Leukemia Virus Type I Tax Induces Expression of the Rel-Related Family of Kappa B Enhancer-Binding Proteins: Evidence for a Pretranslational Component of Regulation. *J Virol.* 1991; 65:6892-6899.

362. Kim JH, Ratto S, Sitz KV, et al. Consequences of Stable Transduction and Antigen-Inducible Expression of the Human Interleukin-7 Gene on Tetanus-Toxoid-Specific T Cells. *Hum Gene Ther.* 1994; 5:1457-1466.

363. Kim JH, Mosca JD, Vahey MT, McLinden RJ, Burke DS, Redfield RR. Consequences of Human Immunodeficiency Virus Type 1 Superinfection of Chronically Infected Cells. *AIDS Res Hum Retroviruses.* 1993; 9:875-882.

364. Kim JH, McLinden RJ, Mosca JD, et al. Transcriptional Effects of Superinfection in HIV Chronically Infected T Cells: Studies in Dually Infected Clones. *J Acquir Immune Defic Syndr Hum Retrovirol.* 1996; 12:329-342.

365. Kim JH, Loveland JE, Sitz KV, et al. Expansion of Restricted Cellular Immune Responses to HIV-1 Envelope by Vaccination: Il-7 and Il-12 Differentially Augment Cellular Proliferative Responses to HIV-1. *Clin Exp Immunol.* 1997; 108:243-250.

366. Michael NL, Chang G, Kim JH, Birx DL. Dynamics of Cell-Free Viral Burden in HIV-1-Infected Patients. *J Acquir Immune Defic Syndr Hum Retrovirol.* 1997; 14:237-242.

367. Bohjanen PR, Johnson MD, Szczech LA, et al. Steady-State Pharmacokinetics of Lamivudine in Human Immunodeficiency Virus-Infected Patients with End-Stage Renal Disease Receiving Chronic Dialysis. *Antimicrob Agents Chemother.* 2002; 46:2387-2392.

368. Raghavan A, Robison RL, McNabb J, Miller CR, Williams DA, Bohjanen PR. Hua and Tristetraprolin Are Induced Following T Cell Activation and Display Distinct but Overlapping RNA Binding Specificities. *J Biol Chem.* 2001; 276:47958-47965.

369. Gottfredsson M, Bohjanen PR. Human Immunodeficiency Virus Type I as a Target for Gene Therapy. *Front Biosci.* 1997; 2:d619-634.

370. Boulware DR, Meya DB, Bergemann TL, et al. Clinical Features and Serum Biomarkers in HIV Immune Reconstitution Inflammatory Syndrome after Cryptococcal Meningitis: A Prospective Cohort Study. *PLoS Med.* 2010; 7:e1000384.

371. Rattenbacher B, Beisang D, Wiesner DL, et al. Analysis of Cugbp1 Targets Identifies Gu-Repeat Sequences That Mediate Rapid mRNA Decay. *Mol Cell Biol.* 2010; 30:3970-3980.

372. Vlasova IA, Tahoe NM, Fan D, et al. Conserved Gu-Rich Elements Mediate mRNA Decay by Binding to Cug-Binding Protein 1. *Mol Cell.* 2008; 29:263-270.

373. Ogilvie RL, Abelson M, Hau HH, Vlasova I, Blackshear PJ, Bohjanen PR. Tristetraprolin Down-Regulates Il-2 Gene Expression through Au-Rich Element-Mediated mRNA Decay. *J Immunol.* 2005; 174:953-961.

374. Raghavan A, Ogilvie RL, Reilly C, et al. Genome-Wide Analysis of mRNA Decay in Resting and Activated Primary Human T Lymphocytes. *Nucleic Acids Res.* 2002; 30:5529-5538.

375. Jackson CR, Vavro CL, Valentine ME, et al. Effect of Influenza Immunization on Immunologic and Virologic Characteristics of Pediatric Patients Infected with Human Immunodeficiency Virus. *Pediatr Infect Dis J.* 1997; 16:200-204.

376. Valentine ME, Jackson CR, Vavro C, et al. Evaluation of Surrogate Markers and Clinical Outcomes in Two-Year Follow-up of Eighty-Six Human Immunodeficiency Virus-Infected Pediatric Patients. *Pediatr Infect Dis J.* 1998; 17:18-23.

377. DiLiberti JH, Jackson CR. Long-Term Trends in Childhood Infectious Disease Mortality Rates. *Am J Public Health.* 1999; 89:1883-1885.

378. Jackson CR. Clinical Experience with Pneumococcal Conjugate Vaccines in Infants and Children. *J Am Osteopath Assoc.* 2002; 102:431-436.

379. Murray D, Jackson C. A Conjugate Vaccine for the Prevention of Pediatric Pneumococcal Disease. *Mil Med.* 2002; 167:671-677.

380. Jackson CR, Turner R. Paediatric Pharmaceutical Medicine: Paediatric Clinical Research and Onsideration for Clinical Trials. *J Clin Stud.* 2010:28-31.

381. Cohen JI, Rosenblum B, Feinstone SM, Ticehurst J, Purcell RH. Attenuation and Cell Culture Adaptation of Hepatitis a Virus (Hav): A Genetic Analysis with Hav Cdna. *J Virol.* 1989; 63:5364-5370.

382. Cohen JI, Feinstone S, Purcell RH. Hepatitis a Virus Infection in a Chimpanzee: Duration of Viremia and Detection of Virus in Saliva and Throat Swabs. *J Infect Dis.* 1989; 160:887-890.

383. Cohen JI, Miller RH, Rosenblum B, Denniston K, Gerin JL, Purcell RH. Sequence Comparison of Woodchuck Hepatitis Virus Replicative Forms Shows Conservation of the Genome. *Virology.* 1988; 162:12-20.

384. Cohen JI, Rosenblum B, Ticehurst JR, Daemer RJ, Feinstone SM, Purcell RH. Complete Nucleotide Sequence of an Attenuated Hepatitis a Virus: Comparison with Wild-Type Virus. *Proc Natl Acad Sci U S A.* 1987; 84:2497-2501.

385. Cohen JI, Ticehurst JR, Feinstone SM, Rosenblum B, Purcell RH. Hepatitis a Virus Cdna and Its RNA Transcripts Are Infectious in Cell Culture. *J Virol.* 1987; 61:3035-3039.

386. Cohen JI, Ticehurst JR, Purcell RH, Buckler-White A, Baroudy BM. Complete Nucleotide Sequence of Wild-Type Hepatitis a Virus: Comparison with Different Strains of Hepatitis a Virus and Other Picornaviruses. *J Virol.* 1987; 61:50-59.

387. Cohen JI. The Biology of Epstein-Barr Virus: Lessons Learned from the Virus and the Host. *Curr Opin Immunol.* 1999; 11:365-370.

388. Cohen JI. Epstein-Barr Virus and the Immune System. Hide and Seek. *JAMA.* 1997; 278:510-513.

389. Cohen JI, Kieff E. An Epstein-Barr Virus Nuclear Protein 2 Domain Essential for Transformation Is a Direct Transcriptional Activator. *J Virol.* 1991; 65:5880-5885.

390. Cohen JI. Genomic Structure and Organization of Varicella-Zoster Virus. *Contrib Microbiol.* 1999; 3:10-20.

391. Cohen JI. Infection of Cells with Varicella-Zoster Virus Down-Regulates Surface Expression of Class I Major Histocompatibility Complex Antigens. *J Infect Dis.* 1998; 177:1390-1393.

392. Cohen JI. Varicella-Zoster Virus. The Virus. *Infect Dis Clin North Am.* 1996; 10:457-468.

393. Karron RA, Daemer R, Ticehurst J, et al. Studies of Prototype Live Hepatitis a Virus Vaccines in Primate Models. *J Infect Dis.* 1988; 157:338-345.

394. Ticehurst JR, Feinstone SM, Chestnut T, Tassopoulos NC, Popper H, Purcell RH. Detection of Hepatitis a Virus by Extraction of Viral RNA and Molecular Hybridization. *J Clin Microbiol.* 1987; 25:1822-1829.

395. Weitz M, Baroudy BM, Maloy WL, Ticehurst JR, Purcell RH. Detection of a Genome-Linked Protein (Vpg) of Hepatitis a Virus and Its Comparison with Other Picornaviral Vpgs. *J Virol.* 1986; 60:124-130.

396. Baroudy BM, Ticehurst JR, Miele TA, Maizel JV, Jr., Purcell RH, Feinstone SM. Sequence Analysis of Hepatitis a Virus Cdna Coding for Capsid Proteins and RNA Polymerase. *Proc Natl Acad Sci U S A.* 1985; 82:2143-2147.

397. Ticehurst JR, Racaniello VR, Baroudy BM, Baltimore D, Purcell RH, Feinstone SM. Molecular Cloning and Characterization of Hepatitis a Virus Cdna. *Proc Natl Acad Sci U S A.* 1983; 80:5885-5889.

398. Tassopoulos NC, Papaevangelou GJ, Roumeliotou-Karayannis A, Ticehurst JR, Feinstone SM, Purcell RH. Detection of Hepatitis B Virus DNA in Asymptomatic Hepatitis B Surface Antigen Carriers: Relation to Sexual Transmission. *Am J Epidemiol.* 1987; 126:587-591.

399. Tassopoulos NC, Papaevangelou GJ, Roumeliotou-Karayannis A, et al. Fulminant Hepatitis in Asymptomatic Hepatitis B Surface Antigen Carriers in Greece. *J Med Virol.* 1986; 20:371-379.

400. Tassopoulos NC, Sjogren MH, Ticehurst JR, et al. Significance of Igm Antibody to Hepatitis B Core Antigen in a Greek Population with Chronic Hepatitis B Virus Infection. *Liver.* 1986; 6:275-280.

401. Arankalle VA, Ticehurst J, Sreenivasan MA, et al. Aetiological Association of a Virus-Like Particle with Enterically Transmitted Non-a, Non-B Hepatitis. *Lancet.* 1988; 1:550-554.

402. Tsarev SA, Tsareva TS, Emerson SU, et al. Elisa for Antibody to Hepatitis E Virus (Hev) Based on Complete Open-Reading Frame-2 Protein Expressed in Insect Cells: Identification of Hev Infection in Primates. *J Infect Dis.* 1993; 168:369-378.

403. Bryan JP, Tsarev SA, Iqbal M, et al. Epidemic Hepatitis E in Pakistan: Patterns of Serologic Response and Evidence That Antibody to Hepatitis E Virus Protects against Disease. *J Infect Dis.* 1994; 170:517-521.

404. Couch RB, Winokur P, Brady R, et al. Safety and Immunogenicity of a High Dosage Trivalent Influenza Vaccine among Elderly Subjects. *Vaccine.* 2007; 25:7656-7663.

405. Ruben FL. Inactivated Influenza Virus Vaccines in Children. *Clin Infect Dis.* 2004; 38:678-688.

406. Ruben FL. Counterpoint to 'Pneumococcal Vaccine after 15 Years of Use'. *Arch Intern Med.* 1995; 155:771-773.

407. Ruben FL. Prevention and Control of Influenza. Role of Vaccine. *Am J Med.* 1987; 82:31-34.

408. Ruben FL, Uhrin M. Specific Immunoglobulin-Class Antibody Responses in the Elderly before and after 14-Valent Pneumococcal Vaccine. *J Infect Dis.* 1985; 151:845-849.

409. Mostow SR, Cate TR, Ruben FL. Prevention of Influenza and Pneumonia. *Am Rev Respir Dis.* 1990; 142:487-488.

410. Ruben FL, Froeschle JE, Meschievitz C, et al. Choosing a Route of Administration for Quadrivalent Meningococcal Polysaccharide Vaccine: Intramuscular Versus Subcutaneous. *Clin Infect Dis.* 2001; 32:170-172.

CITATIONS FOR CHAPTER 8

1. Foege WH. Lecture. 2006.

2. Miller WC, Thielman NM, Swai N, et al. Diagnosis and Screening of HIV/AIDS Using Clinical Criteria in Tanzanian Adults. *J Acquir Immune Defic Syndr Hum Retrovirol.* 1995; 9:408-414.

3. Miller WC, Thielman NM, Swai N, et al. Delayed-Type Hypersensitivity Testing in Tanzanian Adults with HIV Infection. *J Acquir Immune Defic Syndr Hum Retrovirol.* 1996; 12:303-308.

4. Jorgensen AF, Jensen VG, Shao JF, et al. Beta-2-Microglobulin as a Prognostic Marker for Patients with AIDS in Dar-Es-Salaam, Tanzania. *AIDS.* 1990; 4:11-12.

5. Jorgensen AF, Mwakyusa D, Cegielski P, et al. The Effect of Fusidic Acid on Tanzanian Patients with AIDS. *AIDS.* 1990; 4:1037-1038.

6. Cegielski JP. The Epidemiology of AIDS in Africa. *Int Health News.* 1998; 9:9-10.

7. Daley CL, Mugusi F, Chen LL, et al. Pulmonary Complications of HIV Infection in Dar Es Salaam, Tanzania. Role of Bronchoscopy and Bronchoalveolar Lavage. *Am J Respir Crit Care Med.* 1996; 154:105-110.

8. Cegielski JP, Ramiya K, Lallinger GJ, Mtulia IA, Mbaga IM. Pericardial Disease and Human Immunodeficiency Virus in Dar Es Salaam, Tanzania. *Lancet.* 1990; 335:209-212.

9. Cegielski JP, Lwakatare J, Dukes CS, et al. Tuberculous Pericarditis in Tanzanian Patients with and without HIV Infection. *Tuber Lung Dis.* 1994; 75:429-434.

10. Cegielski JP, Devlin BH, Morris AJ, et al. Comparison of PCR, Culture, and Histopathology for Diagnosis of Tuberculous Pericarditis. *J Clin Microbiol.* 1997; 35:3254-3257.

11. Cegielski JP, Msengi AE, Dukes CS, et al. Intestinal Parasites and HIV Infection in Tanzanian Children with Chronic Diarrhea. *AIDS.* 1993; 7:213-221.

12. Cegielski JP, Msengi AE, Miller SE. Enteric Viruses Associated with HIV Infection in Tanzanian Children with Chronic Diarrhea. *Pediatr AIDS HIV Infect.* 1994; 5:296-299.

13. Cegielski JP, Ortega YR, McKee S, et al. Cryptosporidium, Enterocytozoon, and Cyclospora Infections in Pediatric and Adult Patients with Diarrhea in Tanzania. *Clin Infect Dis.* 1999; 28:314-321.

14. Dukes CS, Sugarman J, Cegielski JP, Lallinger GJ, Mwakyusa DH. Severe Cutaneous Hypersensitivity Reactions During Treatment of Tuberculosis in Patients with HIV Infection in Tanzania. *Trop Geogr Med.* 1992; 44:308-311.

15. Hasler T, Handunnetti SM, Aguiar JC, et al. In Vitro Rosetting, Cytoadherence, and Microagglutination Properties of Plasmodium Falciparum-Infected Erythrocytes from Gambian and Tanzanian Patients. *Blood.* 1990; 76:1845-1852.

16. Elliott JF, Albrecht GR, Gilladoga A, et al. Genes for Plasmodium Falciparum Surface Antigens Cloned by Expression in Cos Cells. *Proc Natl Acad Sci U S A.* 1990; 87:6363-6367.

17. Anstey NM, Weinberg JB, Hassanali MY, et al. Nitric Oxide in Tanzanian Children with Malaria: Inverse Relationship between Malaria Severity and Nitric Oxide Production/Nitric Oxide Synthase Type 2 Expression. *J Exp Med.* 1996; 184:557-567.

18. Aguiar JC, Albrecht GR, Cegielski P, et al. Agglutination of Plasmodium Falciparum-Infected Erythrocytes from East and West African Isolates by Human Sera from Distant Geographic Regions. *Am J Trop Med Hyg.* 1992; 47:621-632.

19. Miller WC, Shao JF, Weaver DJ, Shimokura GH, Paul DA, Lallinger GJ. Seroprevalence of Viral Hepatitis in Tanzanian Adults. *Trop Med Int Health.* 1998; 3:757-763.

20. Anstey NM, Tissot Dupont H, Hahn CG, et al. Seroepidemiology of Rickettsia Typhi, Spotted Fever Group Rickettsiae, and Coxiella Burnetti Infection in Pregnant Women from Urban Tanzania. *Am J Trop Med Hyg.* 1997; 57:187-189.

21. Harris JJ, Shao J, Surgarman J. Disclosure of Cancer Diagnosis and Prognosis in Northern Tanzania. *Soc Sci Med.* 2003; 56:905-913.

22. Miller WC, Perkins MD, Richardson WJ, Sexton DJ. Pott's Disease Caused by Mycobacterium Xenopi: Case Report and Review. *Clin Infect Dis.* 1994; 19:1024-1028.

23. Perkins MD, Mirrett S, Reller LB. Rapid Bacterial Antigen Detection Is Not Clinically Useful. *J Clin Microbiol.* 1995; 33:1486-1491.

24. Dietze R, Perkins M, Boulos M, Luz F, Reller B, Corey GR. The Diagnosis of Plasmodium Falciparum Infection Using a New Antigen Detection System. *Am J Trop Med Hyg.* 1995; 52:45-49.

25. Talbot EA, Perkins MD, Silva SF, Frothingham R. Disseminated Bacille Calmette-Guerin Disease after Vaccination: Case Report and Review. *Clin Infect Dis.* 1997; 24:1139-1146.

26. Cannas A, Kalunga G, Green C, et al. Implications of Storing Urinary DNA from Different Populations for Molecular Analyses. *PLoS One.* 2009; 4:e6985.

27. Kunnath-Velayudhan S, Salamon H, Wang HY, et al. Dynamic Antibody Responses to the Mycobacterium Tuberculosis Proteome. *Proc Natl Acad Sci U S A.* 2010; 107:14703-14708.

28. Nahid P, Kim PS, Evans CA, et al. Clinical Research and Development of Tuberculosis Diagnostics: Moving from Silos to Synergy. *J Infect Dis.* 2012; 205 Suppl 2:S159-168.

29. Talbot EA, Pape JW, Sundaram L, Boehme CC, Perkins MD. Transforming Tb Diagnosis: Can Patients and Control Programs Afford to Wait? *Am J Trop Med Hyg.* 2012; 87:202-204.

30. Anstey NM. Artemisinin Compounds in Treatment of Malaria. *Lancet.* 1993; 341:1035.

31. Anstey NM, Currie BJ, Dyer ME. Profound Thrombocytopenia Due to Plasmodium Vivax Malaria. *Aust N Z J Med.* 1992; 22:169-170.

32. Anstey NM, Granger DL, Hassanali MY, Mwaikambo ED, Duffy PE, Weinberg JB. Nitric Oxide, Malaria, and Anemia: Inverse Relationship between Nitric Oxide Production and Hemoglobin Concentration in Asymptomatic, Malaria-Exposed Children. *Am J Trop Med Hyg.* 1999; 61:249-252.

33. Anstey NM, Granger DL, Weinberg JB. Nitrate Levels in Malaria. *Trans R Soc Trop Med Hyg.* 1997; 91:238-240.

34. Anstey NM, Weinberg JB, Wang Z, Mwaikambo ED, Duffy PE, Granger DL. Effects of Age and Parasitemia on Nitric Oxide Production/Leukocyte

Nitric Oxide Synthase Type 2 Expression in Asymptomatic, Malaria-Exposed Children. *Am J Trop Med Hyg.* 1999; 61:253-258.

35. Granger DL, Anstey NM, Miller WC, Weinberg JB. Measuring Nitric Oxide Production in Human Clinical Studies. *Methods Enzymol.* 1999; 301:49-61.

36. Hill AV, Allsopp CE, Kwiatkowski D, et al. Common West African Hla Antigens Are Associated with Protection from Severe Malaria. *Nature.* 1991; 352:595-600.

37. Levesque MC, Hobbs MR, Anstey NM, et al. Nitric Oxide Synthase Type 2 Promoter Polymorphisms, Nitric Oxide Production, and Disease Severity in Tanzanian Children with Malaria. *J Infect Dis.* 1999; 180:1994-2002.

38. Tjitra E, Suprianto S, Dyer M, Currie BJ, Anstey NM. Field Evaluation of the Ict Malaria P.F/P.V Immunochromatographic Test for Detection of Plasmodium Falciparum and Plasmodium Vivax in Patients with a Presumptive Clinical Diagnosis of Malaria in Eastern Indonesia. *J Clin Microbiol.* 1999; 37:2412-2417.

39. Gilbert GL, Clements DA, Broughton SJ. Haemophilus Influenzae Type B Infections in Victoria, Australia, 1985 to 1987. *Pediatr Infect Dis J.* 1990; 9:252-257.

40. Clements DA, Booy R, Dagan R, et al. Comparison of the Epidemiology and Cost of Haemophilus Influenzae Type B Disease in Five Western Countries. *Pediatr Infect Dis J.* 1993; 12:362-367.

41. Clements DA, Weigle KA, Gilbert GL. A Case-Control Study Examining Risk Factors for Invasive Haemophilus Influenzae Type B Disease in Victoria, Australia 1988-90. *J Paediatr Child Health.* 1995; 31:513-518.

42. Clements DA, Gilbert GL. Immunisation for the Prevention of Haemophilus Influenzae Type B Infections: A Review. *Aust N Z J Med.* 1990; 20:828-834.

43. Clements DA, MacInnes SJ, Gilbert GL. Outer Membrane Protein Subtypes of Haemophilus Influenzae Type B Isolates Causing Invasive Disease in Victoria, Australia, from 1988 to 1990. *J Clin Microbiol.* 1992; 30:1879-1881.

44. Benjamin DK, Jr., Miro JM, Hoen B, et al. Candida Endocarditis: Contemporary Cases from the International Collaboration of Infectious Endocarditis Merged Database (Ice-Md). *Scand J Infect Dis.* 2004; 36:453-455.

45. Chirouze C, Cabell CH, Fowler VG, Jr., et al. Prognostic Factors in 61 Cases of Staphylococcus Aureus Prosthetic Valve Infective Endocarditis from the International Collaboration on Endocarditis Merged Database. *Clin Infect Dis.* 2004; 38:1323-1327.

46. Chu VH, Cabell CH, Abrutyn E, et al. Native Valve Endocarditis Due to Coagulase-Negative Staphylococci: Report of 99 Episodes from the International Collaboration on Endocarditis Merged Database. *Clin Infect Dis.* 2004; 39:1527-1530.

47. Anguera I, Miro JM, Evangelista A, et al. Periannular Complications in Infective Endocarditis Involving Native Aortic Valves. *Am J Cardiol.* 2006; 98:1254-1260.

48. Anderson DJ, Olaison L, McDonald JR, et al. Enterococcal Prosthetic Valve Infective Endocarditis: Report of 45 Episodes from the International Collaboration on Endocarditis-Merged Database. *Eur J Clin Microbiol Infect Dis.* 2005; 24:665-670.

49. Cabell CH, Abrutyn E, Fowler VG, Jr., et al. Use of Surgery in Patients with Native Valve Infective Endocarditis: Results from the International Collaboration on Endocarditis Merged Database. *Am Heart J.* 2005; 150:1092-1098.

50. Hoen B, Chirouze C, Cabell CH, et al. Emergence of Endocarditis Due to Group D Streptococci: Findings Derived from the Merged Database of the International Collaboration on Endocarditis. *Eur J Clin Microbiol Infect Dis.* 2005; 24:12-16.

51. Kourany WM, Miro JM, Moreno A, et al. Influence of Diabetes Mellitus on the Clinical Manifestations and Prognosis of Infective Endocarditis: A Report from the International Collaboration on Endocarditis-Merged Database. *Scand J Infect Dis.* 2006; 38:613-619.

52. Lalani T, Kanafani ZA, Chu VH, et al. Prosthetic Valve Endocarditis Due to Coagulase-Negative Staphylococci: Findings from the International Collaboration on Endocarditis Merged Database. *Eur J Clin Microbiol Infect Dis.* 2006; 25:365-368.

53. McDonald JR, Olaison L, Anderson DJ, et al. Enterococcal Endocarditis: 107 Cases from the International Collaboration on Endocarditis Merged Database. *Am J Med.* 2005; 118:759-766.

54. Miro JM, Anguera I, Cabell CH, et al. Staphylococcus Aureus Native Valve Infective Endocarditis: Report of 566 Episodes from the International Collaboration on Endocarditis Merged Database. *Clin Infect Dis.* 2005; 41:507-514.

55. Wang A, Pappas P, Anstrom KJ, et al. The Use and Effect of Surgical Therapy for Prosthetic Valve Infective Endocarditis: A Propensity Analysis of a Multicenter, International Cohort. *Am Heart J.* 2005; 150:1086-1091.

CITATIONS FOR CHAPTER 9

1. Foege WH. *House on Fire.* Los Angeles, CA: The University of California Press; 2011.

2. Plotkin SL, Plotkin SA. A Short History of Vaccination. In: Plotkin SA, Orenstein WA, Offit P, eds. *Vaccines.* 4th ed. Philadelphia, PA: Elsevier Inc; 2004:1-13.

3. Beard JW, Finkelstein H, Sealy WC, Wyckoff RW. The Ultracentrifugal Concentration of the Immunizing Principle from Tissue Diseased with Equine Encephalomyelitis. *Science*. 1938; 87:89-90.

4. Beard JW, Finkelstein H, Sealy WC, Wyckoff RW. Immunization against Equine Encephalomyelitis with Chick Embryo Vaccines. *Science*. 1938; 87:490.

5. Beard JW, Beard D, Finkelstein H. Human Vaccination against Equine Encephalomyelitis Virus with Formolized Chick Embryo Vaccine. *Science*. 1939; 90:215-216.

6. Fothergill LD, Dingle JH, Farber S, Connerly ML. Human Encephalitis Caused by the Virus of Eastern Variety of Equine Encephalomyelitis. *N Engl J Med*. 1938; 219:411.

7. Wesselhoeft CC, Smith EC, Branch CF. Human Encephalitis: Four Fatal Cases Due to Virus of Equine Encephalomyelitis. *JAMA*. 1938; 111:1735-1741.

8. Holden M, Wyckoff RWG. Western Equine Encephalomyelitis in a Laboratory Worker. *JAMA*. 1939; 113:206-207.

9. Howitt BF. Recovery of the Virus of Equine Encephalomyelitis from the Brain of a Child. *Science*. 1938; 88:455-456.

10. Beard JW. Virus of Avian Myeloblastic Leukosis. *Poult Sci*. 1956; 35:203-223.

11. Roberts BW, Jr, London AH. On the Poliomyelitis Epidemic in the Durham North Carolina Area. *Trans Med Soc St N C*. 1936; 83:460-469.

12. Davison WC. Poliomyelitis--a Resume. *Am J Dis Child*. 1936; 52:1158-1178.

13. Feller AE, Enders JF, Weller TH. The Prolonged Coexistence of Vaccinia Virus in High Titre and Living Cells in Roller Tube Cultures of Chick Embryonic Tissues. *J Exp Med*. 1940; 72:367-388.

14. Weller TH, Enders JF. Production of Hemagglutinin by Mumps and Influenza a Viruses in Suspended Cell Tissue Cultures. *Proc Soc Exp Biol Med*. 1948; 69:124-128.

15. Enders JF, Weller TH, Robbins FC. Cultivation of the Lansing Strain of Poliomyelitis Virus in Cultures of Various Human Embryonic Tissues. *Science*. 1949; 109:85-87.

16. Weller TH, Robbins FC, Enders JF. Cultivation of Poliomyelitis Virus in Cultures of Human Foreskin and Embryonic Tissues. *Proc Soc Exp Biol Med*. 1949; 72:153-155.

17. Enders JF, Robbins FC, Weller TH. Classics in Infectious Diseases. The Cultivation of the Poliomyelitis Viruses in Tissue Culture by John F. Enders, Frederick C. Robbins, and Thomas H. Weller. *Rev Infect Dis*. 1980; 2:493-504.

18. Katz SL, Milovanovic MV, Enders JF. Propagation of Measles Virus in Cultures of Chick Embryo Cells. *Proc Soc Exp Biol Med*. 1958; 97:23-29.

19. Enders JF, Katz SL, Holloway A. Development of Attenuated Measles-Virus Vaccines. A Summary of Recentinvestigation. *Am J Dis Child.* 1962; 103:335-340.
20. Enders JF, Katz SL, Milovanovic MV, Holloway A. Studies on an Attenuated Measles-Virus Vaccine. I. Development and Preparations of the Vaccine: Technics for Assay of Effects of Vaccination. *N Engl J Med.* 1960; 263:153-159.
21. Katz SL, Morley DC, Krugman S. Attenuated Measles Vaccine in Nigerian Children. *Am J Dis Child.* 1962; 103:402-405.
22. Offit PA. *Deadly Choices: How the Anti-Vaccine Movement Threatens Us All.* New York, NY: Basic Books; 2011.
23. Katz SL. Poliovaccine Policy--Time for a Change. *Pediatrics.* 1996; 98:116-117.
24. Alpert JJ, Katz SL, Lissauer T. International Exchanges: A Missed Opportunity in Pediatric Graduate Education. *Arch Pediatr Adolesc Med.* 2006; 160:570-571.
25. Cooper LZ, Larson HJ, Katz SL. Protecting Public Trust in Immunization. *Pediatrics.* 2008; 122:149-153.
26. Larson HJ, Cooper LZ, Eskola J, Katz SL, Ratzan S. Addressing the Vaccine Confidence Gap. *Lancet.* 2011; 378:526-535.
27. Becker Y, Joklik WK. Messenger RNA in Cells Infected with Vaccinia Virus. *Proc Natl Acad Sci U S A.* 1964; 51:577-585.
28. Holowczak JA, Joklik WK. Studies on the Structural Proteins of Vaccinia Virus. I. Structural Proteins of Virions and Cores. *Virology.* 1967; 33:717-725.
29. Holowczak JA, Joklik WK. Studies on the Structural Proteins of Vaccinia Virus. II. Kinetics of the Synthesis of Individual Groups of Structural Proteins. *Virology.* 1967; 33:726-739.
30. Joklik WK. Biochemical Studies on Vaccinia Virus in Cultured Cells. II. Localization of Adenine-8-C14 Incorporated in the Course of Infection. *Virology.* 1959; 9:417-424.
31. Joklik WK. Some Properties of Poxvirus Deoxyriboncleic Acid. *J Mol Biol.* 1962; 5:2165-2174.
32. Joklik WK. The Purification of Four Strains of Poxvirus. *Virology.* 1962; 18:9-18.
33. Joklik WK. The Preparation and Characteristics of Highly Purified Radioactively Labelled Poxvirus. *Biochim Biophys Acta.* 1962; 61:290-301.
34. Joklik WK, Abel P, Holmes IH. Reactivation of Poxviruses by a Non-Genetic Mechanism. *Nature.* 1960; 186:992-993.
35. Nevins JR, Joklik WK. Poly (a) Sequences of Vaccinia Virus Messenger RNA: Nature, Mode of Addition and Function During Translation in Vitra and in Vivo. *Virology.* 1975; 63:1-14.

36. Nevins JR, Joklik WK. Isolation and Partial Characterization of the Poly(a) Polymerases from Hela Cells Infected with Vaccinia Virus. *J Biol Chem.* 1977; 252:6939-6947.

37. Nevins JR, Joklik WK. Isolation and Properties of the Vaccinia Virus DNA-Dependent RNA Polymerase. *J Biol Chem.* 1977; 252:6930-6938.

38. Pickup DJ, Bastia D, Stone HO, Joklik WK. Sequence of Terminal Regions of Cowpox Virus DNA: Arrangement of Repeated and Unique Sequence Elements. *Proc Natl Acad Sci U S A.* 1982; 79:7112-7116.

39. Pickup DJ, Bastia D, Joklik WK. Cloning of the Terminal Loop of Vaccinia Virus DNA. *Virology.* 1983; 124:215-217.

40. Pickup DJ, Ink BS, Parsons BL, Hu W, Joklik WK. Spontaneous Deletions and Duplications of Sequences in the Genome of Cowpox Virus. *Proc Natl Acad Sci U S A.* 1984; 81:6817-6821.

41. Pickup DJ, Ink BS, Hu W, Ray CA, Joklik WK. Hemorrhage in Lesions Caused by Cowpox Virus Is Induced by a Viral Protein That Is Related to Plasma Protein Inhibitors of Serine Proteases. *Proc Natl Acad Sci U S A.* 1986; 83:7698-7702.

42. Sarov I, Joklik WK. Studies on the Nature and Location of the Capsid Polypeptides of Vaccinia Virions. *Virology.* 1972; 50:579-592.

43. Sarov I, Joklik WK. Characterization of Intermediates in the Uncoating of Vaccinia Virus DNA. *Virology.* 1972; 50:593-602.

44. Sarov I, Joklik WK. Isolation and Characterization of Intermediates in Vaccinia Virus Morphogenesis. *Virology.* 1973; 52:223-233.

45. Woodson B, Joklik WK. The Inhibition of Vaccinia Virus Multiplication by Isatin-Beta-Thiosemicarbazone. *Proc Natl Acad Sci U S A.* 1965; 54:946-953.

46. Joklik WK, Moss B, Fields BN, Bishop DH, Sandakhchiev LS. Why the Smallpox Virus Stocks Should Not Be Destroyed. *Science.* 1993; 262:1225-1226.

47. Fenner F, Henderson DA, Arita I, Jezek Z, Ladnyi ID. *Smallpox and Its Eradication.* Geneva, Switzerland: World Health Organization; 1988.

48. Mahy BW, Almond JW, Berns KI, et al. The Remaining Stocks of Smallpox Virus Should Be Destroyed. *Science.* 1993; 262:1223-1224.

49. Hamilton JD, Hatch MH, Gutman RA. Serological Evidence of Cross Infection in a Dialysis Unit Hepatitis-B Epidemic. *Kidney Int.* 1974; 6:118-122.

50. Hansen JP, Falconer JA, Hamilton JD, Herpok FJ. Hepatitis-B in a Medical-Center. *J Occup Environ Med.* 1981; 23:338-342.

51. Seeff LB, Wright EC, Zimmerman HJ, et al. Type-B Hepatitis after Needle-Stick Exposure - Prevention with Hepatitis-B Immune Globulin - Final Report of Veterans-Administration Cooperative Study. *Ann Intern Med.* 1978; 88:285-293.

52. Stevens CE, Alter HJ, Taylor PE, Zang EA, Harley EJ, Szmuness W. Hepatitis B Vaccine in Patients Receiving Hemodialysis. Immunogenicity and Efficacy. *N Engl J Med.* 1984; 311:496-501.

53. Szmuness W, Stevens CE, Harley EJ, et al. Hepatitis B Vaccine in Medical Staff of Hemodialysis Units: Efficacy and Subtype Cross-Protection. *N Engl J Med.* 1982; 307:1481-1486.

54. Dukes CS, Street AC, Starling JF, Hamilton JD. Hepatitis B Vaccination and Booster in Predialysis Patients: A 4-Year Analysis. *Vaccine.* 1993; 11:1229-1232.

55. Oddone EZ, Cowper PA, Hamilton JD, Feussner JR. A Cost-Effectiveness Analysis of Hepatitis B Vaccine in Dialysis and Pre-Dialysis Patients. *Clin Res.* 1989; 37:A780-A780.

56. Haynes BF. *HIV Vaccine Research at Duke University: A Continued Quest to End the Epidemic.* Durham, NC: Duke University;2010.

57. Cullen BR, Greene WC. Regulatory Pathways Governing HIV-1 Replication. *Cell.* 1989; 58:423-426.

58. Matthews TJ, Langlois AJ, Robey WG, et al. Restricted Neutralization of Divergent Human T-Lymphotropic Virus Type III Isolates by Antibodies to the Major Envelope Glycoprotein. *Proc Natl Acad Sci U S A.* 1986; 83:9709-9713.

59. Palker TJ, Clark ME, Langlois AJ, et al. Type-Specific Neutralization of the Human Immunodeficiency Virus with Antibodies to Env-Encoded Synthetic Peptides. *Proc Natl Acad Sci U S A.* 1988; 85:1932-1936.

60. Javaherian K, Langlois AJ, LaRosa GJ, et al. Broadly Neutralizing Antibodies Elicited by the Hypervariable Neutralizing Determinant of HIV-1. *Science.* 1990; 250:1590-1593.

61. Bartlett JA, Wasserman SS, Hicks CB, et al. Safety and Immunogenicity of an Hla-Based HIV Envelope Polyvalent Synthetic Peptide Immunogen. DATRI 010 Study Group. Division of AIDS Treatment Research Initiative. *AIDS.* 1998; 12:1291-1300.

62. Clements DA, Armstrong CB, Ursano AM, Moggio MM, Walter EB, Wilfert CM. Over Five-Year Follow-up of Oka/Merck Varicella Vaccine Recipients in 465 Infants and Adolescents. *Pediatr Infect Dis J.* 1995; 14:874-879.

63. Clements DA, Langdon L, Bland C, Walter E. Influenza a Vaccine Decreases the Incidence of Otitis Media in 6- to 30-Month-Old Children in Day Care. *Arch Pediatr Adolesc Med.* 1995; 149:1113-1117.

64. Rennels MB, Hohenboken MJ, Reisinger KS, et al. Comparison of Acellular Pertussis-Diphtheria-Tetanus Toxoids and Haemophilus Influenzae Type B Vaccines Administered Separately Vs. Combined in Younger Vs. Older Toddlers. *Pediatr Infect Dis J.* 1998; 17:164-166.

65. Clements DA, Zaref JI, Bland CL, Walter EB, Coplan PM. Partial Uptake of Varicella Vaccine and the Epidemiological Effect on Varicella Disease in

11 Day-Care Centers in North Carolina. *Arch Pediatr Adolesc Med.* 2001; 155:455-461.

66. Kuter B, Matthews H, Shinefield H, et al. Ten Year Follow-up of Healthy Children Who Received One or Two Injections of Varicella Vaccine. *Pediatr Infect Dis J.* 2004; 23:132-137.

67. Englund JA, Walter EB, Fairchok MP, Monto AS, Neuzil KM. A Comparison of 2 Influenza Vaccine Schedules in 6- to 23-Month-Old Children. *Pediatrics.* 2005; 115:1039-1047.

68. Englund JA, Walter EB, Gbadebo A, Monto AS, Zhu Y, Neuzil KM. Immunization with Trivalent Inactivated Influenza Vaccine in Partially Immunized Toddlers. *Pediatrics.* 2006; 118:e579-585.

69. Walter EB, Neuzil KM, Zhu Y, et al. Influenza Vaccine Immunogenicity in 6- to 23-Month-Old Children: Are Identical Antigens Necessary for Priming? *Pediatrics.* 2006; 118:e570-578.

70. Walter EB, Allred N, Rowe-West B, Chmielewski K, Kretsinger K, Dolor RJ. Cocooning Infants: Tdap Immunization for New Parents in the Pediatric Office. *Acad Pediatr.* 2009; 9:344-347.

71. Walter EB, Allred NJ, Swamy GK, Hellkamp AS, Dolor RJ. Influenza Vaccination of Household Contacts of Newborns: A Hospital-Based Strategy to Increase Vaccination Rates. *Infect Control Hosp Epidemiol.* 2010; 31:1070-1073.

72. Tan W, Viera AJ, Rowe-West B, Grimshaw A, Quinn B, Walter EB. The Hpv Vaccine: Are Dosing Recommendations Being Followed? *Vaccine.* 2011; 29:2548-2554.

73. Chen WH, Winokur PL, Edwards KM, et al. Phase 2 Assessment of the Safety and Immunogenicity of Two Inactivated Pandemic Monovalent H1n1 Vaccines in Adults as a Component of the U.S. Pandemic Preparedness Plan in 2009. *Vaccine.* 2012; 30:4240-4248.

74. Schmader KE, Rahija RJ, Porter KR, Hamilton JD. Murine Cytomegalovirus Gene Amplification and Culture after Submaxillary Salivary Gland Biopsy. *Lab Anim Sci.* 1991; 41:396-400.

75. Schmader KE, Rahija R, Porter KR, Daley G, Hamilton JD. Aging and Reactivation of Latent Murine Cytomegalovirus. *J Infect Dis.* 1992; 166:1403-1407.

76. Schmader KE, Henry SC, Rahija RJ, Yu Y, Daley GG, Hamilton JD. Mouse Cytomegalovirus Reactivation in Severe Combined Immune Deficient Mice after Implantation of Latently Infected Salivary Gland. *J Infect Dis.* 1995; 172:531-534.

77. Henry SC, Schmader K, Brown TT, et al. Enhanced Green Fluorescent Protein as a Marker for Localizing Murine Cytomegalovirus in Acute and Latent Infection. *J Virol Methods.* 2000; 89:61-73.

78. Dworkin RH, Carrington D, Cunningham A, et al. Assessment of Pain in Herpes Zoster: Lessons Learned from Antiviral Trials. *Antiviral Res.* 1997; 33:73-85.

79. Schmader K, George LK, Burchett BM, Hamilton JD, Pieper CF. Race and Stress in the Incidence of Herpes Zoster in Older Adults. *J Am Geriatr Soc.* 1998; 46:973-977.

80. Schmader KE, George LK, Newton R, Hamilton JD. The Accuracy of Self-Reports of Herpes Zoster. *J Clin Epidemiol.* 1994; 47:1271-1276.

81. Coplan PM, Schmader KE, Nikas A, et al. Development of a Measure of the Burden of Pain Due to Herpes Zoster and Postherpetic Neuralgia for Prevention Trials: Adaptation of the Brief Pain Inventory. *J Pain.* 2004; 5:344-356.

82. Oxman MN, Levin MJ, Johnson GR, et al. A Vaccine to Prevent Herpes Zoster and Postherpetic Neuralgia in Older Adults. *N Engl J Med.* 2005; 352:2271-2278.

CITATIONS FOR CHAPTER 10

1. Strauss MB, ed *Familiar Medical Quotations.* Boston, MA: Little, Brown and Company; 1968. Strauss MB, ed.

2. Stead EA, Ebert RV. The Peripheral Circulation in Acute Infectious Diseases. *Med Clin North Am.* 1940; 24:1387-1394.

3. Ebert RV, Stead EA. Circulatory Failure in Acute Infections. *J Clin Invest.* 1941; 20:671-679.

4. Wallace AG, Young WG, Jr., Osterhout S. Treatment of Acute Bacterial Endocarditis by Valve Excision and Replacement. *Circulation.* 1965; 31:450-453.

5. Stead EA, Jr. On Bacterial Endocarditis. *Med Times.* 1966; 94:1250-1252.

6. Durack DT, Beeson PB. Experimental Bacterial Endocarditis. I. Colonization of a Sterile Vegetation. *Br J Exp Pathol.* 1972; 53:44-49.

7. Durack DT, Beeson PB. Experimental Bacterial Endocarditis. II. Survival of a Bacteria in Endocardial Vegetations. *Br J Exp Pathol.* 1972; 53:50-53.

8. Durack DT, Beeson PB, Petersdorf RG. Experimental Bacterial Endocarditis. 3. Production and Progress of the Disease in Rabbits. *Br J Exp Pathol.* 1973; 54:142-151.

9. Durack DT. Experimental Bacterial Endocarditis. IV. Structure and Evolution of Very Early Lesions. *J Pathol.* 1975; 115:81-89.

10. Durack DT, Petersdorf RG, Beeson PB. Penicillin Prophylaxis of Experimental S. Viridans Endocarditis. *Trans Assoc Am Physicians.* 1972; 85:222-230.

11. Durack DT, Petersdorf RG. Chemotherapy of Experimental Streptococcal Endocarditis. I. Comparison of Commonly Recommended Prophylactic Regimens. *J Clin Invest.* 1973; 52:592-598.

12. Pelletier LL, Jr., Durack DT, Petersdorf RG. Chemotherapy of Experimental Streptococcal Endocarditis. IV. Further Observations on Prophylaxis. *J Clin Invest.* 1975; 56:319-330.

13. Southwick FS, Durack DT. Chemotherapy of Experimental Streptococcal Endocarditis. 3. Failure of a Bacteriostatic Agent (Tetracycline) in Prophylaxis. *J Clin Pathol.* 1974; 27:261-264.

14. Durack DT, Pelletier LL, Petersdorf RG. Chemotherapy of Experimental Streptococcal Endocarditis. II. Synergism between Penicillin and Streptomycin against Penicillin-Sensitive Streptococci. *J Clin Invest.* 1974; 53:829-833.

15. Durack DT, Lukes AS, Bright DK. New Criteria for Diagnosis of Infective Endocarditis: Utilization of Specific Echocardiographic Findings. Duke Endocarditis Service. *Am J Med.* 1994; 96:200-209.

16. Dodds GA, Sexton DJ, Durack DT, Bashore TM, Corey GR, Kisslo J. Negative Predictive Value of the Duke Criteria for Infective Endocarditis. *Am J Cardiol.* 1996; 77:403-407.

17. Sexton DJ, Tenenbaum MJ, Wilson WR, et al. Ceftriaxone Once Daily for Four Weeks Compared with Ceftriaxone Plus Gentamicin Once Daily for Two Weeks for Treatment of Endocarditis Due to Penicillin-Susceptible Streptococci. Endocarditis Treatment Consortium Group. *Clin Infect Dis.* 1998; 27:1470-1474.

18. Wilson WR, Karchmer AW, Dajani AS, et al. Antibiotic Treatment of Adults with Infective Endocarditis Due to Streptococci, Enterococci, Staphylococci, and Hacek Microorganisms. American Heart Association. *JAMA.* 1995; 274:1706-1713.

19. Fowler VG, Jr., Li J, Corey GR, et al. Role of Echocardiography in Evaluation of Patients with Staphylococcus Aureus Bacteremia: Experience in 103 Patients. *J Am Coll Cardiol.* 1997; 30:1072-1078.

20. Robinson DL, Fowler VG, Sexton DJ, Corey RG, Conlon PJ. Bacterial Endocarditis in Hemodialysis Patients. *Am J Kidney Dis.* 1997; 30:521-524.

21. Marr KA, Kong L, Fowler VG, et al. Incidence and Outcome of Staphylococcus Aureus Bacteremia in Hemodialysis Patients. *Kidney Int.* 1998; 54:1684-1689.

22. Nettles RE, McCarty DE, Corey GR, Li J, Sexton DJ. An Evaluation of the Duke Criteria in 25 Pathologically Confirmed Cases of Prosthetic Valve Endocarditis. *Clin Infect Dis.* 1997; 25:1401-1403.

23. Conlon PJ, Jefferies F, Krigman HR, Corey GR, Sexton DJ, Abramson MA. Predictors of Prognosis and Risk of Acute Renal Failure in Bacterial Endocarditis. *Clin Nephrol.* 1998; 49:96-101.

24. Gagliardi JP, Nettles RE, McCarty DE, Sanders LL, Corey GR, Sexton DJ. Native Valve Infective Endocarditis in Elderly and Younger Adult Patients: Comparison of Clinical Features and Outcomes with Use of the Duke Criteria and the Duke Endocarditis Database. *Clin Infect Dis.* 1998; 26:1165-1168.

25. Fowler VG, Sanders LL, Sexton DJ, et al. Outcome of Staphylococcus Aureus Bacteremia According to Compliance with Infectious Diseases Specialist Recommendations: Experience with 244 Patients. *Clin Infect Dis.* 1998; 27:478-486.

26. Fowler VG, Jr., Sanders LL, Kong LK, et al. Infective Endocarditis Due to Staphylococcus Aureus: 59 Prospectively Identified Cases with Follow-Up. *Clin Infect Dis.* 1999; 28:106-114.

27. Fowler VG, Jr., Kong LK, Corey GR, et al. Recurrent Staphylococcus Aureus Bacteremia: Pulsed-Field Gel Electrophoresis Findings in 29 Patients. *J Infect Dis.* 1999; 179:1157-1161.

28. Rosen AB, Fowler VG, Jr., Corey GR, et al. Cost-Effectiveness of Transesophageal Echocardiography to Determine the Duration of Therapy for Intravascular Catheter-Associated Staphylococcus Aureus Bacteremia. *Ann Intern Med.* 1999; 130:810-820.

29. McClelland RS, Fowler VG, Jr., Sanders LL, et al. Staphylococcus Aureus Bacteremia among Elderly Vs Younger Adult Patients: Comparison of Clinical Features and Mortality. *Arch Intern Med.* 1999; 159:1244-1247.

30. Piantadosi CA, Fracica PJ, Duhaylongsod FG, et al. Artificial Surfactant Attenuates Hyperoxic Lung Injury in Primates. II. Morphometric Analysis. *J Appl Physiol (1985).* 1995; 78:1823-1831.

31. Welty-Wolf KE, Simonson SG, Huang YC, Fracica PJ, Patterson JW, Piantadosi CA. Ultrastructural Changes in Skeletal Muscle Mitochondria in Gram-Negative Sepsis. *Shock.* 1996; 5:378-384.

32. Huang YC, Fracica PJ, Simonson SG, et al. VA/Q Abnormalities During Gram Negative Sepsis. *Respir Physiol.* 1996; 105:109-121.

33. Simonson SG, Welty-Wolf KE, Huang YC, et al. Aerosolized Manganese Sod Decreases Hyperoxic Pulmonary Injury in Primates. I. Physiology and Biochemistry. *J Appl Physiol (1985).* 1997; 83:550-558.

34. Welty-Wolf KE, Simonson SG, Huang YC, et al. Aerosolized Manganese Sod Decreases Hyperoxic Pulmonary Injury in Primates. II. Morphometric Analysis. *J Appl Physiol (1985).* 1997; 83:559-568.

35. Simonson SG, Huang YC, Fracica PJ, et al. Changes in the Lung after Prolonged Positive Pressure Ventilation in Normal Baboons. *J Crit Care.* 1997; 12:72-82.

36. Carraway MS, Welty-Wolf KE, Kantrow SP, et al. Antibody to E- and L-Selectin Does Not Prevent Lung Injury or Mortality in Septic Baboons. *Am J Respir Crit Care Med.* 1998; 157:938-949.

37. Welty-Wolf KE, Carraway MS, Huang YC, Simonson SG, Kantrow SP, Piantadosi CA. Bacterial Priming Increases Lung Injury in Gram-Negative Sepsis. *Am J Respir Crit Care Med.* 1998; 158:610-619.

38. Gutman LT, Moye J, Zimmer B, Tian C. Tuberculosis in Human Immuno-deficiency Virus-Exposed or -Infected United States Children. *Pediatr Infect Dis J.* 1994; 13:963-968.

39. Marais BJ, Gie RP, Schaaf HS, et al. The Natural History of Childhood Intra-Thoracic Tuberculosis; a Critical Review of Literature from the Pre-Chemotherapy Era. *Int J Tuberc Lung Dis.* 2004; 8:392-402.

40. Wallgren A. Primary Pulmonary Tuberculosis in Childhood. *Am J Dis Child.* 1935; 49:1105-1136.

41. Wallgren A. Pulmonary Tuberculosis--Relation of Childhood Infection to Disease in Adults. *Lancet.* 1938; 1:417-420.

42. Wallgren A. The Time-Table of Tuberculosis. *Tubercle.* 1948; 29 245-251.

43. Brailey M. Prognosis in White and Colored Tuberculous Children According to Initial Chest X-Ray Findings. *Am J Public Health Nations Health.* 1943; 33:343-352.

44. Gedde-Dahl T. Tuberculous Infection in the Light of Tuberculin Matriculation. *Am J Hyg.* 1952; 56:139-214.

45. Bentley FJ, Grzybowski S, Benjamin G. Tuberculosis in Childhood and Adolescence. *The National Association for the Prevention of Tuberculosis.* London, England: Waterlow and Sons Ltd; 1954:1-213 and 238-253.

46. Davies PD. The Natural History of Tuberculosis in Children. *Tubercle.* 1961; 42:1-40.

47. Lincoln EM, Sewell EM. *Tuberculosis in Children.* NY: McGraw-Hill; 1963.

48. Miller FJW, Seal RME, Taylor MD. *Tuberculosis in Children.* London: Churchill; 1963.

49. Starke JR. Childhood Tuberculosis During the 1990s. *Pediatr Rev.* 1992; 13:343-353.

50. Rieder HL, Cauthen GM, Comstock GW, Snider DE, Jr. Epidemiology of Tuberculosis in the United States. *Epidemiol Rev.* 1989; 11:79-98.

51. Centers for Disease Control. Topics in Minority Health Tuberculosis in Blacks--United States. *Morbidity and Mortality Weekly Rep.* 1987; 36:212-214, 219-220.

52. Wilfert CM, Buckley RH, Mohanakumar T, et al. Persistent and Fatal Central-Nervous-System Echovirus Infections in Patients with Agammaglobulinemia. *N Engl J Med.* 1977; 296:1485-1489.

53. McKinney RE, Jr., Katz SL, Wilfert CM. Chronic Enteroviral Meningoencephalitis in Agammaglobulinemic Patients. *Rev Infect Dis.* 1987; 9:334-356.

54. Whitley RJ, Nahmias AJ, Visintine AM, Fleming CL, Alford CA. The Natural History of Herpes Simplex Virus Infection of Mother and Newborn. *Pediatrics.* 1980; 66:489-494.

55. Kimberlin DW, Lin CY, Jacobs RF, et al. Natural History of Neonatal Herpes Simplex Virus Infections in the Acyclovir Era. *Pediatrics.* 2001; 108:223-229.

56. Gutman LT, Wilfert CM, Eppes S. Herpes Simplex Virus Encephalitis in Children: Analysis of Cerebrospinal Fluid and Progressive Neurodevelopmental Deterioration. *J Infect Dis.* 1986; 154:415-421.

57. Rudd C, Rivadeneira ED, Gutman LT. Dosing Considerations for Oral Acyclovir Following Neonatal Herpes Disease. *Acta Paediatr.* 1994; 83:1237-1243.

58. Tiffany KF, Benjamin DK, Jr., Palasanthiran P, O'Donnell K, Gutman LT. Improved Neurodevelopmental Outcomes Following Long-Term High-Dose Oral Acyclovir Therapy in Infants with Central Nervous System and Disseminated Herpes Simplex Disease. *J Perinatol.* 2005; 25:156-161.

59. Kimberlin DW, Whitley RJ, Wan W, et al. Oral Acyclovir Suppression and Neurodevelopment after Neonatal Herjpes. *N Engl J Med.* 2011; 365:1284-1292.

60. Moeller MB, Gutman RA, Hamilton JD. Acquired Cytomegalovirus Retinitis. Four New Cases and a Review of the Literature with Implications for Management. *Am J Nephrol.* 1982; 2:251-255.

61. Klotman ME, Hamilton JD. Cytomegalovirus Pneumonia. *Semin Respir Infect.* 1987; 2:95-103.

62. Kanj SS, Sharara AI, Clavien PA, Hamilton JD. Cytomegalovirus Infection Following Liver Transplantation: Review of the Literature. *Clin Infect Dis.* 1996; 22:537-549.

63. Rubin RH, Tolkoffrubin NE, Oliver D, et al. Multicenter Seroepidemiologic Study of the Impact of Cytomegalo-Virus Infection on Renal-Transplantation. *Transplantation.* 1985; 40:243-249.

64. Kanj S, Hamilton J. The 10 Most Common Questions About Cytomegalovirus Infection in Solid Organ Transplant Recipients. *Infectious Diseases in Clinical Practice.* 1997; 6:29-32.

65. Hamilton JD. *Cytomegalovirus and Immunity.* Basel, Switzerland: Karger; 1982.

66. Hamilton JD, Fitzwilliam JF, Cheung KS, Shelburne J, Lang DJ, Amos DB. Viral Infection-Homograft Interaction in a Murine Model. *J Clin Invest.* 1978; 62:1303-1312.

67. Hamilton JD, Bradley BA, Vanrood JJ. Monolayer Absorption of Human Cytotoxic T Cells: Evidence for Clonality. *Tissue Antigens.* 1979; 13:349-356.

68. Hamilton JD, Fitzwilliam JF, Cheung KS, Lang DJ. Eeffects of Murine Cytomegalovirus-Infection on the Immune-Response to a Tumor Allograft. *Rev Infect Dis.* 1979; 1:976-987.

69. Goulmy E, Hamilton JD, Bradley BA. Anti-Self Hla May Be Clonally Expressed. *J Exp Med.* 1979; 149:545-550.

70. Hamilton JD, Seaworth BJ. Transmission of Latent Cytomegalo-Virus in a Murine Kidney Tissue-Transplantation Model. *Transplantation.* 1985; 39:290-296.

71. Porter KR, Starnes DM, Hamilton JD. Reactivation of Latent Murine Cyto-megalo-Virus from Kidney. *Kidney Int.* 1985; 28:922-925.

72. Klotman ME, Starnes D, Hamilton JD. Tthe Source of Murine Cytomega-lo-Virus in Mice Receiving Kidney Allografts. *J Infect Dis.* 1985; 152:1192-1196.

73. Klotman ME, Henry SC, Hamilton JD. Determinants of the Source of Cy-tomegalovirus in Murine Renal-Allograft Recipients. *Transplantation.* 1987; 44:636-639.

74. Klotman ME, Henry SC, Greene RC, Brazy PC, Klotman PE, Hamilton JD. Detection of Mouse Cytomegalovirus Nucleic-Acid in Latently Infected Mice by in Vitro Enzymatic Amplification. *J Infect Dis.* 1990; 161:220-225.

75. Henry SC, Hamilton JD. Detection of Murine Cytomegalovirus Immediate Early-1 Transcripts in the Spleens of Latently Infected Mice. *J Infect Dis.* 1993; 167:950-954.

76. Yu YH, Henry SC, Xu FJ, Daley GG, Hamilton JD. The Suppressive Effect of Cytomegalovirus on Bone-Marrow - a Murine Model. *Cancer Research Therapy & Control.* 1993; 3:259-267.

77. Yu YH, Henry SC, Xu FJ, Hamilton JD. Expression of a Murine Cytomeg-alovirus Early-Late Protein in Latently Infected Mice. *J Infect Dis.* 1995; 172:371-379.

78. Henry SC, Schmader K, Brown TT, et al. Enhanced Green Fluorescent Protein as a Marker for Localizing Murine Cytomegalovirus in Acute and Latent Infection. *J Virol Methods.* 2000; 89:61-73.

79. Dalod M, Hamilton T, Salomon R, et al. Dendritic Cell Responses to Early Murine Cytomegalovirus Infection: Subset Functional Specialization and Dif-ferential Regulation by Interferon Alpha/Beta. *J Exp Med.* 2003; 197:885-898.

80. Banks TA, Rickert S, Benedict CA, et al. A Lymphotoxin-Ifn-Beta Axis Essen-tial for Lymphocyte Survival Revealed During Cytomegalovirus Infection. *J Immunol.* 2005; 174:7217-7225.

81. Casadevall A, Perfect JR. *Cryptococcus Neoformans.* Washington, DC: ASM Press; 1998.

82. Heitman J, Casadevall A, Kwon-Chung KJ, Kozel T, Perfect JR. *Cryptococcus Neoformans: From Human Pathogen to Model Yeast.* Washington, DC: ASM Press; 2011.

83. Perfect JR, Durack DT. Treatment of Experimental Cryptococcal Meningitis with Amphotericin B, 5-Fluorocytosine, and Ketoconazole. *J Infect Dis.* 1982; 146:429-435.

84. Perfect JR, Savani DV, Durack DT. Comparison of Itraconazole and Fluco-nazole in Treatment of Cryptococcal Meningitis and Candida Pyelonephritis in Rabbits. *Antimicrob Agents Chemother.* 1986; 29:579-583.

85. Savani DV, Perfect JR, Cobo LM, Durack DT. Penetration of New Azole Compounds into the Eye and Efficacy in Experimental Candida Endophthalmitis. *Antimicrob Agents Chemother.* 1987; 31:6-10.

86. Perfect JR. Fluconazole Therapy for Experimental Cryptococcosis and Candidiasis in the Rabbit. *Rev Infect Dis.* 1990; 12:S299-302.

87. Perfect JR, Klotman ME, Gilbert CC, et al. Prophylactic Intravenous Amphotericin B in Neutropenic Autologous Bone Marrow Transplant Recipients. *J Infect Dis.* 1992; 165:891-897.

88. Perfect JR, Rude TH, Penning LM, Johnston SA. Cloning the Cryptococcus Neoformans Trp1 Gene by Complementation in Saccharomyces Cerevisiae. *Gene Amplif Anal.* 1992; 122:213-217.

89. Perfect JR, Toffaletti DL, Rude TH. The Gene Encoding Phosphoribosylaminoimidazole Carboxylase (Ade2) Is Essential for Growth of Cryptococcus Neoformans in Cerebrospinal Fluid. *Infect Immun.* 1993; 61:4446-4451.

90. Denning DW, Lee JY, Hostetler JS, et al. Niaid Mycoses Study Group Multicenter Trial of Oral Itraconazole Therapy for Invasive Aspergillosis. *Am J Med.* 1994; 97:135-144.

91. Lodge JK, Jackson-Machelski E, Toffaletti DL, Perfect JR, Gordon JI. Targeted Gene Replacement Demonstrates That Myristoyl-Coa: Protein N-Myristoyltransferase Is Essential for Viability of Cryptococcus Neoformans. *Proc Natl Acad Sci U S A.* 1994; 91:12008-12012.

92. Perfect JR, Rude TH, Wong B, Flynn R, Chaturvedi V, Niehaus W. Identification of a Cryptococcus Neoformans Gene That Directs Expression of the Cryptic Saccharomyces Cerevisiae Mannitol Dehydrogenase Gene. *J Bacteriol.* 1996; 178:5257-5262.

93. Salas SD, Bennett JE, Kwon-Chung KJ, Perfect JR, Williamson PR. Effect of the Laccase Gene Cnlac1, on Virulence of Cryptococcus Neoformans. *J Exp Med.* 1996; 184:377-386.

94. Odom A, Muir S, Lim E, Toffaletti DL, Perfect J, Heitman J. Calcineurin Is Required for Virulence of Cryptococcus Neoformans. *EMBO J.* 1997; 16:2576-2589.

95. Pappas PG, Bradsher RW, Kauffman CA, et al. Treatment of Blastomycosis with Higher Doses of Fluconazole. The National Institute of Allergy and Infectious Diseases Mycoses Study Group. *Clin Infect Dis.* 1997; 25:200-205.

96. Alspaugh JA, Perfect JR, Heitman J. Cryptococcus Neoformans Mating and Virulence Are Regulated by the G-Protein Alpha-Subunit Gpa1 and Camp. *Genes Dev.* 1997; 11:3206-3217.

97. Yue C, Cavallo L, Alspaugh JA, et al. The Ste12alpha Homolog Is Required for Haploid Filamentation but Largely Dispensable for Mating and Virulence in Cryptococcus Neoformans. *Genetics.* 1999; 153:1601-1615.

98. Mitchell TG, Perfect JR. Cryptococcosis in the Era of AIDS--100 Years after the Discovery of Cryptococcus Neoformans. *Clin Microbiol Rev.* 1995; 8:515-548.

99. Perfect JR. Antifungal Prophylaxis: To Prevent or Not. *Amer J Med.* 1993; 94:233-234.

100. Kirkland KB, Marcom PK, Sexton DJ, Dumler JS, Walker DH. Rocky Mountain Spotted Fever Complicated by Gangrene: Report of Six Cases and Review. *Clin Infect Dis.* 1993; 16:629-634.

101. Kirkland KB, Wilkinson WE, Sexton DJ. Therapeutic Delay and Mortality in Cases of Rocky Mountain Spotted Fever. *Clin Infect Dis.* 1995; 20:1118-1121.

102. Sexton DJ. Rocky Mountain Spotted Fever. *Arch Intern Med.* 1985; 145:2173.

103. Sexton DJ, Corey GR. Rocky Mountain "Spotless" and "Almost Spotless" Fever: A Wolf in Sheep's Clothing. *Clin Infect Dis.* 1992; 15:439-448.

104. Sexton DJ, Dwyer B, Kemp R, Graves S. Spotted Fever Group Rickettsial Infections in Australia. *Rev Infect Dis.* 1991; 13:876-886.

105. Sexton DJ, Banks J, Graves S, Hughes K, Dwyer B. Prevalence of Antibodies to Spotted Fever Group Rickettsiae in Dogs from Southeastern Australia. *Am J Trop Med Hyg.* 1991; 45:243-248.

106. Sexton DJ, Burgdorfer W, Thomas L, Norment BR. Rocky Mountain Spotted Fever in Mississippi: Survey for Spotted Fever Antibodies in Dogs and for Spotted Fever Group Reckettsiae in Dog Ticks. *Am J Epidemiol.* 1976; 103:192-197.

107. Sexton DJ, Muniz M, Corey GR, et al. Brazilian Spotted Fever in Espirito Santo, Brazil: Description of a Focus of Infection in a New Endemic Region. *Am J Trop Med Hyg.* 1993; 49:222-226.

108. Burgdorfer W, Sexton DJ, Gerloff RK, Anacker RL, Philip RN, Thomas LA. Rhipicephalus Sanguineus: Vector of a New Spotted Fever Group Rickettsia in the United States. *Infect Immun.* 1975; 12:205-210.

109. Philip RN, Casper EA, MacCormack JN, et al. A Comparison of Serologic Methods for Diagnosis of Rocky Mountain Spotted Fever. *Am J Epidemiol.* 1977; 105:56-67.

110. Sexton DJ, Kanj SS, Wilson K, et al. The Use of a Polymerase Chain Reaction as a Diagnostic Test for Rocky Mountain Spotted Fever. *Am J Trop Med Hyg.* 1994; 50:59-63.

111. Wilfert CM, MacCormack JN, Kleeman K, et al. The Prevalence of Antibodies to Rickettsia Rickettsii in an Area Endemic for Rocky Mountain Spotted Fever. *J Infect Dis.* 1985; 151:823-831.

112. Wilfert CM, MacCormack JN, Kleeman K, et al. Epidemiology of Rocky Mountain Spotted Fever as Determined by Active Surveillance. *J Infect Dis.* 1984; 150:469-479.

113. Anstey NM, Weinberg JB, Hassanali MY, et al. Nitric Oxide in Tanzanian Children with Malaria: Inverse Relationship between Malaria Severity and Nitric Oxide Production/Nitric Oxide Synthase Type 2 Expression. *J Exp Med*. 1996; 184:557-567.

114. Anstey NM, Granger DL, Hassanali MY, Mwaikambo ED, Duffy PE, Weinberg JB. Nitric Oxide, Malaria, and Anemia: Inverse Relationship between Nitric Oxide Production and Hemoglobin Concentration in Asymptomatic, Malaria-Exposed Children. *Am J Trop Med Hyg*. 1999; 61:249-252.

115. Anstey NM, Weinberg JB, Wang Z, Mwaikambo ED, Duffy PE, Granger DL. Effects of Age and Parasitemia on Nitric Oxide Production/Leukocyte Nitric Oxide Synthase Type 2 Expression in Asymptomatic, Malaria-Exposed Children. *Am J Trop Med Hyg*. 1999; 61:253-258.

116. Granger DL, Anstey NM, Miller WC, Weinberg JB. Measuring Nitric Oxide Production in Human Clinical Studies. *Methods Enzymol*. 1999; 301:49-61.

117. Levesque MC, Hobbs MR, Anstey NM, et al. Nitric Oxide Synthase Type 2 Promoter Polymorphisms, Nitric Oxide Production, and Disease Severity in Tanzanian Children with Malaria. *J Infect Dis*. 1999; 180:1994-2002.

118. Tjitra E, Suprianto S, Dyer M, Currie BJ, Anstey NM. Field Evaluation of the Ict Malaria P.F/P.V Immunochromatographic Test for Detection of Plasmodium Falciparum and Plasmodium Vivax in Patients with a Presumptive Clinical Diagnosis of Malaria in Eastern Indonesia. *J Clin Microbiol*. 1999; 37:2412-2417.

119. Anstey NM, Hassanali MY, Mlalasi JE, Manhenga D, Mwaikambo ED. Elevated Levels of Methaemoglobin in Children with Uncomplicated and Severe Malaria. *Trans R Soc Trop Med Hyg*. 1996; 90:147-151.

120. Dyer ME, Tjitra E, Currie BJ, Anstey NM. Failure of the "Pan-Malarial" Antibody of Ict Malaria P.F/P.V Immunochromatographic Test to Detect Symptomatic Plasmodium Malariae Infection. *Trans R Soc Trop Med Hyg*. 2000; 94:518.

APPENDIX

APPENDIX A
PEDIATRIC INFECTIOUS DISEASES FACULTY IN THE 20TH CENTURY

Current Duke Appointment or

Upon Relocation *

1960's and 1970's

Samuel Katz	Professor emeritus
David Lang	Professor *
Catherine Wilfert	Professor emeritus
Laura Gutman	Associate Clinical Professor emeritus*

1980's

Sandra Lehrman	Assistant Professor*
Ross McKinney	Professor
Robert Drucker	Professor
Dennis Clements	Professor

1990's

Walter Emmanuel	Professor
Jeffrey Snedeker	Assistant Professor*
Nancy Henshaw	Assistant Research Professor
Randall Fisher	Assistant Professor*
Ghassan Dbaibo	Assistant Professor*
Kenneth Alexander	Associate Professor*

APPENDIX B
PEDIATRIC INFECTIOUS DISEASES FELLOWS IN THE 20TH CENTURY

Current Location/ Position if known

1960's and 1970's

Robert Snowe

Robert Greenberg

Mickael Kannan

Ousama Tomeh

Gerald Aronheim   Montreal Children's Hospital/Private Practice

Ziad Idriss   Bethesda, Maryland/Private Practice

James Waller   Deceased

Jeffrey Davis   Madison, Wisconsin/Health Department

Betty Raffin   Saratoga, California/Private Practice

Herbert Lassiter

Cornelia Dekker   Palo Alto, California/Med. Dir.Vaccine Program

Sandra Lehrman   Global Dir. Scientific Affairs/Merck Virology

1980's

John Frank

Ross McKinney   Durham, NC/Duke Professor of Pediatrics

Robert Drucker   Durham, NC/Duke Professor of Pediatrics

Stephen Eppes   Wilmington, DE/Jefferson Med.College Faculty

Jeffrey Snedeker   Syracuse, NY/SUNY Faculty Private Practice

Dennis Clements   Durham, NC/Duke Professor of Pediatrics

Emmanuel Walter   Durham, NC/Duke Professor of Pediatrics

Emilia Rivadeneira   Atlanta, GA/Centers for Disease Control

Ira Dunkel   New York, NY/Sloan Kettering Staff

Randall Fisher   Norfolk, VA/Eastern VA FacultyPrivate Practice

1990's

Ghassan Dbaibo	Beirut, Lebanon/American University Faculty
Gareth Tudor-Williams	London, UK/Imperial College Faculty
Catharine Moffitt	Southeastern Florida/Private Practice
Kenneth Alexander	Univ.Chicago/Chief Pediatric Infect. Dis.
Miguela Caniza	Memphis, TN/Saint Jude Faculty
Cynthia Jackson	Durham, NC/VP Medical.Services Quintiles
Pamela Palasanthiran	Sydney, Australia/Sydney Children's Staff
Paul Adholla	Fayettville, NC/County Health Department
Kenji Cunnion	Norfolk, VA/Eastern VA Medical School Faculty
Alicia Johnston	Springfield, MA/Tufts Faculty
Dan ny Benjamin	Durham, NC/Duke Professor of Pediatrics
Kathleen Vozzelli	Portland, ME/Children's Hospital Staff

APPENDIX C
MEDICINE INFECTIOUS DISEASES FACULTY IN THE 20TH CENTURY

Current Duke Appointment or

<u>Upon Relocation*</u>

1960's and 1970's

Thomas Cate	Houston, TX/Baylor Faculty8
Suydam Osterhout	Deceased
John Hamilton	Durham, NC/Professor emeritus
Harry Gallis	Charlotte, NC/VP Regional Education*
David Durack	Durham, NC/VP Becton Dickinson*
Charles Ellenbogen	Deceased

1980's

John Perfect	Durham, NC/Chief of Duke Infect. Dis.
Ralph Corey	Durham, NC/Professor
Donald Granger	Salt Lake City, UT/Univ. of Utah Prof*
Malcolm McDonald	Darwin, Austra /Menzies School Faculty*
Mary Klotman	Durham, NC/Duke Chair of Medicine
Charles van der Horst	Chapel Hill, NC/UNC ID Faculty*
Michael Cairns	Seattle, WA/Private Practice*
Gunther Lallinger	Chapel Hill, NC/ER Physician*
Kenneth Wilson	Chapel Hill, NC/Professor emeritus
John Bartlett	Durham, NC/Professor
Hetty Waskin	Unknown
James Hathorn	Durham, NC/ Priv. Practice of ID/ONC*

1990's

Kevin Porter	Bethesda, MD/US Navy ID Chief*
Peter Cegielski	Atlanta, GA/TB Branch of CDC*
Mimi Cameron	Washington, DC/Kaiser ID Physician*
Jerome Kim	Washington, DC/Walter Reed Faculty*
Chris Ingram	Raleigh, NC/Private Practice*
Daniel Sexton	Durham, NC/Professor

Kathryn Kirkland	Hanover, NH/Dartmouth Professor*
Souha Kanj	Beirut, Lebanon/AUB ID Chief*
Carol Hamilton	Durham, NC/Dir. FHI 360 Sci. Affairs*
Alison Heald	Seattle, WA/Private Practice*
Richard Frothingham	Durham, NC /Associate Professor
June Almenoff	Durham, NC/Glaxo Smith Klein Staff*
Diego Miralles	LaJolla, CA/VP Johnson and Johnson*
Mark Perkins	Geneva, Switz./Fdn.Innov.New Diagno*.
Charles Hicks	San Diego, CA/UCSD Professor*
Murray Abramson	Directo r of Clinical Trials/Merck*
Gary Cox	Durham, NC/Professor
Andrew Alspaugh	Durham, NC/Professor
Barbara Alexander	Durham, NC/Professor
Vance Fowler	Durham, NC/Professor

APPENDIX D
MEDICINE INFECTIOUS DISEASES FELLOWS IN THE 20TH CENTURY

<u>Current Location/Position if known</u>

1960's and 1970's

Richard McCloskey	Malvern, PA/Industry
John Hamilton	Durham, NC/Duke Professor
David Smith	Kansas City, MO/Private Practice
Harry Gallis	Charlotte, NC/VP Regional Education
Daniel Sexton	Durham, NC/Duke Professor
John Perfect	Durham NC/Duke Professor
John Ticehurst	Baltimore, MD/Johns Hopkins Faculty
Selwyn Lang	Auckland, NZ/Private Practice
Ralph Corey	Durham, NC/Duke Professor
Mark Moeller	Pollicksville, NC/Private Practice
Alan Spanos	Raleigh, NC/Private Practice
Wendy Keitel	Houston, TX/Baylor Faculty
Jonathon Cohen	London, Eng./Dean Brighton Med.Sch.
Barbara Seaworth	SanAntonio, TX/Dir. National TB Center
Ray Smego	Deceased
Robert Brennan	Lynchburg, VA/Private Practice
Malcolm McDonald	Darwin, Austr/Faculty Menzies School
Clark Kerr	Deceased
Mary Klotman	Durham, NC/Chair Duke Medicine Dept.
Joan Drucker	Durham, NC/Industry Consultant
Melinda Wharton	Atlanta, GA/Centers for Disease Control
Gunther Lallinger	Chapel Hill, NC/ER Physician
Michael Cairns	Seattle, WA/Private Practice
John Bartlett	Durham, NC/Duke Professor
Kevin Lee-See	Newcastle, Austr/Private Practice
Glenn Gafford	Wilminton, NC/Private Practice
Sunil Shaunak	London, Eng/Imperial College Faculty
Kevin Porter	Silver Spring, MD/Chief Virology U.S.U.

Alan Street	Melbourne, Austr/Private Practice
Jerome Kim	Washington, DC/Walter Reed Virology
Chris Ingram	Raleigh, NC/Private Practice
Mimi Cameron	Washington, DC/Kaiser Private Practice
Peter Cegielski	Atlanta, GA/Centers for Disease Control
Carol Dukes Hamilton	Durham, NC/FHI 360 Dir. Scien. Affairs
Claire Beiser	Bellingham, WA/Private Practice
David Tanner	Charlotte, NC/Private Practice
Michael Towns	Sparks, MD/Industry
Karen Welty -Wolfe	Durham, NC/Duke Professor
Kathryn Kirkland	Hanover, NH/Dartmouth Professor
Richard Frothingham	Durham, NC/Duke Professor
Souha Kanj	Beirut, Lebanon/AUB Professor
William Miller	Chapel Hill, NC/Univ of NC Professor
James Schwartz	Columbia, MO/Univ.of Missouri Faculty
Janet Hammond	Fairfield, CT/ Industry
Chris Hahn	Boise, ID/Dir. State Health Department
Lennox Archibald	Atlanta, GA/Centers for Disease Control
Gary Cox	Durham, NC/Duke Professpr
David Wininger	Columbus, OH/Ohio State Professor
Igor Abolnik	Salt Lake City, UT/Private Practice
Nicholas Anstey	Darwin, Austr/Menzies School Professor
Murray Abramson	Philadelphia, PA/Industry
Allison Heald	Seattle, WA/Private Practice
Scott Letendre	San Diego, CA/UCSD Professor
Fran Meredith	Cary, NC/Private Practice
Andrew Alspaugh	Durham, NC/Duke Professor
Sanjay Desai	Bethesda, MD/NIH Research Faculty
Elizabeth Talbot	Hanover, NH/Dartmouth Professor
Barbara Alexander	Durham, NC/Duke Professor
Vance Fowler	Durham, NC/Duke Professor
Magnus Gottfredsson	Reykjavik, Ice./Medical School Prof.
Jamie Whitehouse	Asheville, NC/Private Practice

Paul Bohjanen	Minneapolis, MN/Univ. of Minn. Prof.
David Murdoch	Christchurch, NZ/Univ. of Otago Prof.
Steven Wilson	Indianapolis, IN/Private Practice
Dan Wray	Charleston, SC/Med Univ of SC Prof.
Debra Miller	Spartanburg, SC/Private Practice
Jason Stout	Durham, NC/Duke Professor
Chris Woods	Durham, NC/Duke Professor
Debra Friedman	Geelong, Austr/Private Practice

INDEX

North Carolina TB Association 28

O

Oas 74, 232
Occupational 87, 90, 95, 99, 242, 244
Oden 101, 160
Official accreditation 157
Official Duke Hospital policies 96
Opportunistic Infections/OI 49–50,
    63–64, 78, 82, 105, 109, 113,
    174, 228, 247, 271
Osler 1, 16, 24, 27
Osterhout 54, 58, 91, 96, 99, 117, 131,
    133–134, 149, 259–260, 296
Other contributors 84
Other transplant programs were
    established at Duke 109

P

PACTG 076 72–73, 168
Palker 67, 74, 78, 140, 194–195, 227–
    228, 231–232, 294
Para-amino-salicylic acid 30
Pasteur 4, 60, 171, 187
Paul McCain 11
Pediatric AIDS Clinical Trials Group/
    PACTG 72–74, 144, 168–169,
    205, 230–231, 267
Pediatric infectious diseases faculty/
    Table 1 57, 142, 305
Pediatric infectious diseases/Table
    2 38, 55, 57–58, 73–74, 97,
    112–113, 142–145, 147–148, 154,
    163–164, 166, 168–169, 305–306
Perfect 58, 62, 79, 84, 114–116, 119,
    125, 138, 210, 227, 247–253, 257,
    263–264, 272–274, 278, 301–303
Perkins 41, 43–45, 155, 163, 179–180,
    223–224, 251, 280, 288
Perlzweig 18

Peters 122–123, 255
Phase I 69, 74, 146–147, 230, 266–
    267, 279
Phase II 69–70
Pickard 97–98
Pickup 138, 145, 191–192, 226, 260–
    262, 267, 293
Plasma viral load 72
Pneumocystis carinii Pneumonia/
    PCP 63, 228
Polio 18, 34, 141, 190
Popovic 67, 228
Post-herpetic neuralgia 197
Prior fellows wrote brief
    summaries 160
Public Health Movement 26
Purcell 170–171, 284–285

R

Rammelkamp 37, 150, 171
Rankin 3, 17–20, 24, 27–28, 31, 218–
    219, 280
Reed 4, 17, 71, 81, 167, 171, 223
Reller 38, 58, 117, 150, 154, 169, 183,
    224, 251, 280–281, 288
Research Center on AIDS and HIV
    Infection/RCAHI 76, 78–79,
    140, 157
Research emphasis 155
Research Themes
    Pediatrics 16, 31, 33, 36, 38–39,
        43–44, 50, 55–59, 62–63,
        72–74, 84, 97, 109, 111–113,
        116–117, 121–122, 129, 131,
        136, 141–150, 154, 162–164,
        166–170, 174, 176, 178, 181,
        190–191, 196, 203, 205, 210,
        212, 222, 230–231, 245–
        246, 251, 264, 266–269,
        275, 281, 284, 287, 292, 295,
        299, 305–306

www.ingramcontent.com/pod-product-compliance
Lightning Source LLC
Chambersburg PA
CBHW031820170526
45157CB00001B/128